T0211493

How Transistor Area Shrank by 1 Million Fold

Howard Tigelaar

How Transistor Area Shrank by 1 Million Fold

 Springer

Howard Tigelaar
Tigelaar Consulting, LLC
Allen, TX, USA

ISBN 978-3-030-40023-1   ISBN 978-3-030-40021-7   (eBook)
https://doi.org/10.1007/978-3-030-40021-7

This Springer imprint is published by the registered company Springer Nature Switzerland AG
The registered company address is: Gewerbestrasse 11, 6330 Cham, Switzerland

Preface

I worked in semiconductor research, development, and manufacturing for over 30 years. I saw the number of transistors on a computer chip go from about two thousand per chip to several billion (see Table 1). I know how it is done—I helped do it. It still seems like magic to me. My common sense tells me no way should the integrated circuits in my cell phone and laptop, as complex as they are, continue to work flawlessly year after year. No way should over a billion (1,000,000,000) electrical switches (transistors) fit on an IC about the size of my thumbnail and all work perfectly. No way should three wires (electrical signal in, electrical signal out, flip the switch), 1000 times smaller than a strand of my hair, be able to operate these switches and convey synchronized signals flawlessly between these switches. No way should these switches be able to turn on and off over 3 billion times a second, 24 h a day, 7 days a week, for years and years and still keep going. In fact, every day, millions of these integrated circuits are manufactured, packaged, and used to manufacture a host of other goods. Each day, millions of us make phone calls, video calls, listen to music, surf the internet, and drop our phones but they keep working. Day after day these miniscule electronic switches run our cell phones, our computers, our TVs, electronic toys and the myriad integrated circuit chips in almost every electrical device and appliance we own. To my thinking this is absolutely amazing—it borders on the miraculous.

At Texas Instruments (TI) my job was process integration. I led a team of engineers that collaborated closely with many other teams of engineers and talented people to produce the manufacturing flows for the next-generation integrated circuit technology nodes (see Fig. 1). At each technology node, our goal was to shrink the area required to build an integrated circuit by 50%. By cutting the area in half but keeping the chip (die) size about the same, we could pack more transistors and therefore more computing power on each chip. TI introduced a new technology node every 2–4 years. Over the span of about 40 years, the number of transistors on a computer chip skyrocketed from about two thousand to over two billion (Table 1) [1]. Over the span of about 40 years, the area occupied by an integrated circuit got smaller by a million-fold.

Table 1 Summary of CPU scaling at Intel

Node nm	Approx. year	Power supply	Metal levels	Intel CPU	Approx. transistor count	CPU speed MHz
10,000	1971	5.0	1	4004	2300	.2
6000	1976	5.0	1	8080	6000	2
3000	1979	5.0	1	8088	29,000	8
1500	1982	5.0	1	80286	134,000	12
1000	1987	5.0	2	386SX	275,000	33
800	1989	4.0	3	486DX	1,200,000	33
500	1993	3.3	4	IntelDX4	1,600,000	100
350	1995	2.5	4	Pentium	3,300,000	133
250	1998	1.8	5	Pentium2	7,500,000	300
180	1999	1.6	6	Pentium3	28,000,000	700
130	2001	1.4	6	Celeron	44,000,000	1330
90	2003	1.2	7	Pentium4	125,000,000	3060
65	2005	1.0	8	Pentium4	184,000,000	3800
45	2007	1.0	9	Xenon	410,000,000	3000
32	2009	0.75	10	Core i7 G	1,170,000,000	3460
22	2011	0.75	11	Core i7 IB	1,400,000,000	3900
14	2014	0.70	12	Core i7 BU	1,900,000,000	4500

1 nm is one billionth of a meter
1 MHz is one million cycles per second [1]
https://www.intel.com/pressroom/kits/quickrefyr.htm

It was an exciting time to work in semiconductors. The technology was in a constant state of invention and renewal. At each scaled technology node, new equipment, new manufacturing processes, and new materials were invented, developed, and introduced into manufacturing. Equipment, materials, and IC manufacturing companies invested hundreds of millions of dollars every year in research and development. Billions of dollars each year went to support the collaboration between industry and universities to keep this scaling going. Multiple approaches to equipment and materials were attempted. Only the ones that proved to be manufacturable survived.

What the teams (hundreds of teams working together) accomplished to produce the first integrated circuit chip at each new node is flat out amazing (Fig. 1). Thousands of engineers with advanced degrees (bachelors, masters, doctorate) plus a lot of other really smart people without advanced degrees invented solutions to thousands of highly technical problems and generated thousands of patents in the process. At each new technology node, thousands of complex technical problems in almost every scientific field of chemistry, math, and physics were solved and integrated into a manufacturable process flow to produce ICs. (Developing a process that a Ph.D. can run in a lab is far easier than developing a process that manufacturing personnel can run three shifts a day, seven days a week.) In parallel, as my team was developing the IC manufacturing flow, teams of design engineers were

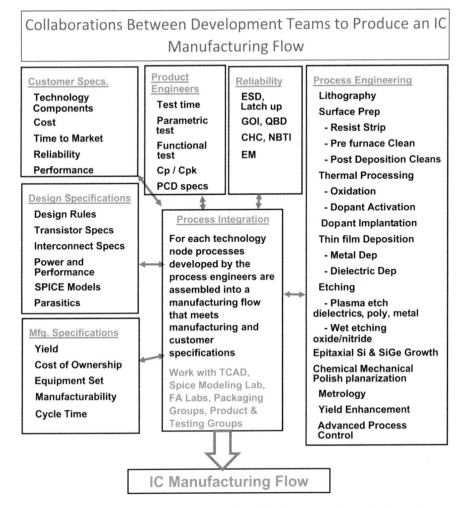

Fig. 1 Summary of collaborations between various development teams required to produce a manufacturing flow for an integrated circuit

designing new circuits, teams of equipment engineers were designing and developing new manufacturing equipment, teams of process engineers were developing new manufacturing processes, teams of testing engineers were developing new testing programs and testing equipment, teams of reliability engineers were developing new reliability tests and reliability equipment, and teams of manufacturing engineers were designing and building the new factory that would manufacture the new IC using the new manufacturing flow. All these activities progressed in parallel with a target completion date a couple of years down the road. At each and every node, all these development activities had to come together and had to seamlessly work together to manufacture IC chips and to generate the profits necessary to fund the next new technology and keep IC scaling on track with Moore's law.

Throughout this process the IC industry and academia collaborated closely. Academic research funded directly by the IC industry or funded with government tax dollars from the IC industry evaluated new materials, new processes, and new transistor structures. Ones that looked promising were either proven or disproven in industrial labs. If proven, they were incorporated into manufacturing flows. If disproven, they were discarded. The feasibility of many of the new processes for thin film deposition and etching were first demonstrated in university labs and consortia and then developed into manufacturable processes at industrial labs and R&D centers. The enormous expense of semiconductor equipment prevented academia and most research institutes from building prototyping manufacturing lines capable of producing state-of-the-art integrated circuits. Professors at universities collaborated closely with industrial scientists to develop extremely sophisticated modeling programs for many IC structures and processes. Simulated experiments using these modeling programs saved enormous amounts of time and helped to guide the industry research, development, and circuit design teams.

I lucked into my job at TI just when the bulk of IC chips manufactured at TI was transitioning from bipolar ICs and NMOS ICs to CMOS ICs. I watched and participated as CMOS IC transistor gate lengths shrank by 1000-fold. For a tech nerd like me, it was the perfect job. I had ready access to scores of really smart people with wide-ranging technical expertise. Instead of having to research a topic for hours, days, or weeks, I could satisfy my curiosity in a few minutes with a phone call or conversation. Some of the engineers, especially a few brilliant ones, were a bit quirky. I learned to work with and around their quirks to benefit from what they uniquely could offer.

In this book, I hope to convey to the reader an understanding of how integrated circuits are manufactured, an appreciation for the constant invention and innovation that kept the scaling going, and an appreciation for some of the incredible new technologies that were invented and made manufacturable along the way. I also hope to convey an appreciation for the amazing collaboration and teamwork that occurred between thousands of scientists, engineers, and others to produce each new technology node.

This book covers just a small fraction of the innovations that helped ICs scale. Whole books are written on the technical challenges of scaling each type of memory cell, or high voltage switch, or ESD protection transistor or ... Whole books are written on the innovations and technical challenges of designing, laying out, and debugging complex ICs with billions of transistors. Whole books are written on the innovations and technical challenges of designing, building, and debugging each of the many pieces of manufacturing equipment needed to build an IC. This book gives a glimpse into one small part of the marvelous IC technology revolution.

In Chapter 1, I attempt to convey to the reader the incredible wealth generated by IC technology that spawned and funded new technologies and industries and raised the standard of living for people across the globe. I conclude that IC scaling has virtually stopped, and the IC industry is in the process of transitioning to a new paradigm.

Chapter 2 is an overview of IC manufacturing. A cross-sectional view of a manufactured IC is described. Descriptions of some of the transistors and memory cells found in ICs is presented. The packaging of individual IC dies after they complete manufacturing is described.

Chapter 3 is an introduction to semiconductors, diodes, metal-oxide-silicon field effect transistors (MOSFET or MOS transistors), and bipolar transistors. A section introducing the inverter circuit and why it is so important for low-power devices and long battery life is included. Analog transistors are compared with digital transistors.

Chapter 4 describes the engineering that goes in to designing CMOS transistors to go fast.

Chapter 5 describes parasitic MOS and bipolar transistors that must be engineered to never turn on simultaneously with engineering the CMOS transistors to go fast.

Chapter 6, describes step-by-step a manufacturing flow from wafer start through transistor source and drain construction for building an inverter circuit. The purpose of each step is explained.

Chapter 7 continues the manufacturing flow from transistor source and drain construction through the first layer of metal wiring.

Chapter 8 continues the manufacturing flow with the addition of more levels of metal wiring and completion of IC manufacture.

Chapter 9 lists a few of the amazing technologies that are enablers for transistor scaling.

Chapter 10 describes some of the inventions and innovations by node that were implemented during the scaling of transistors and transistor isolation.

Chapter 11 describes some of the inventions and innovations by node that were implemented during the scaling of pre-metal dielectric, contacts, and aluminum alloy wires.

Chapter 12 describes some of the inventions and innovations by node that were implemented during the scaling of multilevel copper metallization.

Chapter 13 is a compilation of anecdotes and stories from my career in IC manufacturing.

Chapter 14 Wrapup and what is next?

Reference

1. Intel Microprocessor Quick Reference Guide, 1971—2008. https://www.intel.com/pressroom/kits/quickrefyr.htm

Allen, TX   Howard Tigelaar

Acknowledgements

To my late wife who, in spite of being a TI widow, gave me unwavering support, unconditional love, bore our two sons, and almost single handedly raised them into the amazing young men they are today.

To my sons, their wives, and five granddaughters who are a constant joy to me.

To my many former colleagues at Texas Instruments who supported me and patiently answered my constant barrage of questions. I especially acknowledge the engineers in my groups, men and women of many different ages and nationalities, who formed teams that focused on solving problems and getting the job done. Very especially to the technicians in my groups who worked night and day processing and inspecting our lots. Innumerable times they saved months of work by spotting and fixing my and my engineers mistakes.

To my friend and colleague, Patrick Fernandez, who shares my philosophy of avoiding problems at all cost. Together we developed and instituted a number of systems and procedures at Texas Instruments still in use today. Patrick's vast knowledge of semiconductor manufacturing equipment, manufacturing processes, and the yield problems they cause was key to rapidly building and qualifying semiconductor factories, rapidly qualifying new technologies, rapidly ramping yield, and controlling semiconductor factories for maximum yield and profit.

To the late Dr. W. H. Flygare, my graduate research professor at U. of Illinois, who inspired me with his enthusiastic pursuit of a wide range of sciences and technologies. He dove head first without hesitation and with great enthusiasm into new areas of science that he was unfamiliar with such as radio astronomy and biological light scattering.

To my parents whose sacrifice and unwavering support enabled me to be the first from our family to attend and graduate from Hope College.

Special thanks to my sister-in-law, Susan Gaunt, for patiently tolerating me going on and on about things of which she had little understanding and even less interest while I organized my thoughts when writing this book. And also for her keen insights and her unique ability to deliver cutting criticism with humor so I'd get her point and still walk away with a smile.

Contents

1 Integrated Circuit Revolutions 1
  1.1 Introduction 1
  1.2 Integrated Circuit Scaling and Technological/Workplace
      Revolutions 2
  1.3 Integrated Circuit Scaling Impact on Universities,
      Industrial Labs, and Consortia. 3
  1.4 Integrated Circuit Scaling and the Worldwide Rise
      in the Standard of Living 4
  1.5 Integrated Circuits: An Industry in Transition 4
  References. 5

2 Overview of Integrated Circuit Manufacturing. 7
  2.1 Integrated Circuit Manufacturing. 7
      2.1.1 Photolithography (Printing Patterns in Photoresist) 9
      2.1.2 Plasma Etch 9
      2.1.3 Thermal Oxidation of Silicon to Grow Silicon
            Dioxide Dielectric 10
      2.1.4 Chemical Vapor Deposition (CVD) 10
      2.1.5 Physical Vapor Deposition (Sputter Deposition) 12
  2.2 Integrated Circuit Transistors. 14
      2.2.1 Memory Transistors in Integrated Circuits 14
  2.3 Packaged Integrated Circuits 16
      2.3.1 Introduction to Packaged ICs. 16
      2.3.2 Packaging Individual IC Dies 16
      2.3.3 Packaging Multiple Integrated Circuit Dies. 18
  References. 19

3 Semiconductors, Diodes, Transistors, and Inverters 21
  3.1 Insulators, Conductors, and Semiconductors. 21
  3.2 Chemistry of Silicon and Carbon. 21
  3.3 Single Crystal Carbon Insulator (Diamond) 23
  3.4 Single Crystal Silicon Semiconductor 23

3.5   Low Resistance Doped Single Crystal Silicon   24
       3.5.1   N-Type Doped Single Crystal Silicon Semiconductor   24
       3.5.2   P-Type Doped Single Crystal Silicon Semiconductor.   25
3.6   PN-Junction Diodes [2]   26
       3.6.1   Forward-Biased Diodes   26
       3.6.2   Reverse-Biased Diodes   27
3.7   Metal on Silicon (MOS) Transistors [3]   30
       3.7.1   NMOS Transistors   30
       3.7.2   PMOS Transistors   32
3.8   CMOS Inverter [4]   33
       3.8.1   Inverter Circuit: Two Switches in Series   34
       3.8.2   CMOS Inverter Structure in Silicon.   35
3.9   Logic Functions "AND" and "OR"   38
3.10   Digital and Analog MOS Transistors and Signals   39
       3.10.1   Converting Analog Signals to Digital Signals [5]   40
       3.10.2   Digitizing Analog Signals   40
3.11   Digital MOS Transistors   41
       3.11.1   Halos on Digital MOS Transistors.   43
3.12   Analog MOS Transistors [6]   44
3.13   Bipolar Transistors [7]   46
       3.13.1   Bipolar Transistor Structure and Circuit Diagram.   46
       3.13.2   Bipolar Transistors in CMOS Integrated Circuits   48
References.   50

4   **Engineering MOS Transistors for High Speed and Low Power**   51
4.1   Water Flow to Current Flow Analogy   51
4.2   MOS Transistor Figure of Merit (FOM) [1]   53
4.3   MOS Transistor Resistance and Capacitance   55
       4.3.1   MOS Transistor Series Resistance.   55
       4.3.2   MOS Transistor Capacitance   57
       4.3.3   MOS Transistor Conventional Capacitors   59
       4.3.4   MOS Transistor Diode Capacitors.   61
4.4   MOS Transistor OFF Leakage Current   62
       4.4.1   Test Structures to Characterize MOS Transistor
                OFF Leakage Current   62
       4.4.2   Source to Drain Leakage Current   63
       4.4.3   Area Diode Leakage Current   64
       4.4.4   Gated Diode Leakage Current   64
       4.4.5   Band-to-Band Tunneling Leakage Current   64
       4.4.6   Gate Dielectric Leakage Current   65
       4.4.7   Silicon on Insulator (SOI) Substrates for Low
                MOS Transistor OFF Current   65
Reference   66

5   Parasitic MOS and Bipolar Transistors in CMOS ICs   67
   5.1   Introduction   67
   5.2   Design Rules [1, 2]   68
   5.3   Parasitic MOS Transistors Between Core MOS Transistors   69
           5.3.1   Parasitic MOS Transistors Between Adjacent Core MOS
                     Transistors   69
           5.3.2   Parasitic MOS Transistors Between Core MOS
                     Transistors and Adjacent WELLS   70
   5.4   Parasitic Bipolar Transistors in Integrated Circuits   70
           5.4.1   Parasitic Bipolar Transistors Under Core
                     MOS Transistors   70
           5.4.2   Parasitic Bipolar Transistors Between Core
                     Transistors and Wells in Integrated Circuits   72
   References   72

**6   CMOS Inverter Manufacturing Flow: Part 1 Wafer
       Start Through Transistors**   73
   6.1   Introduction to CMOS Inverter Manufacturing   73
   6.2   Shallow Trench Isolation (STI)   75
           6.2.1   Active Photoresist Pattern   75
           6.2.2   Shallow Trench Etch   76
           6.2.3   Trench Dielectric Fill   77
           6.2.4   Chemical Mechanical Polish Planarization   78
   6.3   Transistor WELL and Transistor Turn ON Voltage Doping   79
           6.3.1   NWELL and PWELL Formation   79
           6.3.2   Turn ON Voltage of NMOS and PMOS Transistors   82
   6.4   NMOS and PMOS Transistor Gates   84
           6.4.1   Introduction   84
           6.4.2   Transistor Gate Dielectric   84
           6.4.3   Transistor Polysilicon Gate   85
           6.4.4   Transistor Gate Patterning   86
           6.4.5   Transistor Gate Etching   86
           6.4.6   Gate Poly Oxidation   88
   6.5   Transistor Source and Drain Extensions   88
           6.5.1   NMOS Extension and Halo Dopant Pattern
                     and Implants   89
           6.5.2   NMOS Halo Dopant Implants   89
           6.5.3   PMOS Extension Pattern and Implant   92
           6.5.4   PMOS Halo Pattern and Implant   93
   6.6   Transistor Deep Source and Drain Diodes   95
           6.6.1   Transistor Dielectric Sidewalls   95
           6.6.2   NMOS Deep Source/Drain Dopant Pattern
                     and Implant   97

6.6.3   PMOS Transistor Deep Source/Drain Dopant
         Pattern & Implant.   99
6.6.4   Source/Drain Extension and Deep Source/Drain
         Diode Anneal   100
6.7   Silicided Source, Drains, and Gates.   100
6.7.1   Introduction to Silicides.   100
6.7.2   Nickel Silicide Formation   101
6.7.3   Nickel Metal Strip   101
6.7.4   Nickel Silicide Resistance Lowering Anneal.   102
References.   102

7   **CMOS Inverter Manufacturing Flow: Part 2 Transistors
    Through Single-Level Metal**   103
7.1   Introduction   103
7.2   Premetal Dielectric Layers (PMD)   104
7.2.1   Contact Etch-Stop Layer (CESL)   104
7.2.2   Gap Fill Dielectric: PMD1.   105
7.2.3   PMD1 Planarization.   106
7.2.4   PMD2.   107
7.3   Contacts   107
7.3.1   Contact Pattern.   107
7.3.2   Contact Etch.   108
7.3.3   Contact Barrier Layer   109
7.3.4   Contact Hole Metal Fill   111
7.3.5   Contact CMP   111
7.4   Metal1 Wiring Layer   112
7.4.1   Overview   112
7.4.2   First Intermetal Dielectric (IMD1).   114
7.4.3   Metal1 Trench Pattern and Etch.   115
7.4.4   Metal1 Trench Copper Fill.   116
7.4.5   Metal1 Copper CMP   120
7.5   Protective Overcoat and Bond Pads.   120
References.   121

8   **CMOS Inverter Manufacturing Flow: Part 3 Additional
    Levels of Metal Through PO**   123
8.1   Overview   123
8.2   Metal2 Wiring Layer   123
8.2.1   Metal2/Metal1 Intermetal Dielectric: IMD2   125
8.2.2   Via1 Pattern and Etch.   125
8.2.3   Metal2 Trench Pattern and Etch.   127
8.2.4   Via1 and Metal2 Trench Fill   128
8.2.5   Metal2 Copper CMP   129
8.3   Additional Metal Wiring Levels.   130
8.4   Protective Overcoat (PO) and Bond Pad   131

8.4.1 PO Layer Deposition, Pattern, and Etch 131
8.4.2 Bond Pad Deposition, Pattern, and Etch 132
8.5 Forming Gas Anneal (Sintering) 135
8.6 Example Packaged ICs 135
References. 136

9 The Incredible Shrinking IC: Part 1 Enabling Technologies 137
9.1 Introduction 137
9.2 Clean Room and Wafers. 139
9.2.1 Introduction 139
9.2.2 Clean Room Attire 139
9.2.3 Evolution of Clean Rooms. 140
9.2.4 Increasing Wafer Size 143
9.2.5 Wafer Carriers 145
9.2.6 Wafer Handling Tools 147
9.3 MMST Program: IC Chips Manufacture in 3 Days! 149
9.4 Lithography 151
9.4.1 Introduction 151
9.4.2 Printing IC Dies on a Wafer. 151
9.4.3 Photolithography Printers: Introduction 153
9.4.4 Printers: Light Wavelength as ICs Scaled 155
9.4.5 Printers: Intel's 2002 Lithography Roadmap. 156
9.4.6 Printers: Steppers and Scanners. 157
9.4.7 Printers: 193 nm Immersion Scanner. 158
9.4.8 Printers: The 157-nm Photolithography System Saga. 159
9.4.9 Photoresist: Introduction 161
9.4.10 Photoresist: Coating Wafers. 161
9.4.11 Photoresist: Kodak Thin Film Photoresist 161
9.4.12 Photoresist: Novolak Positive Photoresist 162
9.4.13 Photoresist: Chemically Amplified Photoresist 162
9.4.14 Photoresist: Bottom Antireflective Coating 163
9.4.15 Photoresist: Top Antireflective Coating 164
9.4.16 Photoresist: Bilayer Resists 165
9.4.17 Ultraviolet Hardened Resists 166
9.4.18 Photomasks (Reticles) 166
9.4.19 Photomasks: Phase Shift Masks 167
9.4.20 Photomasks: Alternating Phase Shift Masks 167
9.4.21 Photomasks: Attenuated Phase Shift Masks 167
9.4.22 Photomasks: Compensating for Light Interference
and Etch Loading. 168
9.4.23 Photomasks: Optical Proximity Correction (OPC) 170
9.4.24 Dummy Fill Geometries 173
9.5 Planarization: Chemical Mechanical Polish (CMP) 176
9.5.1 Shallow Trench Isolation 176
9.5.2 Pre-metal Dielectric Planarization. 177

	9.5.3	Multi-level Metal	177
	9.5.4	Copper Metallization	178
	9.5.5	Metal Replacement Gate Transistors	178
9.6	Rapid Thermal Processing		179
	9.6.1	Introduction	179
	9.6.2	Rapid Thermal Anneal	179
	9.6.3	Spike Anneal	179
	9.6.4	Flash Anneal	179
	9.6.5	Laser Anneal	180
9.7	Defect Detection and Defect Analysis		180
	9.7.1	Introduction	180
	9.7.2	Killer Particles	181
	9.7.3	Automated Killer Particle Detection	183
	9.7.4	Killer Particle Analysis	183
9.8	Functional and Reliability Testing		184
	9.8.1	Introduction	184
	9.8.2	Functional Testing	185
	9.8.3	Reliability Testing: Introduction	185
	9.8.4	Accelerated Reliability Testing	186
	9.8.5	Accelerated Reliability Modeling	186
	9.8.6	Qualification Reliability Testing	187
9.9	Computer Technology; Computer Simulations; Data Analysis		188
	9.9.1	Introduction	188
	9.9.2	Computer Simulation Programs: Transistor Processing	189
	9.9.3	Computer Simulation Programs: IC Processing	192
	9.9.4	Computer Simulation Programs: Integrated Circuit Design	192
	9.9.5	Computer Simulation Programs: Thermal Heating in IC Dies and IC Packages	193
	9.9.6	Computer Simulation Programs: Stress in IC Dies and IC Packages	194
	9.9.7	Computer Simulation Programs: Reliability Simulation Programs	194
	9.9.8	Computer Simulation Programs: Photo Simulation Programs	195
	9.9.9	Computer Simulation Programs: Other Simulation Programs	195
	9.9.10	Data Analysis Programs: Probe Data	195
	9.9.11	Data Analysis Programs: Sensor Data and Fault Detection and Classification (FDC) Data	197
References			197

10 The Incredible Shrinking IC: Part 2 FEOL Isolation
   Scaling and Transistor Scaling. 201
   10.1   Introduction 201
   10.2   Shrinking Transistor Isolation 201
           10.2.1   Overview 201
           10.2.2   LOCOS Isolation 202
           10.2.3   Shallow Trench Isolation (STI) 202
           10.2.4   Void-Free Gap Fill 205
   10.3   Shrinking MOS Transistor. 208
           10.3.1   Overview 208
           10.3.2   Scaling Source/Drain Junction Diodes:
                      Implant Angle 211
           10.3.3   Scaling Source/Drain Junction Diodes:
                      Source/Drain Extensions 212
           10.3.4   Engineering Source/Drain Extension
                      Resistance and Capacitance. 213
           10.3.5   Gate Etch 215
           10.3.6   Gate Dielectric Scaling 217
           10.3.7   Silicide 219
   10.4   Transistor Performance Enhancement Using Stress 221
           10.4.1   High-Performance PMOS Transistors
                      with Compressive Channel Stress 221
           10.4.2   High-Performance NMOS Transistors
                      Using Tensile Stress Memorization. 221
           10.4.3   High-Performance NMOS Transistors
                      Using Contact Etch-Stop Applied Stress. 223
           10.4.4   Single Contact Etch-Stop Liner Stress Technology. 225
           10.4.5   Dual Contact Etch-Stop Liner Stress Technology:
                      DSL Technology 225
   References. 226

11 The Incredible Shrinking IC: Part 3 BEOL Aluminum
   Alloy Single and Multilevel Metal. 227
   11.1   Introduction 227
   11.2   Pre-Metal Dielectric (PMD) 229
           11.2.1   Overview 229
           11.2.2   Pre-metal Dielectrics and PMD Stack 230
           11.2.3   Pre-metal Dielectric Planarization: Reflow 232
           11.2.4   Pre-metal Dielectric Planarization: Resist Etch back . . . 234
           11.2.5   Pre-metal Dielectric Planarization:
                      Chemical Mechanical Polish. 234
           11.2.6   Pre-metal Dielectric Gap Fill. 235
   11.3   Contact Process 236
           11.3.1   Overview 236
           11.3.2   Contact Pattern. 236
           11.3.3   Contact Etch. 237

        11.3.4   Contact Barrier: Overview........................ 238
        11.3.5   Contact Barrier Material and Deposition............ 238
        11.3.6   Contact Fill and Metal1 239
        11.3.7   Contact Plugs: Tungsten Chemical Vapor Deposition... 241
    11.4 Single-Level Metal................................... 242
        11.4.1   Single-Level CVD-W Metal (SLM) 242
        11.4.2   Single-Level Aluminum Alloy Metal............... 243
        11.4.3   Aluminum Metal1: Electromigration............... 243
    11.5 Multilevel Aluminum Metal............................ 245
        11.5.1   Multilevel Aluminum Metal: IMD1 Resist Etch Back
                  Planarization 246
        11.5.2   Multilevel Aluminum Metal: Tungsten
                  Filled Via Plugs 246
        11.5.3   Multilevel Aluminum Metal: Spin-on-Glass
                  Etch Back Planarization 248
        11.5.4   Multilevel Aluminum Metal: CMP Planarization 250
    References.. 251

12 **The Incredible Shrinking IC: Part 4 BEOL Low-K**
    Intermetal Dielectrics and Multilevel Copper Metallization 253
    12.1 Introduction 253
    12.2 Low-K Intermetal Dielectrics (Low-K IMDs)............... 254
    12.3 Damascene Copper Metalization....................... 255
        12.3.1   Single Damascene Copper Metalization 256
        12.3.2   Dual Damascene Copper Metalization 256
    12.4 180 nm Copper Multilevel Metal......................... 256
        12.4.1   180 nm Single Damascene Copper Metal1 256
        12.4.2   180 nm Dual Damascene Copper Metal2 257
    12.5 130 nm Copper Multilevel Metal......................... 258
    12.6 90 nm Copper Multilevel Metal.......................... 259
        12.6.1   90 nm Single Damascene Copper Metal1 259
        12.6.2   90 nm Dual Damascene Copper Metal2 260
    12.7 65 nm Copper Multilevel Metal.......................... 261
        12.7.1   65 nm Single Damascene Copper Metal1 261
        12.7.2   65 nm Dual Damascene Copper Metal2 261
    12.8 45 nm Copper Multilevel Metal.......................... 263
        12.8.1   45 nm Single Damascene Copper Metal1 263
        12.8.2   45 nm Dual Damascene Copper Metal2 264
    12.9 32 nm Copper Multilevel Metal.......................... 265
        12.9.1   32 nm Single Damascene Copper Metal1 266
        12.9.2   32 nm Dual Damascene Copper Metal2 266
    12.10 Copper Metallization Challenges Going Forward............. 267
    References.. 268

13 Anecdotes 271
    13.1 High-Voltage Transistors for DLP 271
    13.2 TI Widows and Orphans 271
    13.3 Priority Zero Lots 273
    13.4 When Nerds at Play Paid Off Big Time 273
    13.5 Super Tech Nerds 275
    13.6 Performance of PDSOI Circuits Versus Conventional CMOS ... 276
    13.7 Semiconductor Industry Job Insecurity 277
    13.8 A Very Short Career at Texas Instruments 277
    13.9 Still Needed Test Equipment 278
    13.10 Exploited Technical Artists 279
    13.11 Antifuse Technology Development Successful and Sold 281
    13.12 Managing Tech Nerds 281
    13.13 One Tech Nerd I Failed to Manage 282
    13.14 Rapid Learning and Change Control 282
    13.15 Engineer versus Technician 283
    13.16 Customer Crisis Averted 284
    13.17 Reliability Failures on a Qualified Manufacturing Flow 285
    13.18 Texas Instruments Training Institute
          and Semiconductors in Asia 286
    13.19 GUI Interface to Run Split Lots in Manufacturing 286
    References 288

14 Wrap Up and What Is Next? 289
    14.1 Wrap Up 289
    14.2 What Is Next: AI 290
    14.3 What Is Next: ICs and Electronic Devices 291
    14.4 What Is Next: Photonic Computers 292
    14.5 What Is Next: 3-D Integration in Silicon 293
    14.6 What Is Next: More Integration in Packaging—Multichip
          Packages 295
    References 295

Appendix A: References by Topic 297

Index 311

About the Author

Howard Tigelaar began his professional career as a polymer chemist in plastics R&D at Rohm and Haas Co. At Micromedic Systems, a subsidiary at Rohm and Haas, he switched into clinical chemistry, developed and manufactured immunoassay kits before spending several years as a clinical chemist at Abbott Laboratories in North Chicago. Howard then took a job at Texas Instruments, becoming a thin film deposition and etch engineer before transitioning into a process integration engineer. He spent the bulk of his career at TI managing groups that developed manufacturing flows for next-generation integrated circuit technologies. He was named TI Fellow. While at TI, he was a named author on over 40 technical papers and is a named inventor on over 65 issued patents. While working for TI, he spent a year at LETI in Grenoble, France, studying silicon on insulator technology and 2 years at IMEC in Leuvin, Belgium, learning hi-k, metal gate transistor technology.

After retiring from TI, Howard worked for several years at PDF Solutions as a senior yield enhancement engineer before starting his own consulting business. As a consultant, Howard has written over 200 technical patent applications for IP Law Firms as well as yield enhancement consulting at major semiconductor companies. At one company, yield improvement led to an increase in profits of over $1 million per day.

Howard earned his PhD at University of Illinois in microwave spectroscopy. He was a member of the University of Illinois radio astronomy team that discovered a cloud of interstellar formamide—the first interstellar molecule discovered which contained all the atoms of life—carbon, hydrogen, oxygen, and nitrogen ($HCONH_2$). He did a postdoctorate in physics and quantum optics at the University of Arizona. While at Texas Instruments, he took graduate courses in device physics, circuit design, and computer architecture taught on campus by professors from University of Texas at Dallas.

Howard was born on a farm in Michigan and lived his early years in a house with no running water, no indoor bathroom, and no central heating. He attended Zutphen Elementary School (two room school) and had the same teacher for all subjects in grades 4 through 8. Howard played football, basketball, and golf at Hudsonville Public High school. He earned a BA at Hope College in Holland, MI. He worked his way through Hope College with various jobs such as a hired laborer on dairy farms, a pig farm, fruit orchards, muck fields, construction, and as a janitor.

Chapter 1
Integrated Circuit Revolutions

1.1 Introduction

The space program demonstrated that amazing things can be accomplished when clear technical goals are set, and teams of tech-nerd scientists and engineers are showered with money. In less than 10 years, the space program evolved from a goal (May 25, 1961—President Kennedy) of landing a man on the moon and bringing him safely back to actually landing a man on the moon and bringing him safely back (July 24, 1969). The need for complex real-time computation on the Apollo rockets and on the Lunar Rover drove huge investments into research and development of transistors, integrated circuits, and portable computers. Advances in solid-state transistors and computers throughout the 1950s paved the way for the incredible integrated circuit technology revolution that began in the 1960s and continues today.

Severe cutbacks in the space program after the last Apollo 17 mission in 1972 flooded the market with out of work, highly educated, highly skilled scientists and engineers. Fortunately, the growing demand for more sophisticated, low-power consumer products (such as digital watches, handheld calculators, portable radios, and business computers) created a demand for highly trained engineers across the broad spectrum of the sciences and engineering. Within a few years, the flood of scientists and engineers from the space program was absorbed, and a growing demand for even more technically educated workers was created by the rapidly growing semiconductor industry.

Over a period of 8 years, absolutely incredible new technologies were invented and developed that put men on the moon. Over a period of 40+ years, even more incredible new technologies were invented and developed to scale the area of IC chips by over a million-fold.

© Springer Nature Switzerland AG 2020
H. Tigelaar, *How Transistor Area Shrank by 1 Million Fold*,
https://doi.org/10.1007/978-3-030-40021-7_1

1.2 Integrated Circuit Scaling and Technological/ Workplace Revolutions

Increasingly cheaper and more powerful ICs spawned the creation of new electronic devices that revolutionized technologies, birthed new industries, and greatly increased productivity. Numerous technological/workplace revolutions enabled by IC scaling evolved in parallel with IC scaling.

A few of the technology/workplace revolutions include the following:

1. The invention/development of CMOS technology and the low-power CMOS inverter circuit enabled the change from plugged in electronic equipment to wireless, portable, and battery-powered electronic equipment.
2. The invention/development of fast, multibit analog-to-digital and digital-to-analog converters plus digital signal processing technology revolutionized how information is stored, transmitted, and processed by computers.
3. As ICs scaled smaller, computers became smaller and more powerful. When it became cost-effective to put a desk top computer on the desk of every professional, thousands of secretarial jobs vanished. Many secretaries were retrained as data entry clerks. Professionals who relied upon secretaries to help them prepare papers for publication and slides for presentations now had to do it themselves on their personal desk top computer.
4. The world wide web combined with more powerful personal computers and electronic devices revolutionized communication, revolutionized how we do business, and revolutionized how we shop.
5. A first robotic revolution produced computerized machine tools that transformed many manufacturing sectors. Items such as automobiles, steel, and integrated circuits transitioned from being manufactured solely by humans to being manufactured by humans with the aid of robots. Robots transformed banking with automatic teller machines (ATMs) and transformed fast food restaurants with food order kiosks. Bar code technology transformed retail store checkouts reducing errors and eliminating the need for checkout clerks that could memorize and enter the prices of hundreds of items and could calculate correct change.
6. A second robotic revolution is now taking place as artificial intelligence is being combined with robotics to perform jobs that require intelligent decisions as new situations arise. In some jobs, humans are totally eliminated.

ASIDE:
I and a group of colleagues went to a restaurant in a shopping mall in Beijing, China, in 2018. When we arrived, a greeter asked us to bring up their website on our cell phones, review their menu, and place our orders. They told us to browse in the mall and they would text us when our table was ready. About half an hour later we received the text. Within a few minutes after we were seated, our first course was served. When our meal was finished, we could select to electronically pay individually or with one check.
END ASIDE

7. The wireless revolution and the internet of things allow people to communicate with each other and with other electronic machines whenever and wherever.
8. Wireless pagers, personal digital assistants (PDA), and smart phones are revolutionizing personal communications, revolutionizing social interactions, revolutionizing work, revolutionizing shopping, and revolutionizing entertainment.
9. Cybercrime is huge. Cyber security is now a major industry. Whole new industries and new governmental departments have been generated to conduct cyber war.
10. Smart weapons are revolutionizing the war industry and revolutionizing how wars are fought.
11. The IC industry has and is revolutionizing the medical industry (MRIs, CAT scans, DNA sequencing, diagnostic testing, implantable devices such as heart pacers and insulin pumps).
12. The IC revolution transformed the gaming industry from pinball machines in bars and one-armed bandits in casinos to computer game consoles and portable computer games virtually every child and child-at-heart adult owns.
13. Social media is transforming the news industry and revolutionizing marketing. Many magazines and newspapers have closed. Targeted advertisements are being delivered directly to consumers on their personal social media devices.
14. Photography is transformed. Film is for the most part a thing of the past. Expensive film cameras have been largely replaced with cell phone cameras. Motion pictures are recorded digitally using CCD cameras and projected onto screens in theaters using digital light projectors (DLPs). Anyone can now produce high-quality video using a handheld camera. Amazing special effects are now generated electronically. Animated cartoons are now created by digital artists using animation software instead of by thousands of artists drawings on cells.
15. Heavy cathode ray televisions with screens limited to about 40 in. and weighing several hundred pounds are replaced by LCD and LED televisions with screens exceeding 200 in. and weighing less than 100 pounds [1]. https://www.businessinsider.com/samsung-219-inch-the-wall-tv-2019-1

1.3 Integrated Circuit Scaling Impact on Universities, Industrial Labs, and Consortia

The increased demand for IC chips and the race to smaller and smaller ICs that kept Moore's law on track for over 40 years created a huge demand for scientists and engineers with advanced degrees. In response to this demand, physical science departments including math, physics (quantum, solid state, optics, plasma, information), chemistry (polymer, plasma, cleanup, chemical vapor deposition, thin film deposition, crystal growth, photochemistry), and materials science (solid state materials, crystal doping, thin film dielectrics, thin film metals, polymers for ICs and packaging) ballooned. In addition, new departments (computer programming, computer modeling, robotics, information technology, digital signal processing, wireless technology, internet technology, internet security, etc.) were established to support these evolving new technologies. Many private and government research labs funded by IC chip profits were established to support the semiconductor industry both within the United States and worldwide [2–6]. (*Americas* [2]; *European Union A-J* [3]; *European Union K-Z* [4]; *Europe Outside EU* [5]; *Asia and Australia* [6]).

1.4   Integrated Circuit Scaling and the Worldwide Rise in the Standard of Living

The increase in the standard of living across the world enabled by integrated circuits is truly mind boggling. Cell phones and color TVs can be found in even the most remote villages. The development of modern medical equipment such as MRIs, EKGs, CAT scans, and implanted pacemakers would not have been possible without ICs. ICs run most modern household appliances. LEDs and LCDs provide far more light using far less energy than incandescent light bulbs. The agricultural revolution that has reduced hunger in the world to a level that is lower than ever before in recorded history was enabled by ICs. Without the increase in human productivity enabled by ICs, the dramatic increase in the standard of living across the world would not have been possible.

1.5   Integrated Circuits: An Industry in Transition

Integrated circuit scaling has virtually stopped. The number of transistors that can be crammed on an IC chip and switched simultaneously is pushing physical limits. A silicon atom is about 0.2 nm. Transistor gates have now scaled to less than 10 nm. The resistance and capacitance of the highly scaled, closely spaced wires is limiting how fast ICs can run. The heat generated by the billions of transistors is pushing reliability limits. The thickness of transistor gate dielectrics has scaled to just a few layers of dielectric molecules.

Despite the end of IC scaling, the semiconductor industry is still going strong. The demand for ICs able to perform an even wider variety of functions is still on the rise. Integrated circuit factories, each costing ten billion dollars or more, are still being built—most of them in Asia. In the USA, the focus of semiconductor companies is changing from research and development of equipment and processes that can manufacture smaller transistors to a focus on research and development (R&D) of transistors that can switch higher voltages and more power, R&D into new sensors and other new electronic components, and R&D into new packaging methods and procedures that reduce cost, reduce size, and increase the number of functions packaged ICs can perform. Some of the new packaging methods include wafer scale packages where IC chips skip the usual packaging processing and are mounted directly onto circuit boards after the wafer is diced. Other new packaging methods combine multiple dies with widely different capabilities into one package avoiding mounting the multiple dies separately on a circuit board thus saving area and reducing cost.

Other industries such as biomedical, aerospace, oil and gas, transportation, agriculture and a host of others are taking advantage of new wide-ranging IC capabilities to enhance the existing products and to develop and offer new products. Research into areas such as the internet of things, robotics, artificial intelligence, bio transistors, wearable electronics, smart prosthetics, self-driving vehicles, and mega data ensures the demand for ICs with new and increased capabilities will continue for some time.

References

1. Samsung's absurd 219-inch TV takes up an entire wall—thus its name, The Wall, Business Insider (2019). https://www.businessinsider.com/samsung-219-inch-the-wall-tv-2019-1.
2. Semiconductors-Industries Website, Semiconductor Research Laboratories of North (USA & CANADA), Central and South America. http://www.semiconductors.co.uk/research_laboratories/americas.htm. Accessed 4 Mar 2020.
3. Semiconductors-Industries Website, Semiconductor Research Laboratories within the European Union Countries A to J. http://www.semiconductors.co.uk/research_laboratories/europe-eu_a-to-j.htm. Accessed 4 Mar 2020.
4. Semiconductors-Industries Website, Research Laboratories within the European Union Countries K to Z. http://www.semiconductors.co.uk/research_laboratories/europe-eu_k-to-z.htm. Accessed 4 Mar 2020.
5. Semiconductors-Industries Website, European Semiconductor Research Laboratories Outside the European Union. http://www.semiconductors.co.uk/research_laboratories/europe-noneu.htm. Accessed 4 Mar 2020.
6. Semiconductors-Industries Website, Semiconductor Research Laboratories in Asia and Australasia. http://www.semiconductors.co.uk/research_laboratories/asia&australasia.htm. Accessed 4 Mar 2020.

Chapter 2
Overview of Integrated Circuit Manufacturing

2.1 Integrated Circuit Manufacturing

To manufacture an integrated circuit, a silicon wafer is typically processed 24 h a day, 7 days a week, for 2–3 months. One high-volume integrated circuit fab costs upwards of 10 billion dollars to build and equip. The fab costs over five million dollars per day just to operate. Most of the machines in an IC fab cost over one million dollars each. Multiple machines are required for each manufacturing step. The most expensive machine is the immersion scanner which costs over 100 million dollars for just one. These machines print the extremely small and complicated patterns on the wafer [1]. https://en.wikipedia.org/wiki/Semiconductor_fabrication_plant.

Figure 2.1 is a cross-sectional view of a simple complementary metal-oxidesemiconductor (CMOS) integrated circuit built in single crystal silicon substrate. The CMOS IC uses both NMOS (n-type channel where current is negatively charged electrons) and PMOS (p-type channel where current is positively charged holes) transistors. The NMOS and PMOS transistors are isolated from each other by a shallow trench (STI) filled with silicon dioxide (SiO_2) dielectric. The NMOS and PMOS transistors are covered with silicon dioxide (SiO_2) premetal dielectric (PMD). Metallic contact plugs electrically connect the NMOS and PMOS transistors to the overlying copper metal1 wiring layer. A first layer of intermetal dielectric (IMD1) electrically isolates the metal1 wires from each other. A second layer of intermetal dielectric (IMD2) electrically isolates a second layer of copper wiring (metal2) from the metal1 copper wires. Copper filled holes (via1) electrically connect metal2 wires to the underlying metal1 wires. A third layer of metal wiring (metal3) is covered with a protective overcoat (PO) dielectric such as SiO_2 or polyimide. Three levels of copper wiring are shown. Some ICs have a dozen or more levels of copper wiring. An aluminum bond pad, shorted to the top layer of metal3, is formed through an opening in the PO to provide electrical connection to the underlying transistors. A wire attached to the aluminum bond pad (not shown)

© Springer Nature Switzerland AG 2020
H. Tigelaar, *How Transistor Area Shrank by 1 Million Fold*,
https://doi.org/10.1007/978-3-030-40021-7_2

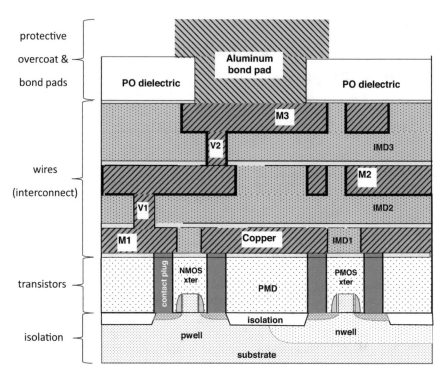

Fig. 2.1 Cross section of a CMOS integrated circuit with three layers of metal wires (interconnect) plus a metal bond pad layer

communicates electrical signals to and from this IC to other ICs, to other electrical devices, and to the power supply.

The integrated circuit manufacturing process starts with a bare silicon wafer (substrate). The integrated circuit is manufactured by repeatedly applying a photoresist coating on the wafer, printing a pattern in the photoresist, and then performing manufacturing steps such as plasma etching to transfer the photoresist pattern onto electronic structures on the wafer. For example, after the PMD dielectric layer is deposited it is coated with photoresist and a contact hole pattern is printed in the photoresist with contact holes penetrating through the photoresist. An etching step is then performed that etches contact holes through the PMD layer. The contact photoresist pattern is then removed, and the contact holes are filled with metallic contact plugs. The IMD1 dielectric layer is then deposited covering the contact plugs and the PMD layer. A photoresist coating is then applied on the IMD1 layer and a metal1 trench pattern is printed in the photoresist with metal1 trenches penetrating through the photoresist. The metal1 trenches are then etched through the IMD1 layer and filled with copper to form the metal1 copper wires. These steps of IMD_x deposition, forming a $metal_x$ trench pattern in photoresist, etching the $metal_x$ trenches and refilling them with copper are repeated for additional wiring layers.

2.1.1 *Photolithography (Printing Patterns in Photoresist)*

Before digital photography, a camera was used to capture pictures on film. The film was then developed to form a negative. The negative was put into a projection printer and light was projected through the negative to expose photo reactive chemicals on photographic paper. The exposed photographic paper was then developed with chemicals to produce the photograph. In semiconductor processing the surface of the wafer is coated with a photoactive polymer (photoresist). Portions of the photoresist are exposed by light that is projected onto the photoresist through a pattern (photomask or reticle). Photoresist polymer that is exposed by the light undergoes a chemical reaction that renders it soluble to a solvent (developer). After exposure, the developer washes away exposed photoresist leaving the unexposed photoresist pattern on the wafer. The photoresist geometries protect the underlying surface of the wafer from being changed by subsequent manufacturing processes. Some photo resist geometries in ICs manufactured today can be 20 nm or smaller. Over 1000 times smaller than a human hair.

The metal trench pattern openings in a photoresist pattern expose the underlying IMD layer. Where the IMD silicon dioxide (SiO_2) layer is exposed, reactive gases that are produced in a plasma react with the exposed SiO_2 and etch it away forming trenches through the IMD layer.

2.1.2 *Plasma Etch*

When an electric current flows through a gas, the electric current can dislodge electrons from the gaseous atoms forming highly reactive atoms. The electric current that flows through neon gas inside a neon tube knocks electrons from neutral (uncharged) neon gas atoms forming highly reactive positively charged neon gas ions (charged atoms). These highly reactive neon ions scavenge electrons as soon as possible to return the neons ion back into uncharged neon atoms. As the scavenged electrons neutralize the charge on the reactive neon ions, they emit light (photons) with the characteristic neon color. Florescent light is also produced by a plasma in the florescent tube.

In plasma etching, electric current flows through an etchant gas forming highly reactive etchant ions. One process that plasma etches the silicon dioxide (SiO_2) PMD and IMDx layers passes electric current through chloroform (CHF_3) and oxygen (O_2) to form highly reactive atomic fluorine, hydrogen, and oxygen atoms. These highly reactive atoms react with exposed silicon dioxide surface removing silicon and oxygen atoms. SiO_2 etch products are silicon tetrafluoride (SiF_4), carbon dioxide (CO_2), and water (H_2O) gas molecules. These gasses are pumped away through a vacuum pump.

Gases such as sulfur hexafluoridez(SF_6) or carbon tetrafluoride (CF_4) decompose in a plasma releasing highly reactive fluorine atoms that etch single crystal silicon and polysilicon. The reaction forms silicon tetrafluoride (SiF_4) gas which is pumped away.

2.1.3   Thermal Oxidation of Silicon to Grow Silicon Dioxide Dielectric

Silicon dioxide (SiO_2) dielectric is grown on exposed silicon surfaces by the reaction with oxygen (O_2) or water vapor (H_2O) at temperatures in the range of 800 –1200 °C.

In early technology nodes, SiO_2 was simultaneously grown on multiple wafers (batch processing) in quartz furnace tubes with oxygen or water vapor flowing through.

In advanced technology nodes, SiO_2 is grown on one wafer at a time (single wafer processing) in rapid thermal processing tools (RTP) using oxygen or water vapor. Gate dielectrics in advanced technologies are so thin (a few nanometers) that the gate dielectric can be grown in a few seconds or less. RTP tools can raise the temperature of a wafer from room temperature to 900 °C or more and return the wafer back to room temperature in less than a second (without it shattering due to thermal stresses!)

2.1.4   Chemical Vapor Deposition (CVD)

In a CVD process, gaseous molecules are decomposed, usually thermally or with the help of a plasma, to deposit dielectric or metallic thin film layers. Many different types of dielectric and metal thin films are deposited using a variety of CVD methods.

In thermal activated CVD, gaseous molecules hit the hot surface of a wafer, decompose, and react depositing a dielectric or metal thin film. For example, the gas tetraethyl orthosilicate (TEOS) decomposes thermally when it hits the hot (~700 C) surface of the wafer and deposits silicon dioxide (SiO_2) dielectric. TEOS decomposes into silicon dioxide (SiO_2) and diethyl ether. The SiO_2 solid deposits on the wafer. The diethyl ether gas is pumped away.

Silane (SiH_4) plus ammonia (NH_3) or dichlorosilane (SiH_2Cl_2) plus NH_3 thermally decompose to deposit silicon nitride dielectric (Si_3N_4) films.

High aspect ratio (narrow and deep) contact holes are filled with tungsten metal using CVD. Tungsten hexafluoride gas decomposes to form tungsten metal (W) plus fluorine gas (F_2). Tungsten metal fills the contact holes and fluorine gas is pumped away.

Vias are lined with CVD tantalum nitride (CVD-TaN) to prevent copper diffusion into the surrounding intermetal dielectric (IMD).

Many variations on the CVD process have been invented and developed to deposit various thin films over wide-ranging underlying topographies.

To enable engineers to build smaller and smaller transistors, chemists developed various CVD reactant gases to deposit thin films at lower and lower temperatures. For example, CVD depositions of chlorinated silanes such as SiH_3Cl, SiH_2Cl_2, and $SiHCl_3$ occur at lower temperatures than silane (SiH_4). CVD depositions of organosilanes such as tetraethyl orthosilicate (TEOS) occur at an even lower temperature.

Equipment scientists and engineers repeatedly redesigned CVD reactors to fill smaller and smaller trenches void-free as IC area scaled. Reactant gases have a longer mean free path (distance they travel before a collision) at lower pressures. Reactant gasses can better fill narrow trenches and have improved step coverage at lower pressure.

Some of the various CVD deposition methods are explained in more detail in Chap. 10. Included are low pressure CVD (LPCVD), sub atmospheric CVD (SACVD), atmospheric CVD (APCVD), high-density plasma CVD (HDP), high aspect ratio process CVD (HARP), and atomic layer deposition CVD (ALD).

New CVD variations with improved gap filling (such as flowable CVD) are being researched and developed.

APCVD (Atmospheric Pressure Chemical Vapor Deposition) Operating Pressure 1 atm (760 Torr). Temperatures 100–1000 °C. Wafers on a heated moving belt pass under the gas source. Equipment is inexpensive. Deposition rate is fast. Films tend to be lower quality and less uniform.

LPCVD (Low Pressure Chemical Vapor Deposition) Operating pressure between 0.25 and 2 Torr. Operating temperatures 300–900 °C. Many wafers are simultaneously processed in a batch in LPCVD furnace tube. Slow deposition rate is compensated by the large batch size. Good step coverage and good gap fill.

SACVD (Sub atmospheric Pressure Chemical Vapor Deposition) Operating pressure between 6 and 600 Torr. Good step coverage and good gap fill. Higher dep. rate than LPCVD. Denser dielectric film than LPCVD.

PECVD (Plasma Enhanced Chemical Vapor Deposition) Operating pressure 0.1–5 Torr. Low temperature deposition—usually about 400 °C. Plasma is used to lower the CVD deposition temperature. Higher deposition rates than thermal CVD techniques are achieved.

HDP (High Density Plasma Chemical Vapor Deposition) Ionizes argon atoms and accelerates them to sputter etch the dielectric surface as the dielectric film is depositing. Top flat surface sputters faster than bottom of trenches. Trenches fill void-free. Deposited dielectric film is dense and is largely planarized. A negative is that gate dielectrics can be degraded due to charge from the argon ions.

HARP™ (High Aspect Ratio Process Chemical Vapor Deposition) No plasma. Very good void-free gap fill of narrow trenches. Concentrations of TEOS vs ozone are adjusted throughout process to first fill small trenches and then speed up deposition rate to shorten deposition time.

ALD (Atomic Layer Chemical Vapor Deposition) First reactant gas decomposes on hot surfaces on the wafer to deposit a monolayer of atoms or molecules. Reaction is self-limiting. Second reactant gas decomposes on hot surface to deposit a second monolayer on top of the first monolayer. Monolayers are repeatedly deposited to build up the desired film thickness. https://web.stanford.edu/class/ee311/NOTES/Deposition_Planarization. pdf [2].

2.1.5 Physical Vapor Deposition (Sputter Deposition)

Metal thin films are usually deposited with sputtering. An argon (Ar) ion plasma is formed in the sputter deposition tool. Electric fields are used to accelerate the argon ions causing them to impinge on a metal target. High electric fields impart enough energy to the argon ions for them to dislodge (sputter) metal atoms and clusters of metal atoms from the target. The metal atoms and clusters rain down and deposit on the surface of the wafer forming a layer of the desired metal. Mirrors are formed by placing glass plates under an aluminum target in a sputter deposition tool and sputter depositing an aluminum metal film on them.

As ICs scaled scientists and engineers developed number of different sputter deposition processes to provide improved metal film uniformity and improved step coverage.

Sputter Deposition

I began my career at Texas Instruments as a sputter deposition engineer. When I hired into TI, I had not heard of sputter deposition. I did some reading prior to my start date so I knew what a sputter machine was but had never seen one. My first day I was put in charge of the sputter machine in the Houston Process Development Lab where TI was prototyping a 16 MEG DRAM. Luckily, I was assigned a very good and experienced technician. He introduced me to the Perkin Elmer 2400 and taught me how to run it. My job was to either quickly become a sputter deposition expert or to find another job.

I describe various manufacturing methods for sputter deposition used over the years to give you a feel for the incredible research and development (R&D) expense and effort that provided tools to keep transistor scaling moving forward. The equipment and other semiconductor manufacturing processes have similar or even more amazing stories to tell.

DC sputtering: A DC voltage accelerates argon ions (Ar+) into a plate (target) of the metal to be deposited. The argon ions are given sufficient energy to dislodge individual metal atoms and small clusters of metal atoms from the surface of the target. These atoms and clusters rain down and deposit on the wafer coating it with a thin film of the desired metal.

DC magnetron sputtering: Magnets behind the target confine and accelerate electrons in the plasma ionizing the Ar atoms more efficiently. Deposition rates are increased anywhere from 10 to 100 times depending upon the material.

RF sputtering: To sputter deposit insulating materials, RF energy from an RF generator behind the insulating target is coupled through the insulator target to create a plasma in the sputtering chamber. The RF provides sufficient energy to ionize the argon gas and sputter deposit the insulator.

Reactive sputtering: A reactive gas is introduced into the sputtering chamber along with the argon gas during sputter deposition. The reactive gas reacts with the target metal being sputtered to form the desired molecular compound. For example, nitrogen (N_2) gas is introduced into the sputtering chamber during sputter deposition of titanium to form titanium nitride (TiN) thin films (or during the deposition of tantalum to form tantalum nitride TaN).

Columnated sputtering: In columnated sputtering, a collimator (looks like a honeycomb) is placed under the target (between the target and the wafer). The collimator prevents atoms and clusters sputtered at oblique angles from reaching the wafer. Columnated sputtering improves bottom and sidewall coverage of high aspect ratio contact and via holes (small and deep).

IMP Sputtering: Sputtered metal atoms are given a positive charge and are accelerated with an electric field toward the wafer. Electric fields are also used to confine the charged metal atoms into a column directed toward the surface of the wafer. The high directionality of the IMP metal atoms enables them to reach the bottom of very deep and narrow contact and via holes and the bottom of deep and narrow damascene metal trenches.

SIP Sputtering: In self-ionized plasma sputtering (SIP) a specially designed magnetron increases the ionization in the plasma and increases resputtering of the metal from the surface. The resputtering results in a more uniform layer covering the bottom and sidewalls of vias and trenches.

IBS Sputtering: In ion beam sputtering an ion beam with mono energetic ions is created and accelerated toward the target. The ion beam is neutralized before it hits the target so that the metal or insulator is sputtered with neutral, high-energy atoms. IBS sputtering produces very dense, high-quality films. Handbook of Physical Vapor Deposition (PVD) Processing [3].

2.2   Integrated Circuit Transistors

In addition to low-voltage core CMOS transistors that perform the bulk of the digital logic, most integrated circuits also have high-voltage input/output (I/O) transistors. The I/O transistors translate the low-voltage signals from the core CMOS logic to the higher voltage levels needed for the IC chip to communicate with other IC chips such as those that run keyboards, video displays, speakers, and motors.

In addition to core CMOS transistors and I/O transistors, many integrated circuits also have bipolar transistors, analog transistors, various versions of high-voltage transistors, various versions of high-power transistors, electrostatic discharge (ESD) transistors, and memory transistors such as SRAMs, DRAMs, EPROMS, EEPROMS, and FLASH. The manufacturing processes required for these other types of transistors are similar to the manufacturing processes for core CMOS transistors. SRAM memories are built with the same process flow as the core CMOS transistors. Only a few additional steps are required to manufacture SRAM CMOS transistors. EPROM, FLASH, and DRAM memories on the other hand add about one-third more process steps to the core CMOS manufacturing flow. In some cases, specialized equipment and processes used only for the manufacture of integrated circuits with these memory cells are required.

2.2.1   Memory Transistors in Integrated Circuits

Static Random-Access Memory (SRAM)

The vast majority of SRAM memory cells are six transistor or 6 T SRAM cells. SRAM transistors are identical to core CMOS transistors except for their turn on voltages (V_T). An SRAM cell stores a logic state 1 or a logic state 0 while drawing very little power for as long as the SRAM memory is connected to a battery. Once disconnected, all logic information is lost (Volatile memory).

Dynamic Random-Access Memory (DRAM)

A DRAM memory cell is simply a capacitor connected to the source of an NMOS transistor. The DRAM capacitor can be either charged (logic state = 1) or discharged (logic state = 0). When the DRAM capacitor is charged to a logic state = 1, the source diode is reverse-biased and a small reverse-biased diode leakage current flows. The DRAM capacitor slowly discharges losing the logic state. In a DRAM memory cell, the logic information is dynamically refreshed (rewritten) every few milliseconds. When disconnected from the battery all logic information is lost (Volatile memory).

Erasable Programmable Read Only Memory (EPROM and FLASH)

EPROM and FLASH memories are nonvolatile memories. When the battery is disconnected, the logic states are preserved.

A FLASH memory is a version of an EPROM memory. An EPROM memory transistor is a NMOS transistor with a piece of polysilicon that is electrically isolated between the NMOS gate and the NMOS channel (see Figs. 2.2 and 2.3). During programming a high voltage is applied to the control gate and drain of the EPROM transistor. This turns the EPROM transistor on hard and generates channel hot carriers (CHC) (extremely energetic electrons) in the EPROM transistor channel. High voltage on the control gate diverts the path of some of these CHC electrons from the drain toward the floating gate. Some of the CHC electrons have enough energy to penetrate the gate dielectric between the floating gate and the EPROM transistor channel. These electrons get trapped on the floating gate giving the floating gate a negative charge (logic state 1). An EPROM transistor that has not been programmed (no charge trapped on the floating gate) has a logic state 0. These logic states remain programmed for 10 years or more even when no battery is attached! (This is how your car radio remembers your presets when your car battery goes dead.) To deprogram or erase the negative charge off the floating gate of an EPROM or FLASH transistor, a positive voltage is applied to the substrate and a negative voltage applied to the control gate with sufficient strength to create an electrical field that forces the trapped electrons to quantum mechanically tunnel off the floating gate.

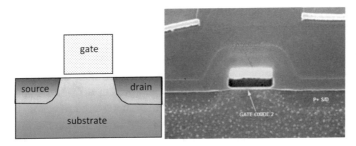

Fig. 2.2   Core NMOS transistor cross-sectional drawing (left) and SEM (right) [4]. http://smithsonianchips.si.edu/ice/cd/9504_407.pdf

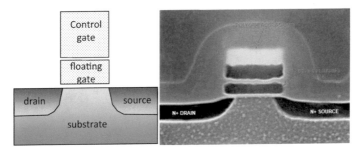

Fig. 2.3   EPROM transistor cross-sectional drawing (left) and SEM (right) [4]. http://smithsonianchips.si.edu/ice/cd/9504_407.pdf

2.3   Packaged Integrated Circuits

After the integrated circuit chips (dies) are manufactured on a wafer, the wafer is cut up into individual IC dies. The dies are then mounted on a lead frame and encased in molding compound to protect it.

2.3.1   Introduction to Packaged ICs

After the wafer is processed through the protective overcoat (PO) layer in the fab, it is sent to multiprobe testing where each die is electrically tested to determine if the dies are fully functional good electrical dies (GEDs). Nonfunctional and partially functional dies (non GEDs) are identified with an ink dot to indicate they are bad. The inked wafers are then sent to a packaging house (factory), where the wafers are cut apart into individual dies using saws or lasers. Inked dies are trashed.

2.3.2   Packaging Individual IC Dies

The packaging of good electrical IC dies (GEDs) is illustrated in Figs. 2.4 through 2.10. A GED (Fig. 2.4) is mounted on a lead frame (Fig. 2.5). Lead frames are typically made of copper or brass coated with a solderable metal such as nickel. Wire bonds (usually gold) electrically connect bond pads on the IC die to leads on the lead frame (Fig. 2.6). The lead frame mounted IC die is then placed into an injection mold and a plastic packaging or mold compound such as a filled epoxy resin is injected into the mold to encapsulate the IC and protect it from the environment. (Fig. 2.7). Portions of the lead frame leads remain uncovered to enable them to be soldered to leads on a circuit board. The final step in the packaging process is to remove the support frame from the lead frame (Fig. 2.8). A cross-sectional view of the packaged IC die is shown in Fig. 2.9. Figure 2.10 is a projection view of a commercial dual inline packaged (DIP) IC die.

Examples of other individual IC die packages are shown in Fig. 2.11.

Fig. 2.4 Good IC die

Fig. 2.5 Lead frame

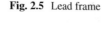

Fig. 2.6 GED mounted on
lead frame with bond wires

Fig. 2.7 GED and portion
of lead frame encapsulated
with plastic molding
compound

Fig. 2.8 Support frame
removed from
packaged IC die

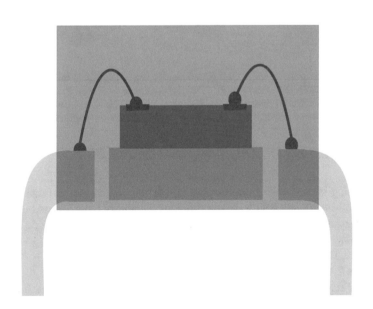

Fig. 2.9 Cross-sectional view of dual inline packaged IC die (DIP)

Fig. 2.10 Projection view of a commercial DIP

Fig. 2.11 Single die packaged ICs. QFP (Quad Flat Package), DIP (Dual Inline Package), BGA (Ball Grid Array), PGA (Pin Grid Array), SOP (Small Outline Package), PLCC (Plastic Leaded Chip Carrier), SSOP (Shrink Small Outline Package), TSOP (Thin Small Outline Package), and PQFP (Plastic Quad Flat Package) [5]. https://www.alibaba.com/showroom/rtl8019as.html

2.3.3 Packaging Multiple Integrated Circuit Dies

IC Packaging is evolving at an amazing rate to provide increased functionality at lower cost. Multiple dies with widely differing capability are packaged together in a single package [4]. These multi-die packages reduce area and provide increased functionality.

Figure 2.11 are examples of a few of the single die packages.

Figures 2.12 and 2.13 are examples of multiple die packages with the molding compound removed. A packaging machine stacks the dies on top of each other and then attaches a bond wire from each bond pad on the dies to a lead on the lead frame. Encapsulating the dies and the bond wires in plastic without shorting the wires together is just one of the packaging challenges.

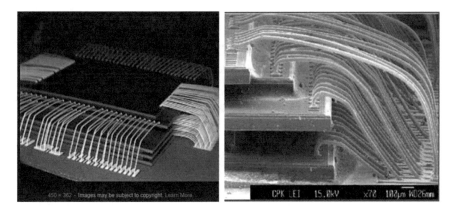

Fig. 2.12 Wire bonding complexity in multiple stacked die packages [6–8]. https://www.toshiba.
co.jp/about/press/2004_01/pr2101.htm.   https://www.palomartechnologies.com/applications/
stacked-die

Fig. 2.13 Packaged stacked IC dies in a Low-profile Quad Flat Package (LQFP) [9]. https://
c44f5d406df450f4a66b-1b94a87d576253d9446df0a9ca62e142.ssl.cf2.rackcdn.com/2018/02/
3DStacked_Die_TS104.pdf

References

1. Semiconductor Fabrication Plant, Wikipedia, https://en.wikipedia.org/wiki/Semiconductor_
   fabrication_plant. Accessed 4 Mar 2020
2. Prof. K. Saraswat, Thinfilm Deposition and Planarization, EE3111 Class Notes, Stanford
   University,   https://web.stanford.edu/class/ee311/NOTES/Deposition_Planarization.pdf.
   Accessed 4 Mar 2020
3. D. Mattox, *Handbook of Physical Vapor Deposition (PVD) Processes* (Elsevier Inc.,
   Amsterdam,   2010).   https://www.elsevier.com/books/handbook-of-physical-vapor-deposi-
   tion-pvd-processing/mattox/978-0-8155-2037-5?countrycode=US&format=print&campa
   ign_source=google_ads&campaign_medium=paid:search&campaign_name=usashopping&g-
   clid=CjwKCAiA1fnxBRBBEiwAVUouUiKqASl7dd4MZIH4lXImIYC5fMX97lDnOfMr3tZ
   Kq6uq2zNZGN_pOhoCz8UQAvD_BwE

4. Construction Analysis—ISSI IS27HC010 1Mbit UVEPROM, ICE Report SCA9504–407, http://smithsonianchips.si.edu/ice/cd/9504_407.pdf
5. Alibaba.com Website, https://www.alibaba.com/showroom/rtl8019as.html. Accessed 4 Mar 2020
6. F. Carson, G. Narvaez, HC Choi, DW Son, Stack Die CSP Interconnect Challenges, ChipPAC, Inc. IEEE/CPMT Seminar, http://www.ewh.ieee.org/soc/cpmt/presentations/cpmt0210b.pdf
7. Toshiba's New Multi Chip Package Stacks Nine Layers in a Package Only 1.4mm High, Toshiba Press Release (2004), https://www.toshiba.co.jp/about/press/2004_01/pr2101.htm
8. Stacked Die, Palomar Technologies, Website, https://www.palomartechnologies.com/applications/stacked-die. Accessed 4 Mar 2020
9. Technology Solutions, 3D and Stacked Die, Amcor Technology (2018), https://c44f5d406df450f4a66b-1b94a87d576253d9446df0a9ca62e142.ssl.cf2.rackcdn.com/2018/02/3DStacked_Die_TS104.pdf

Chapter 3
Semiconductors, Diodes, Transistors, and Inverters

3.1 Insulators, Conductors, and Semiconductors

Insulators do not conduct electricity. Put a battery across and insulator and no current flows. Insulators have extremely high resistance to the flow of electric current. Excellent insulators include glass, diamond, wood, and most plastics.

Current flows through conductors with little resistance. Excellent conductors include heavily doped silicon and metals such as aluminum, gold, copper, and silver. Only a very small amount of current flows when a battery is connected across a semiconductor. Semiconductors include single crystal silicon, single crystal gallium arsenide (GaAs), single crystal indium phosphide (IP), single crystal gallium nitride (GaN), and others.

3.2 Chemistry of Silicon and Carbon

The atomic elements used in silicon-based integrated circuits are single crystal silicon (Si) and dopant atoms boron (B), phosphorus (P), and arsenic (As). Single crystal silicon is the semiconductor substrate in which all silicon-based transistors and integrated circuits are built. Conductive paths for electrical current are formed in the single crystal silicon replacing some of the silicon atoms in the silicon crystal lattice with dopant atoms. Single crystal silicon with some of the silicon atoms replaced with phosphorus and/or arsenic dopant atoms (silicon doped n-type) provide conductive channels through which negative electrical current (negatively charged electrons) flow. Single crystal silicon with some of the silicon atoms replaced with boron dopant atoms provides conductive channels through which positive electrical current (positively charged holes) flow.

As shown in Fig. 3.1, carbon (C) and silicon (Si) are in the Group IV column of the periodic table. Elements in the Group IV column have four electrons in their

© Springer Nature Switzerland AG 2020
H. Tigelaar, *How Transistor Area Shrank by 1 Million Fold*,
https://doi.org/10.1007/978-3-030-40021-7_3

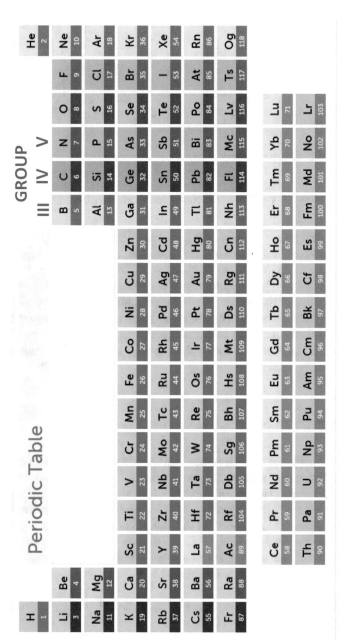

Fig. 3.1 Periodic table of the elements [1] https://www.acs.org/content/acs/en/education/whatischemistry/periodictable/periodic-table-gallery.html

outer shell. These four electrons can form covalent (shared electron) bonds either with themselves to form carbon crystals and silicon crystals or can form bonds with other atoms to form molecules such as methane (CH_4) and silane (SiH_4). In methane (CH_4), one carbon atom is covalently bonded to four hydrogen atoms. Each hydrogen atom (hydrogen is in Group I of the periodic table) has one electron to share and form one covalent bond. In carbon dioxide (CO_2), one carbon atom is covalently bonded to two oxygen atoms. Each oxygen atom shares two electrons and forms a double bond (two covalent bonds) with the carbon atom.

3.3   Single Crystal Carbon Insulator (Diamond)

Diamond (single crystal carbon) is formed of pure carbon. Each carbon atom shares its four electrons with four other carbon atoms forming four shared carbon-to-carbon covalent bonds (see Fig. 3.2). Carbon-to-carbon covalent bonds are very strong. When a strong electric field is applied across a diamond, no electrons are dislodged from the bonds and no electrical current flows. Single crystal carbon (diamond) is an excellent electrical insulator (dielectric).

3.4   Single Crystal Silicon Semiconductor

Like carbon, silicon (Si) is in the group IV column and has four electrons in its outer shell available for forming covalent bonds. Single crystal silicon is formed when each silicon atom forms covalent bonds to four other silicon atoms (see Fig. 3.3).

Fig. 3.2 Diamond crystal

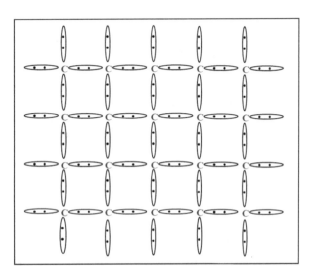

Fig. 3.3 Silicon crystal

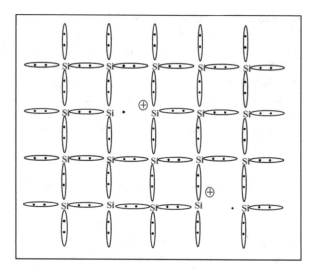

   Silicon-to-silicon covalent bonds are not as strong as carbon-to-carbon covalent
bonds. When a strong electric field is applied across single crystal silicon, a few of
the covalent bonds break and release electrons. When a strong electric field is
applied across a silicon crystal, a weak electrical current flows. Single crystal silicon is classified as a semiconductor because it conducts electricity poorly.

3.5   Low Resistance Doped Single Crystal Silicon

3.5.1   N-Type Doped Single Crystal Silicon Semiconductor

The element phosphorus (P) is in the Group V column of the periodic table. It has
five electrons in the outer shell available to form covalent bonds with other elements.
By replacing some of the silicon atoms in a silicon crystal with phosphorus atoms,
the silicon can be n-type doped and made electrically conductive for negatively
charged electrons (see Fig. 3.4). Silicon is doped by implanting phosphorus atoms
into the single crystal silicon. An ion implanter accelerates a beam of phosphorus
atoms to near the speed of light and shoots (implants) them into the silicon crystal.
These implanted phosphorus atoms are then heated to 900+ °C causing some of them
to replace silicon atoms in the silicon single crystal (dopant activation). Four of the
five electrons in the outer shell of phosphorus form covalent bonds with adjacent silicon atoms in the single crystal lattice. The fifth electron is relatively free to move
about. When an electric field is applied across phosphorus doped silicon, these electrons flow relatively freely (electron current flow). Higher concentrations of phosphorus atoms in the single crystal silicon lattice lower the electrical resistance
allowing the electrons to move more freely. Arsenic (As), which is also in the Group
V column, and has five electrons in the outer shell, is another n-type dopant commonly used to form n-type conductive pathways in silicon based integrated circuits.

Fig. 3.4 Phosphorus
doped single crystal silicon

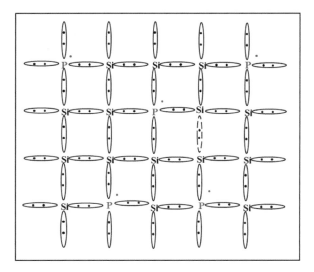

Fig. 3.5 Boron doped
single crystal silicon

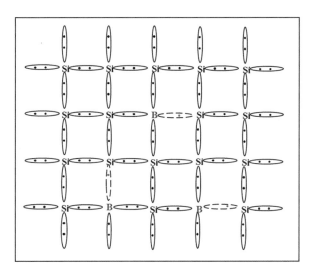

3.5.2 P-Type Doped Single Crystal Silicon Semiconductor

The element Boron (B) is in the Group III column of the periodic table. It has three electrons in the outer shell available for forming covalent bonds with other elements. Boron atoms can be implanted into single crystal silicon and then annealed at 900+ °C to replace silicon atoms in the silicon single crystal. The three electrons in the outer shell of boron form covalent bonds to adjacent silicon atoms (see Fig. 3.5). A "hole" is left in the silicon lattice where an electron is missing between the boron atom and the fourth silicon atom. This hole acts as a positive charge carrier (p-type). When an electric field is applied across boron doped silicon, an electron from an adjacent covalent bond moves to fill the vacant hole creating a new hole where the electron vacated the adjacent covalent bond. Under the influence of the electric field, this positively charged hole moves through the single crystal silicon lattice. Holes move at about one-third the speed (have one-third the mobility) of electrons. The higher the concentration of boron dopant atoms, the lower the electrical resistance to hole current.

3.6 PN-Junction Diodes [2]

When n-type doped single crystal silicon is brought into contact with p-type doped single crystal silicon, a pn-junction diode is formed. A pn-diode (diode) readily passes current when forward-biased (negative pole of battery connected to the n-type side of the diode and positive pole of the battery connected to the p-type side of the diode), but passes only a small leakage current when reverse-biased (negative pole of battery connected to the p-type side of the diode and positive pole of the battery connected to the n-type side of the diode).

Figure 3.6 is the electrical symbol for a pn-junction diode. Figure 3.7 is a cartoon of a pn-junction diode. Figure 3.8 is the electrical symbol for a voltage source (battery).

3.6.1 Forward-Biased Diodes

Figure 3.9 shows a forward-biased pn-diode. The electric field drives the electrons (negative carriers) and the holes (positive carriers) from the battery terminals to the interface between the n-type and p-type silicon (pn-junction). The electrons and holes recombine at the pn-junction eliminating both the electron and the hole charge

Fig. 3.6 PN-diode symbol

Fig. 3.7 PN-diode cartoon

Fig. 3.8 Battery cartoon

Fig. 3.9 Forward-biased
pn-diode

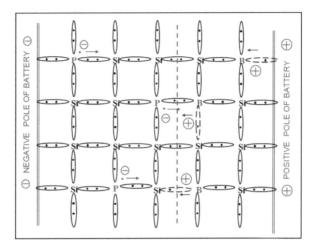

carriers. The electrons and holes keep recombining and electric current keeps flowing as long as the battery terminals remains connected and as long as the battery is not completely drained of electrons and holes. As shown in the Fig. 3.10 graph, a small forward bias voltage (positive voltage) across the pn-junction results in a substantial current flow.

3.6.2 Reverse-Biased Diodes

Figure 3.11 shows a reverse-biased pn-junction diode where the negative pole of the battery is connected to the p-type doped single crystal silicon side of the pn-junction diode, and the positive pole of the battery is connected to the n-type doped single crystal side of the pn-junction diode. The electric field drives electrons (negative carriers) away from the pn-junction to the positive pole of the battery. These electrons (from the n-type single crystal silicon) combine with holes from the positive pole of the battery. The electric field also drives the holes (positive carriers from the p-type silicon) away from the pn-junction to the negative pole of the battery. These holes combine with electrons from the negative pole of the battery. The n-type and p-type doped single crystal silicon adjacent to the pn-junction becomes depleted of electron and hole carriers (depletion region). Once this region is depleted of carriers, little current flows. As shown by the blue line in the Fig. 3.12 graph, only a small leakage current flows as the voltage is raised negative across a reverse-biased pn-junction.

The reverse-biased voltage is dropped across the depletion region which is devoid of carriers. This is analogous to the voltage dropped across a capacitor dielectric which also is devoid of carriers. In integrated circuits, reverse-biased diodes are capacitors. The depletion region is the capacitor dielectric of the diode capacitor.

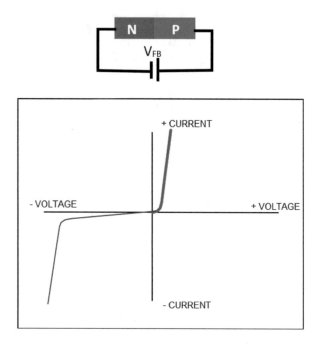

Fig. 3.10 Current vs. Voltage curve for forward-biased diode

Fig. 3.11 Reversed-biased pn-diode

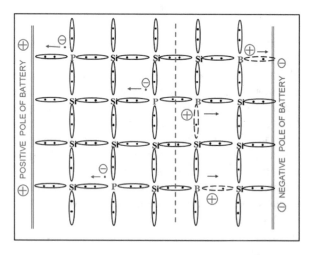

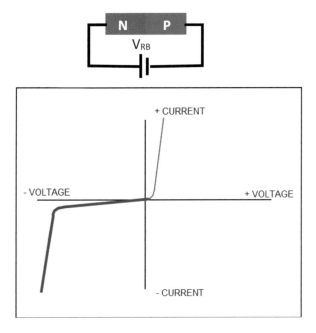

Fig. 3.12   Current vs. Voltage curve for reverse-biased diode

The charge that is depleted from the n-type side of the pn-junction is equal to the charge depleted from the p-type side of the pn-junction. If the n-type doping is the same concentration as the p-type doping, then the width of the n-type depletion region is equal to the width of the p-type depletion region. If the n-type doping is heavier than the p-type doping in the pn-junction, then the width of the n-type depletion region is narrower than the width of the p-type depletion region (number of n-type carriers depleted from n-type region must equal the number of p-type carriers depleted from the p-type region). It is desirable for the drain diodes of NMOS and PMOS transistors to be as heavily doped as possible to provide low series resistance to the MOS transistor. The source and drains are doped as heavily as possible. It is also desirable for the drain diodes of NMOS and PMOS transistors to have as low capacitance as possible (fewer electrons are needed to charge low capacitance). For this reason, the source and drain diodes are graded with a high concentration of dopant at the surface for low series resistance and light doping below the surface for reduced capacitance.

Substantial current eventually flows when the reverse-biased voltage is raised so high that the diode breaks down (Fig. 3.12 at high negative voltage). Breakdown occurs when the reverse biased electric field is so intense that it breaks covalent bonds creating electron-hole pairs (electrons and holes that are free to move). The free electrons, accelerated by the electric field, hit and break additional covalent bonds creating additional electron-hole pairs. This chain reaction breaks even more covalent bonds causing high current to flow that destroys the diode (avalanche breakdown).

3.7   Metal on Silicon (MOS) Transistors [3]

3.7.1   NMOS Transistors

The electrical symbol for the NMOS transistor is given in Fig. 3.13. A cartoon of an NMOS transistor that is turned OFF is shown in Fig. 3.14. The substrate of the NMOS transistor is lightly doped p-type single crystal silicon (semiconductor silicon wafer). In the Figures, N− indicates lightly doped n-type silicon and N+ indicates heavily doped n-type silicon. A transistor gate dielectric such as silicon dioxide is grown on the substrate. A conductive transistor gate (historically doped polysilicon but now sometimes metal) is formed on the gate dielectric. Polysilicon gates (composed of randomly oriented silicon mini crystals) are formed by depositing and etching doped polysilicon. A capacitor is formed across the gate dielectric between the conductive transistor gate and conductive p-type substrate. An n-type dopant such as phosphorus and/or arsenic is implanted adjacent to both sides of the gate to form the n-type doped source (current comes to the transistor) and drain (current leaves the transistor) of the NMOS transistor. The n-type source and drains are pn-diodes in the p-type substrate.

NMOS Transistor Turned OFF

When a positive power supply voltage is applied to the drain and 0 V to the source, gate, and substrate, the NMOS transistor is turned OFF (Fig. 3.14). Since the gate and substrate are at the same potential, there is no electric field across the gate dielectric and no channel is formed in the substrate under the gate. Consequently, no current flows between the source and drain. Only a small reverse-biased diode leakage current flows across the pn-diode formed between the n-type drain and the p-type substrate.

NMOS Transistor Turned ON

The NMOS transistor is turned on when a positive power supply voltage is applied to the gate and drain, and the source and substrate are at 0 V (Fig. 3.15). When a positive voltage (+1.5 V) is applied to the transistor gate, the voltage drop between

Fig. 3.13 NMOS
transistor symbol

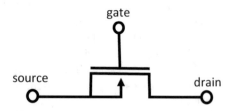

Fig. 3.14 NMOS
transistor cartoon

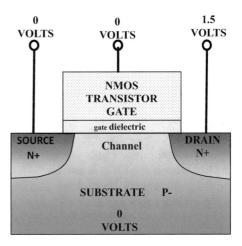

Fig. 3.15 NMOS
transistor cartoon with
NMOS transistor
turned ON

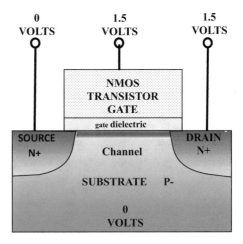

the gate and substrate creates an electric field across the gate dielectric. This electric field attracts mobile electrons in the p-substrate to the channel region under the gate (Fig. 3.15). When a sufficiently high positive voltage is applied to the gate, a greater number of electrons than holes are gathered into the channel region. This turns the channel region from p-type silicon to n-type silicon. With the channel region n-type, a continuous n-type single crystal silicon channel exists between the source and drain. Electron current freely flows from source to drain through this channel. The gate voltage at which source to drain current (transistor current) starts flowing is the turn ON voltage (V_{TN}) of the NMOS transistor. Small changes in the NMOS gate voltage cause large changes in NMOS transistor current. A small signal on the gate is amplified by the NMOS transistor.

3.7.2   PMOS Transistors

The electrical symbol for the PMOS transistor is given in Fig. 3.16. A cartoon of a PMOS transistor is shown in Fig. 3.17. The PMOS transistor structure is the same as the NMOS transistor structure except the source, drain, and substrate have opposite doping. The result is that PMOS transistors are turned ON by negative voltage and OFF by positive voltage, whereas NMOS transistors are turned ON by positive voltage and OFF by negative volts. The PMOS transistor is formed in a lightly phosphorus doped n-type single crystal silicon substrate (Fig. 3.17). A gate dielectric such as silicon dioxide is grown on the n-type substrate. A transistor gate is formed on the gate dielectric by depositing and etching doped polysilicon. (The doping of the polysilicon gate can be either n-type or p-type depending upon the desired turn on voltage V_{TP} of the PMOS transistor.) A p-type dopant such as boron (B) and/or boron difluoride (BF_2) is implanted adjacent to both sides of the PMOS gate to form the p-type source and drain. The p-type source and drain diffusions form pn-diodes with the n-type substrate.

PMOS Transistor Turned OFF

The PMOS transistor is turned OFF when positive voltage, is on the drain, gate, and substrate and 0 V is on the source (Fig. 3.17) Because the gate and the substrate are at the same positive potential, there is no electric field across the gate dielectric. Consequently, there is no channel formed under the gate and no current flowing from the drain to the source. Only a small reverse-biased diode leakage current flows through the pn-diode formed between the p-type drain and the n-type substrate.

PMOS Transistor Turned ON

The PMOS transistor is turned ON when a voltage more negative than the substrate, in this case 0 V is applied to the PMOS transistor gate (Fig. 3.18). When 0 voltage is applied to the transistor gate, the voltage drop between the 0 V on the gate and the +1.5 V on the substrate creates an electric field across the gate dielectric. This negative electric field repels mobile electrons from the channel region under the gate dielectric. When a sufficiently high negative voltage (less positive than the +1.5 V on the NWELL) is applied to the gate, enough electrons are repelled from the

Fig. 3.16 PMOS
transistor symbol

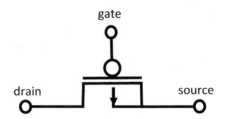

Fig. 3.17 PMOS
transistor cartoon

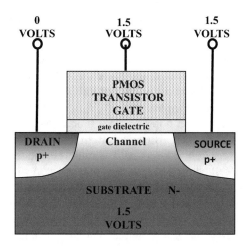

Fig. 3.18 PMOS transistor
cartoon with PMOS
transistor turned ON

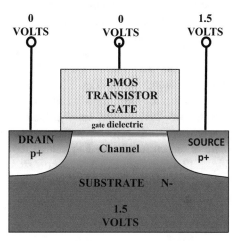

channel region to turn the channel from n-type to p-type single crystal silicon
(Fig. 3.18). The continuous p-type channel region between the p-type source and
p-type drain enables hole current to freely flow. The gate voltage at which source to
drain current starts flowing is the PMOS transistor turn ON voltage (V_{TP}) of the
transistor. A small signal on the PMOS gate is amplified by the PMOS transistor.

3.8 CMOS Inverter [4]

The CMOS inverter is the heart of low-power integrated circuits. The CMOS
inverter stores and switches logic states while using very little power. The CMOS
inverter is the key to the long battery life in digital electronics. Only reverse-biased
diode leakage current flows when an inverter is storing a logic state. Only the current required to charge or discharge the transistor drain capacitor flows when an
inverter switches its logic state.

3.8.1 Inverter Circuit: Two Switches in Series

The circuit symbol for a CMOS inverter is shown in Fig. 3.19.

The circuit diagram for a CMOS inverter is given in Fig. 3.20. It consists of one NMOS transistor and one PMOS transistor connected in series between the power supply (V_{DD}) and ground (V_{SS}). The importance of the CMOS inverter is that it can store either a logic state = 0 or logic state = 1 using very little power. The CMOS inverter can also switch from logic state = 0 to logic state = 1 or vice versa using very little power. CMOS circuits are the reason we have portable electronic devices (cell phones, fit bits, mp3 players, etc.) with long battery lives.

Figures 3.21 and 3.22 show a schematic of an inverter that is storing a logic state 0 (0 V) and storing a logic state 1 (1.5 V). The inverter consists of two switches wired in series between a 1.5 V power supply and ground (0 V). The two switches are coupled together, so that both switches flip at the same time. When one of the switches is ON and the other switch is always OFF.

In Fig. 3.21, the lower switch is closed shorting the storage node to ground (0 V). Logic state = 0 (0 V) is stored by the inverter. Since the upper switch is open, no current flows while the logic state = 0 is being stored.

When the two switches are simultaneously flipped, the upper switch closes and the lower switch opens (Fig. 3.22). The lower switch opens removing the connection between the storage node and the ground. The upper switch closes shorting the storage node to the power supply. The power supply charges the storage node to the power supply voltage (1.5 V). The only current that flows during switching the logic state is the current needed to charge the storage node capacitance. As shown in Fig. 3.22, logic state 1 (1.5 V) is now stored by the inverter. Since the lower switch is open, no current flows while the logic state = 1 is being stored. Since no current

Fig. 3.19 CMOS inverter circuit symbol

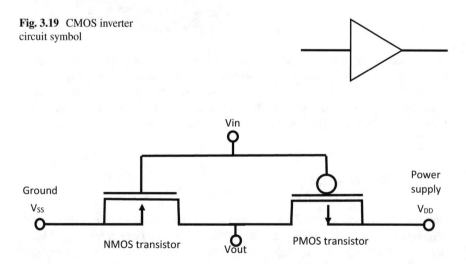

Fig. 3.20 CMOS inverter circuit diagram

Fig. 3.21 Inverter storing
logic state = 0 (0 V)

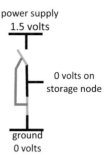

power supply
1.5 volts

0 volts on
storage node

ground
0 volts

Fig. 3.22 Inverter storing
logic state = 1 (1.5 V)

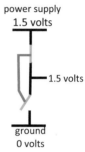

power supply
1.5 volts

1.5 volts

ground
0 volts

(except for diode leakage current) flows when the logic states are stored and very little current flows when the logic state is switched, CMOS circuits can run for a long time on battery power.

3.8.2   CMOS Inverter Structure in Silicon

Figure 3.23 shows a cross section of a CMOS inverter built in a p-type silicon substrate. The CMOS inverter consists of one PMOS transistor and one NMOS transistor connected in series between the positive terminal of a power supply (V_{DD}) and the negative terminal (Vss) which usually is ground. The source of the PMOS transistor is connected to V_{DD}. The source of the NMOS transistor is connected to Vss or ground. The gates of the NMOS and PMOS transistors are shorted together. The shorted gates are the inverter input node (V_{IN}). The drains of the NMOS and PMOS transistors are also shorted together. The shorted drains are the inverter output node (V_{OUT}) where the logic state (1 or 0) is stored (inverter storage node). The drain diode capacitance is charged to V_{DD} when the inverter stores logic state = 1 and is discharged to Vss or ground when the inverter stores logic state = 0.

In order to build the PMOS transistor on the p-type silicon substrate, the p-type silicon where the PMOS transistor is to be built is counter doped with sufficient phosphorus to overwhelm the light p-type doping and change the doping of the substrate from p-type silicon to n-type silicon (NWELL).

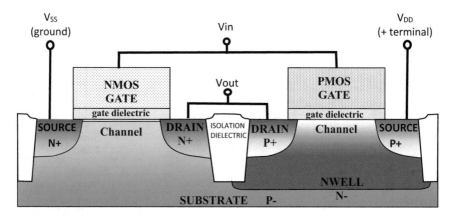

Fig. 3.23 CMOS inverter

Fig. 3.24 Inverter storing
logic state = 1

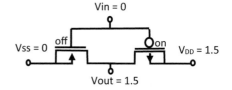

CMOS Inverter Storing a Logic State = 1

Figure 3.24 is a circuit diagram and Fig. 3.25 is a cross section of a CMOS inverter
storing a logic state 1 (1.5 V). The gates of the NMOS and PMOS transistors in the
CMOS inverter are shorted together. An input signal of 0 V (V_{IN} = 0 V) to the
CMOS transistor gates turns the PMOS transistor ON and the NMOS transistor
OFF. The drains of the NMOS and PMOS transistors are shorted together and form
the storage node of the CMOS inverter. The inverter output signal (V_{OUT}) is read
from the storage node. The ON PMOS transistor connects the storage node to V_{DD}
(power supply = 1.5 V). The storage node (V_{OUT}) is charged up to the power supply
voltage. When V_{IN} = 0 V (logic state 0), V_{OUT} = 1.5 V (logic state = 1). The output
signal (V_{OUT}) is opposite the input signal (V_{IN}). The CMOS inverter inverts the
logic signal.

CMOS Inverter in Silicon Storing a Logic State = 0

Figure 3.26 is a circuit diagram and Fig. 3.27 is a cross section of a CMOS inverter
storing a logic state = 0 (0 V). The gates of the NMOS and PMOS transistors in the
CMOS inverter are shorted together. An input signal of 1.5 V (V_{IN} = 1.5 V) to the
CMOS transistor gates turns the NMOS transistor ON and the PMOS transistor
OFF. The ON NMOS transistor connects the storage node to V_{SS} (ground = 0 V).

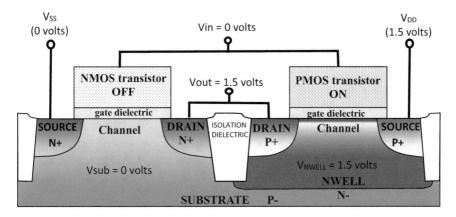

Fig. 3.25   CMOS inverter with $V_{IN} = 0$ V and $V_{OUT} = 1.5$ V

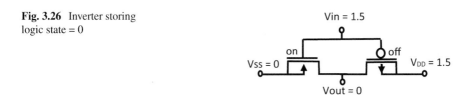

Fig. 3.26   Inverter storing logic state = 0

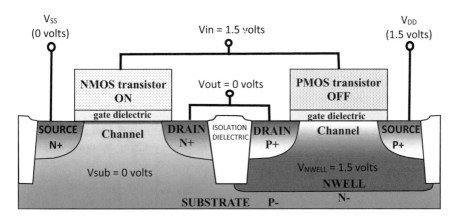

Fig. 3.27   CMOS inverter with $V_{IN} = 1.5$ V and $V_{OUT} = 0$ V

Any voltage on the storage node (V_{OUT}) is discharged to ground = 0 V. When $V_{IN} = 1.5$ V (logic state 1), $V_{OUT} = 0$ V (logic state = 0). The output signal (V_{OUT}) of a CMOS inverter circuit is opposite the input signal (V_{IN}). The CMOS inverter inverts logic signals.

As long as a power supply is connected to the CMOS inverter, the logic state is stored with no direct current path from power supply to ground. The only current that flows while the CMOS inverter is storing a logic state = 0 is diode leakage current through the reverse-biased NMOS transistor drain. The only current that flows out of the power supply while the CMOS inverter is storing logic state = 1 is diode leakage current through the reverse-biased PMOS transistor drain.

3.9   Logic Functions "AND" and "OR"

Two of the most commonly used logic functions in digital circuits are "AND" and "OR" logic functions. The circuit symbols for the "AND" logic and "OR" logic functions are given in Figs. 3.28 and 3.30.

The "AND" logic circuit symbol is shown in Fig. 3.28. The AND logic function is implemented by wiring two or more transistors in series as is illustrated in Fig. 3.29. The storage node on the CMOS inverter in Fig. 3.29 does not switch from logic state 1 (power supply voltage) to logic state 0 (ground) unless NMOS transistor 1 and NMOS transistor 2 are both ON. Transistors wired in series must all be ON for a signal to be propagated. Series transistors 1 and transistor 2 and transistor 3 ... must all be ON for a signal to be propagated.

Fig. 3.28 Logic AND circuit symbol

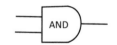

Fig. 3.29 CMOS inverter plus AND logic gate

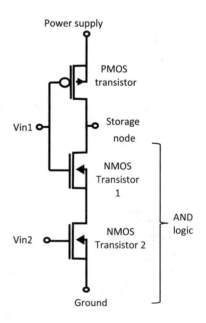

Fig. 3.30 Logic OR
circuit symbol

Fig. 3.31 CMOS inverter
plus OR logic gate

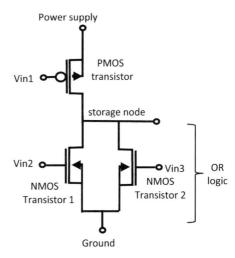

The "OR" logic circuit symbol is shown in Fig. 3.30. The OR logic function implemented by wiring two or more transistors in parallel as is illustrated in Fig. 3.31. The storage node on the CMOS inverter in Fig. 3.31 does not switch from logic state 1 (power supply voltage) to logic state 0 (ground) unless at least one of NMOS transistor 1 or NMOS transistor 2 are ON. For the logic state to change either transistor 1 or transistor 2 must be ON (or both can be ON). At least one of transistors wired in parallel must be ON for a signal to be propagated. One of the parallel transistors 1 or transistor 2 or transistor 3 ... must be ON for a signal to be propagated.

3.10   Digital and Analog MOS Transistors and Signals

Most signals in nature are analog. Analog signals are continuous. Analog signals can have any value. For example, an analog sound signal can have any value from total silence to full volume. An analog light signal can have any value from total darkness to full sun. Color can have any hue from total blackness to total whiteness. Before the advent of digital radios, analog radio signals were received and amplified in transistor radios using NMOS transistors. The problem with the amplifying the analog signals is that the NMOS transistors were constantly ON. An NMOS transistor that is constantly ON, constantly draws power from the battery. In the 1960s going to the beach meant one person in the group brought their transistor radio and others in the group each brought sets of D-cell batteries to keep the music going throughout the afternoon.

Analog audio signals are stored in grooves of vinyl records or on audio tapes such as eight-track tapes and cassette tapes. Analog video signals are stored on film such as 8 mm and 16 mm film. Prior to the digital age, some radio stations were difficult to hear because of static in the analog signal, and some TV stations were difficult to see because of snow and jitter in the analog signal.

3.10.1 Converting Analog Signals to Digital Signals [5]

The invention of digital signal processing integrated circuits (DSPs) revolutionized the radio, television, movie, and communications industry. We no longer had to contend with static and interference when listening to radio stations. We could now get clear, stable pictures on our new digital TVs. Digital Signal Processors manipulate the digital signals to correct transmission errors, to remove unwanted interfering audio and visual signals, and amplify only the desired audio and visual signals. DSPs do such things as remove scratches from digitized vinyl recordings, remove static from broadcast radio signals, remove snow and flicker from broadcast TV signals, brighten digitized photographs, and help create amazing animations and movie visual effects.

Analog signals such as from microphones and cameras need to be converted to digital signals before they can be manipulated by a DSP. Specialized integrated circuits called analog-to-digital converters (ADCs) were invented and perfected to perform this task. After analog signals are converted to digital signals in the ADC, the DSP can manipulate the digitized signals. After the DSP performs its magic on the digital signals, they must be converted back to analog signals to power such analog devices as speakers in radios and TVs, and the screens in our laptops, TVs, and printers. Specialized integrated circuits called digital-to-analog converters (DACs) were invented and perfected to perform this task. DSP, ADC, and DAC circuitry and software relies heavily upon information technology, sampling theory, statistical physics, and computation physics.

3.10.2 Digitizing Analog Signals

Depending upon how closely you want the digitized music to replicate the music coming out of the soundboard (analog music) or how closely you want the digitized picture to duplicate the photograph, you can choose the number of bits in the digitized version of the analog audio or analog visual signal. Points on the analog signal can be converted to digital numbers with 8 bits, 16 bits, 32 bits, 64 bits, resolution.

Once the analog signal is converted to a series of digital numbers, these digital numbers can be stored in digital memories such as thumb drives and DVDs and can be processed in DSPs.

The drain current from a transistor is shown in Fig. 3.32. The current continuously changes from 0 to 4.5 mA as the drain voltage is ramped from 0 to 1 V. In Fig. 3.33 the analog drain current signal is approximated by a 5-bit digital signal. The 5-bit digital signal in Fig. 3.33 is the result of feeding the analog signal in Fig. 3.32 through a 5-bit ADC. In Fig. 3.34 the analog drain current signal is approximated by an 11-bit digital signal. The 11-bit digital signal is the result of feeding the analog signal through a higher resolution 11-bit ADC.

Digital signals are stored in bytes. Each byte (also called an 8-bit word) has 8 bits, that is, eight 1s and 0s. Digital audio players are typically 8-bit (2^8), 16-bit (2^{16}), or 24 (2^{24}) bits. One 8-bit word can represent any number from 0 to 255. One 16-bit word can represent any number from 0 to 65,535. Digital signals are stored on CDs, DVDs, and in solid-state memories such as are found in solid-state hard drives, thumb drives, and memory cards.

3.11   Digital MOS Transistors

The majority of MOS transistors in central processing units (CPUs), microprocessors, microcontrollers, and other integrated circuits are digital CMOS transistors. Digital transistors are designed to switch between 1 and 0 logic states (switch between power supply voltage = V_{DD} and ground = $V_{SS} = 0$ V) as fast as possible. An inverter in a 4-GHz computer CPU switches logic states four billion times in each second! The current/voltage curves (I/V) of a digital MOS transistor are shown in Fig. 3.35. The gate voltage is stepped from 0 to 1.0 V in 0.25-V increments. For each gate voltage (0.25, 0.50, 0,75, 1.0), the source to drain current (I_{DS}) is plotted

Fig. 3.32 Transistor output current (analog)

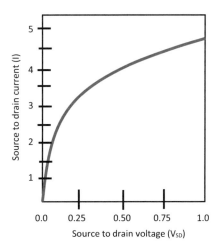

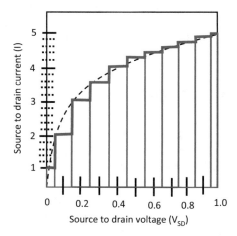

Fig. 3.33 Transistor output current (digitized with 5 bits)

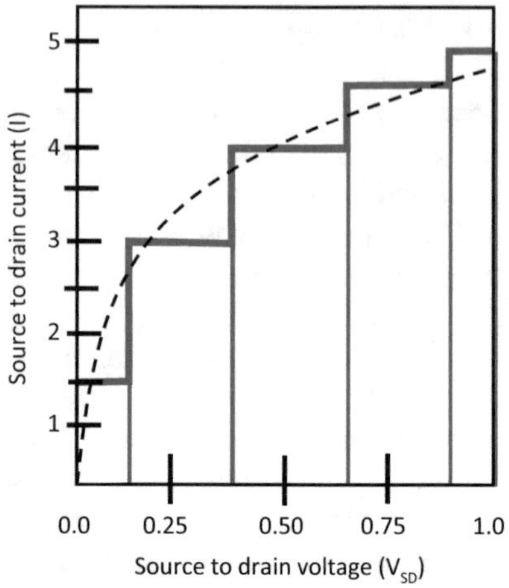

Fig. 3.34 Transistor output current (digitized with 11 bits)

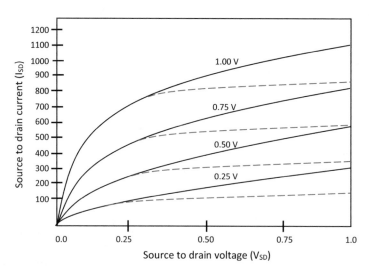

Fig. 3.35 Current vs. Voltage curves of a core digital NMOS transistor. Dashed lines indicate what the transistor current would be without channel resistance modulation (without halo implants)

as the drain voltage is ramped from 0 to 1.0 V (horizontal axis). In most digital logic integrated circuits, the source on an NMOS transistor is 0 V, the drain is either 0 V or V_{DD}, and the gate is switched between 0 V and V_{DD}. The gate voltage is switched from logic state zero (0 V) to logic state 1 (V_{DD}). The digital NMOS transistor is usually either OFF (substrate = source = gate = 0 V and drain = V_{DD}) or ON in full saturation (substrate = source = 0 V and gate = drain = V_{DD}). To enable the digital MOS transistor to switch as fast as possible, the transistor gate length is made as short as possible (minimum possible resistance between source and drain when the transistor is ON but still able to stop current from flowing between source and drain when transistor is OFF). At the 180 nm node, the transistor channel had scaled so short that the gate was no longer able to keep the transistor turned OFF. At the 180 nm node, the turn on voltage of the transistor (V_{TN}) dropped rapidly as the transistor channel got shorter (less voltage on the gate allowed current to flow between source and drain.) See the no halo curve in Fig. 3.36 and no halo transistor cartoon in Fig. 3.37.

3.11.1   Halos on Digital MOS Transistors

To reduce this short channel effect (V_{TN} roll off), engineers added halo (pocket) doping around the ends of the source and drain extensions (Fig. 3.38). Halo's are the same doping type as the substrate. Halo's are the opposite doping of the extensions. Halo implants overlap as the channel length gets shorter. Since halos are the same doping type as the substrate, this increases the doping in the channel under the gate. Increased substrate doping in the channel raises the V_T. The rising V_T caused by the overlapping halos counters the falling V_T caused by short channel effects. This enables the transistor engineers to scale the digital MOS transistors to a shorter minimum gate length. See halo V_{TN} roll off curve in Fig. 3.36 and transistor cartoon with pockets in Fig. 3.38.

The halos cause a modulation of the channel resistance as the drain voltage is ramped. This results in a significant positive slope on the I/V curves (called Early

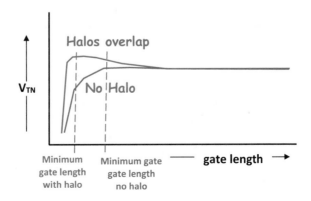

Fig. 3.36 V_{TN} roll off without and with a halo

Fig. 3.37 NMOS
transistor with no halos

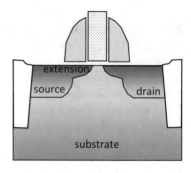

Fig. 3.38 NMOS
transistor with halos

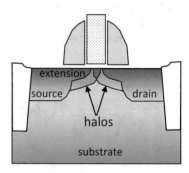

effect) as the drain voltage is ramped up (Fig. 3.35). The I/V curves of MOS transistors without pocket doping are relatively flat as indicated by the dashed curves in Fig. 3.35.

3.12   Analog MOS Transistors [6]

Analog MOS transistors are used to amplify weak analog signals and are used for circuits such as current mirrors, voltage references, and operational amplifiers in integrated circuits. The current vs voltage (I/V) curves of an analog MOS transistor are shown in Fig. 3.39. Linearity and transistor matching are the main concern of circuit design engineers for analog transistors that these designers use in digital IC circuits. Switching speed is less of a concern. To reduce the effect of noise in analog transistors and to provide better matching between analog transistors (so both analog transistors amplify the signal by the same amount), the gate length used for analog MOS transistors is longer than the gate length used for digital MOS transistors. The region of the I/V curve between 0 V on the drain and about 0.1 V on the drain (linear region) is primarily used for analog amplifiers. In this region a small change in the drain voltage produces a linear change in the drain to source current. Signals can be amplified with little distortion using this region.

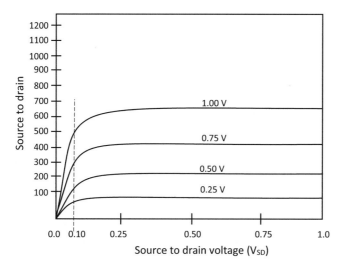

Fig. 3.39 Current vs. Voltage curves of an analog MOS transistor as gate voltage is stepped from 0 to 1 V in 0.25-V increments. The source to drain voltage is ramped from 0 to 1 V

For analog integrated circuits such as current mirrors, it is important for transistor current to be independent of the drain voltage in the saturation region. While the slope on the I/V curve of the digital MOS transistor in Fig. 3.35 is not a problem for digital integrated circuits, it is a killer problem for analog integrated circuits.

Analog MOS transistors can be manufactured using the same patterns and doping implants as the digital MOS transistor. Analog MOS transistors have longer gate lengths and lack halo doping. The I/V curves of an analog MOS transistor are shown in Fig. 3.39. Because the gate length is longer, the drive current is lower. Because the analog MOS transistor lacks halos, the slope of the IV curves in the saturation region is virtually flat.

180 nm Node Analog

At the 180 nm transistor node, halo implants were required for the first time to facilitate scaling to shorter gate lengths (Fig. 3.36). The halo implants caused a slope on the I/V curves (Fig. 3.35) making them unusable for analog designers. Until the 180 nm node, analog designers used long channel digital MOS transistors. Until the 180 nm node halo doping was not used. Analog designers could add their analog circuits to digital circuits with zero added cost. At the 180 nm node engineers in my group were tasked with providing transistors for the analog designers with no added manufacturing cost. For 3 months we ran experiments trying to adjust the halos and to flatten the I/V curves sufficiently for the analog design engineers. We held weekly meetings to review our progress. Every week we said: "You have to allow us to add two additional patterning steps to block the pocket implants from core NMOS and PMOS transistors to make your analog transistors. The extra patterning steps cost very little." Every week the analog designers said: "The additional cost will price us out of the market. The transistors you are trying to stick us with are crap. Our designs cannot work with your crappy digital transistors." For 3 months we were at loggerheads. Then one of my engineers said: "We simply can't build high performance digital transistors without halos. The channel length is just too short." One of the analog engineers then replied: "We don't want your high-performance transistors. The channel length on the transistors we use is always a half micron or more. You've got to get rid of those damn halos." It took 3 months of weekly meetings for that critical piece of information to come out. I and my digital CMOS transistor engineers did not know that analog CMOS designers did not care about transistor speed. Until that moment we did not know analog designers only used long channel transistors. Once we understood this, the solution was immediately obvious. We built high-performance digital core transistors with halos and also built high-voltage input/output (I/O) transistors without halos in our manufacturing flow. For analog transistors, our transistor engineers simply put the I/O extensions without halos on long channel core transistors. They accomplished this by swapping photoresist patterns for the core and I/O extension implants on the NMOS and PMOS extension reticles. The analog designers got exactly the transistors they wanted, and it did not cost them a dime. It only took us 3 months of bickering back and forth for us to hit upon the obvious solution.

3.13 Bipolar Transistors [7]

3.13.1 Bipolar Transistor Structure and Circuit Diagram

Bipolar transistors are formed with two closely spaced back-to-back pn-junction diodes. A cartoon of an NPN bipolar transistor is shown in Fig. 3.40. A cartoon of a PNP bipolar transistor is shown in Fig. 3.41. Circuit diagrams of an NPN and a PNP bipolar transistor are presented in Figs. 3.42 and 3.43, respectively.

Bipolar transistors consist of two back-to-back pn-diodes. As is illustrated in Fig. 3.40, in an NPN bipolar transistor, the collector/base diode is reverse-biased (positive voltage on n-type collector and negative voltage on p-type base) and the emitter/base diode is slightly forward-biased. The depletion region formed around the reverse-biased collector/base junction is in close proximity to the slightly forward-biased emitter/base junction. Emitter current I_E from the slightly

Fig. 3.40 NPN bipolar
transistor cartoon

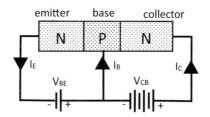

Fig. 3.41 PNP bipolar
transistor cartoon

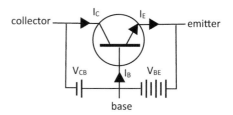

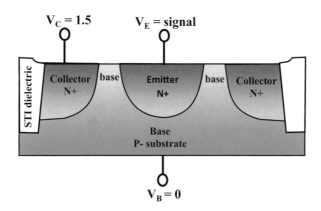

Fig. 3.42 NPN bipolar transistor circuit diagram

forward-biased emitter/base diode is collected by the reverse-biased collector/base diode. Small changes in forward bias of the emitter/base diode cause small changes in the emitter forward bias current I_E. This current is then collected by the collector/base diode. The collector/base and emitter/base diodes are connected in series. Small changes in emitter current I_E cause large changes in the collector current I_C. A small signal applied to the emitter/base circuit gets amplified in the collector circuit of the NPN bipolar transistor.

The PNP transistor in Figs. 3.43, 3.44, and 3.45 operate the same way as the NPN transistor in Figs. 3.40, 3.41, and 3.42 but with the doping reversed and with the voltages reversed.

Fig. 3.43 PNP bipolar
transistor circuit diagram

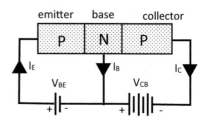

Fig. 3.44 NPN bipolar
transistor. P-type base is a
ring around the
n-type emitter

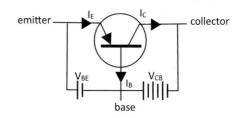

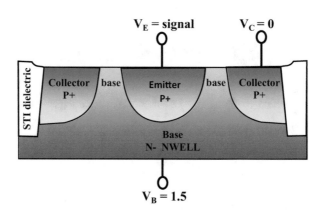

Fig. 3.45 PNP bipolar transistor. N-type base is a ring around the p-type emitter

3.13.2   Bipolar Transistors in CMOS Integrated Circuits

BICMOS (bipolar plus CMOS) integrated circuits combine the advantages of CMOS transistors (low power and high packing density) with the advantages of BIPOLAR transistors (high current and fast switching speed). Various useful bipolar transistors can be built with no added cost in MOS transistor manufacturing flows.

The circuit diagram in Fig. 3.41 and the cross section of a horizontal NPN bipolar transistor that can be formed with no additional cost in a CMOS integrated circuit manufacturing flow is shown in Fig. 3.42. The n-type emitter and collector diodes in the horizontal NPN bipolar are formed using the NMOS transistor source and drain photo pattern and n-type dopant implants. The base of the NPN bipolar transistor is the p-type silicon substrate. The NPN bipolar transistor is isolated from the NMOS transistors and other NPN bipolar transistors with dielectric filled shallow trench isolation (STI) trenches. In this example, the base (p-type substrate) forms a ring around the n-type emitter and the n-type collector forms a ring around the p-type base.

Figure 3.45 is a cross section of a horizontal PNP bipolar transistor that can be formed with no additional cost while making PMOS transistors. The p-type emitter and collector diodes are formed using the PMOS transistor source and drain photo pattern and p-type dopant implants. The base of the PNP bipolar transistor is the n-type silicon NWELL. This is the same NWELL in which PMOS transistors are built. In this example, the base (n-type NWELL) forms a ring around the p-type emitter and the p-type collector forms a ring around the n-type base.

Figure 3.46 is an example of vertical PNP bipolar transistor that can be formed with no additional cost in a CMOS integrated circuit process flow. The emitter can be formed using the PMOS source and drain photoresist pattern and p-type dopant implants. The base of the vertical PNP bipolar transistor is an NWELL formed at the same time as the PMOS transistor NWELLs are formed. The collector is the p-type substrate. Frequently a separate bipolar transistor emitter pattern and p-type implant and/or a separate base pattern and n-type implant are added with increased cost to improve the gain of the vertical PNP bipolar transistor.

Figure 3.47 is an example vertical NPN bipolar transistor that can be formed in a CMOS integrated circuit process flow by adding one additional photoresist

Fig. 3.46 Vertical PNP bipolar

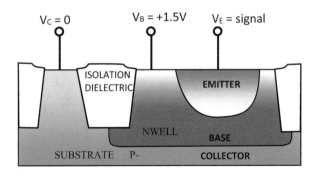

Fig. 3.47 Vertical NPN bipolar

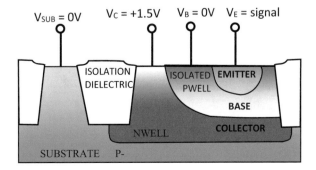

patterning step and one additional dopant implantation step to form an electrically isolated PWELL inside the NWELL. The emitter can be formed using the NMOS source and drain photoresist pattern plus the NMOS source and drain n-type dopant implants. The base of the vertical PNP bipolar transistor is the isolated PWELL formed by adding an isolated PWELL photo patterning and dopant implanting steps. The collector is for the vertical NPN bipolar transistor, the NWELL.

These are just a few illustrative examples. Many other vertical and horizontal NPN and PNP bipolar transistor configurations can be engineered into CMOS manufacturing flows to meet the needs of specific BICMOS integrated circuit requirements.

References

1. Gallery or Periodic Tables, ACS Chemistry for Life, https://www.acs.org/content/acs/en/education/whatischemistry/periodictable/periodic-table-gallery.html. Assessed 4 Mar 2020
2. G.W. Neudeck, *Volume II the PN Junction Diode* (Addison-Wesley Publishing Company, Inc., Boston, 1983)
3. R.F. Pierret, *Volume IV Field Effect Devices* (Addison-Wesley Publishing Company, Inc., Boston, 1983), pp. 81–103
4. D.K. Schroder, *Advanced MOS Devices* (Addison-Wesley Publishing Company, Inc., Boston, 1987), pp. 167–174
5. Analog-to-Digital Converter, Wikipedia, The free encyclopedia, https://en.wikipedia.org/wiki/Analog-to-digital_converter. Assessed 4 Mar 2020
6. List of MOSFET Applications, Wikipedia, The free encyclopedia, https://en.wikipedia.org/wiki/List_of_MOSFET_applications. Assessed 4 Mar 2020
7. G.W. Neudeck, *The Bipolar Junction Transistor* (Addison-Wesley Publishing Company, Inc., Boston, 1983)

Chapter 4
Engineering MOS Transistors for High Speed and Low Power

4.1 Water Flow to Current Flow Analogy

A convenient analogy to help understand the roles that resistance and capacitance play in high-speed inverters and integrated circuits is to compare the transistor to a faucet, to compare electron flow to water flow, to compare the resistance of electrons flowing through a wire to the resistance of water flowing through a pipe, and to compare the ability of a capacitor to store electrons to the ability of a bucket to store water. Figure 4.1 is a circuit diagram of a CMOS inverter. The CMOS inverter is composed of a PMOS and NMOS electrical switches in series between a power supply and ground. The gates of the PMOS and NMOS switches are connected so that they both switch at the same time. When the PMOS transistor switch turns on, the NMOS transistor switch turns off and vice versa. Figure 4.2 is a corresponding water inverter composed of water switches (faucets) and a water capacitor (bucket). The water source of the water inverter corresponds to the power supply terminal of the CMOS inverter, and the drain of the wafer inverter corresponds to the ground terminal of the CMOS inverter. The challenge is to design the CMOS inverter to charge and discharge the storage node capacitance with electrons as quickly as possible. Similarly, the challenge is designing the water inverter to fill and empty the bucket with water quickly as possible.

   To fill the bucket fast, you want faucet #1 to allow as much water as possible to flow when faucet #1 is turned on (high PMOS transistor drive current). To empty the bucket fast, you want faucet #2 to allow as much water as possible to flow when faucet #2 is turned on (high NMOS transistor drive current). For high water flow (high drive current), you want the faucets and the water pipes to and from the faucets to be large (series resistance of source and drains low). To fill and empty the bucket quickly, you want the bucket to be small (small capacitance). To keep the water logic state = 1 (bucket full of water), you want your bucket (low reverse diode leakage current) and faucet #2 not to leak (low transistor off current). Water leaking out of the bucket unintentionally changes the logic state from 1 (full bucket) to logic state

© Springer Nature Switzerland AG 2020
H. Tigelaar, *How Transistor Area Shrank by 1 Million Fold*,
https://doi.org/10.1007/978-3-030-40021-7_4

Fig. 4.1 CMOS inverter

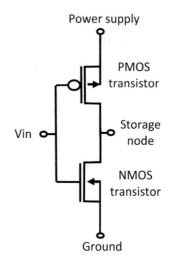

Fig. 4.2 Water inverter

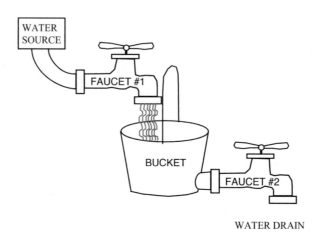

0 (empty bucket). Likewise, when the water inverter is storing logic state = 0 (bucket empty), you do not want faucet #1 to leak water into the bucket. Any water leaking through the faucets or leaking from the bucket wastes water. Any electrons leaking through the CMOS transistors or from the storage node capacitance wastes electrons (shorter battery life.)

High-speed CMOS inverter design requires transistors that pass as much current as possible when turned ON and requires resistances and capacitances to be as small as possible. Low-power CMOS inverter design requires low leakage current through the CMOS transistors when they are turned OFF, low leakage current through reverse-biased diodes, and low leakage current through the thin gate dielectric.

4.2 MOS Transistor Figure of Merit (FOM) [1]

The switching speed (frequency) of a MOS transistor is given by the transistor Figure of Merit (FOM).

$$\text{FOM} = \frac{\text{DRIVE CURRENT}}{\text{CAPACITANCE} * \text{VOLTAGE}} = \frac{\dfrac{dQ}{dt}}{C * V = Q}$$

$$= \frac{d}{dt} = \text{frequency} = \text{switches per second.}$$

The number of electrons (dQ) that flow through a transistor in a second (dt) is given by the ratio dQ/dt. The number of electrons needed to charge capacitances in the transistor (diode and gate dielectric capacitance) is given by the capacitance times the voltage charging the capacitance, $C * V$. Dividing the drive current dQ/dt by the $C * V$ gives the Figure of Merit (FOM) of the transistor. FOM has units of inverse seconds (per second) and is the calculated frequency (speed) at which the transistor can switch. Since drive current is in the numerator of the equation, the larger the drive current, the larger the frequency (more water flow fills the bucket faster). Since capacitance C is in the denominator of the equation, smaller capacitance results in higher transistor switching speed (smaller water bucket fills and empties faster).

Figure 4.3 is a high-resolution scanning electron microscope (SEM) image of an NMOS transistor.

Figure 4.4 is a cartoon of the NMOS transistor. A step-by-step description of how CMOS transistors and how a CMOS inverter are manufactured is presented in Chaps. 6, 7, and 8. The most important structure in the NMOS transistor is the channel under the gate. The channel must be high resistance when the transistor is turned OFF, so no leakage current flows from source to drain which would shorten battery life. The channel must be low resistance when the transistor is turned ON to provide high current for fast transistor switching. The capacitor formed by the gate dielectric between the gate and channel turns the NMOS transistor ON and OFF when the voltage on the gate (V_G) switches between power supply voltage (V_{DD}) and ground ($V_{SS} = 0$ V). The metal contact plugs between an overlying metal wiring layer (metal1) and the deep source and drain diodes connect the NMOS transistor drain to the power supply voltage (V_{DD}) and the NMOS transistor source to ground ($V_{SS} = 0$). A silicide layer between the metal contact plug and the source and drain diodes lowers the contact resistance between the metal plug and the heavily n-type doped single crystal silicon. N-type doped extensions electrically connect the n-type deep source and drain diodes to the transistor channel. P-type halos covering the ends of the extensions reduce transistor turn ON voltage roll-off at short transistor gate lengths as described in Sect. 3.11.1. Dielectric filled trenches in the substrate electrically isolate the NMOS transistor from other transistors and other electrical devices in the substrate. The dielectric transistor sidewalls, contact etch stop layer, and premetal dielectric (PMD) prevent the transistor gate from shorting to the metal contact plug and to metal1. The intermetal dielectric layer (IMD1) prevents metal1 leads from shorting to each other and from shorting to overlying layers of metal wiring.

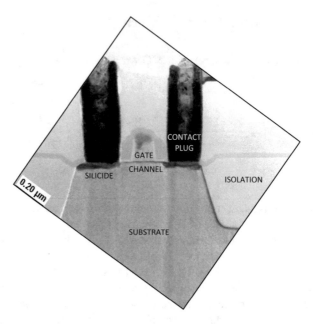

Fig. 4.3 Scanning electron microscope image of an NMOS transistor. SEM is courtesy of Texas Instruments

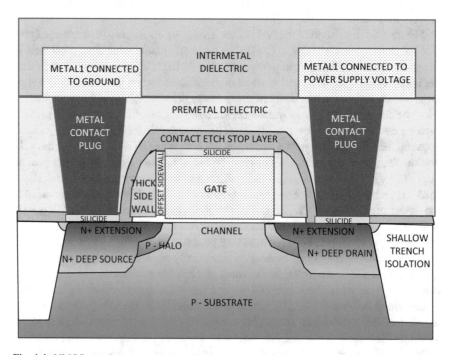

Fig. 4.4 NMOS transistor cartoon

4.3   MOS Transistor Resistance and Capacitance

The MOS transistor structure introduces a number of resistances between the input and output of the transistor. Transistor engineers must minimize the resistance of each of these resistors for maximum transistor performance. The MOS transistor structure introduces a number of capacitances between the input and output of the transistor. The input transistor current must fully charge each of these capacitances to turn the transistor ON and discharge each of these capacitances to turn the transistor OFF. Transistor engineers must minimize the capacitance of each of these capacitors for maximum transistor performance.

The final value chosen by the transistor engineers for the resistors and capacitors is always a compromise between maximum integrated circuit performance and longest battery life.

4.3.1   MOS Transistor Series Resistance

Figure 4.5 is the circuit symbol for a resistor.

Figure 4.6 is a copy of the NMOS transistor shown in Fig. 4.4 with the various resistances in the NMOS transistor labeled a through f. Descriptions of each of these resistances are listed in Table 4.1.

When engineering these transistor structures, the objective is to make these series resistances as low as possible.

The most important resistance for switching speed is resistance of the transistor channel (region "a"). The length and the doping level of the transistor channel are the

Fig. 4.5 Circuit symbol for a resistor

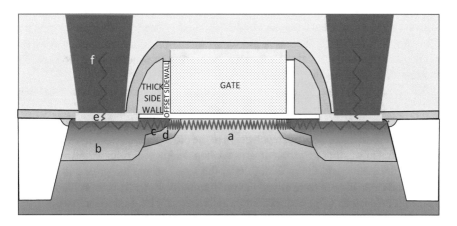

Fig. 4.6 Series resistance in an NMOS transistor

Table 4.1 Series resistance in an NMOS transistor

Resistor	Resistance description	Details
a	Transistor channel	Resistance is controlled by the transistor gate
b	Deep source/drain	Highly doped silicon at the surface.—Graded to light doping near the bottom
c	Source/drain extension	Shallow highly doped silicon
d	Halo	Pocket with higher channel doping around the ends of the extensions
e	Silicide contact	Doped silicon/nickel silicide/contact plug
f	Contact plug	Tungsten metal

primary factors that determine the channel resistance. Over a period of about 40 years, the transistor channel length scaled about 1000 times shorter (From about 10 μm at the 10 μm transistor node to about 10 nm at the 14 nm transistor node). A brief summary of this incredible feat is given in Chap. 10. To keep the transistor channel turned OFF as the transistor gate length got shorter, the thickness of the gate dielectric was reduced to increase the coupling of the transistor gate to the transistor channel. As the gate dielectric thickness was reduced, the power supply voltage had to be reduced to avoid breaking down the thinner gate dielectric. At the 10 μm technology node, the power supply voltage was 5 V. At the 10 nm technology node, the power supply voltage is less than 1 V. As transistors scaled smaller, commonly used power supply voltages were 3.3, 2.5, 1.5, and 1.25 V.

The resistance of the deep source and drain diodes (regions "b") can be made lower by increasing the doping and by increasing the depth of the diode. The problem encountered with increased doping is the increase of diode capacitance. The problem with increasing the diode depth is the increase of leakage current from source to drain under the gate when the transistor is OFF. As a compromise between resistance and capacitance, the surface of the source and drain diodes are doped as heavily as possible and the doping concentration is graded to lighter doping as the diode gets deeper. This is accomplished by a series of n-type dopant implants with less and less concentration and higher and higher energy. The lighter doping at the bottom of the diode increases the width of the depletion region between the source and drain diodes and the substrate. Increasing the width of the diode depletion region reduces the diode capacitance.

The depth of the deep source and drain diodes is determined by how deep they must be to prevent contact leakage when the contact plugs are formed. The contact etch can damage the surface of the source and drain diodes. The diode must be deep enough so this damage does not reach down to the diode junction and cause increased leakage. A show stopper problem arose when leakage current (transistor OFF) between the source and drain diodes under the gate became too great and the deep source and drain diodes were already at minimum depth. To fix this problem source and drain extensions were invented and implemented. This enabled transistor engineers to space the deep source and diffusions a sufficient distance apart to avoid this under the channel source to drain current leakage. The source and drain extensions form a shallow conductive path between the deep source and drain diodes and the channel of the transistor.

Source and drain extension diodes (region "c") are doped as heavily as possible, are made as shallow as possible, and are kept as short as possible to reduce the resistance of the source and drain extensions. More details of how the extensions were engineered as transistors scaled smaller are covered in Sect. 10.3.4 (extension doping).

The spacing of the source and drain extensions to the transistor channel is determined by the thickness of the offset dielectric sidewall. The source and drain dopants are implanted self-aligned to the offset dielectric sidewall.

The spacing of the deep source and drains (region "b") to the transistor channel is determined by the thickness of the offset sidewall plus the thickness of the thick dielectric sidewall. The deep source and drain diode dopants are implanted self-aligned to the thick dielectric sidewall (Deep source/drain doping).

Halos (region "d") are explained in Sect. 3.11.1. Halos are the same doping as the substrate. Increased substrate doping increases transistor channel resistance. Halo doping is a compromise between controlling V_{TN} roll-off and increased channel resistance. Halo doping is engineered to be just enough to counter V_{TN} roll-off with minimum increase in channel resistance.

Contact resistance (region "e") between the contact metal plug and the deep source and drain diodes is greatly reduced by siliciding the surface of the deep source and drain diodes. The silicide is formed on the exposed silicon surfaces of deep source and drain diodes and on the transistor gate by depositing a refractory metal such as nickel (Ni) and then annealing it to cause the refractory metal silicide to form. Silicides have orders of magnitude lower resistance than heavily doped single crystal silicon. As the contact size continued to get smaller as transistor area shrunk, to counter the increasing resistance, contacts went from no silicide to titanium silicide to cobalt silicide to nickel silicide.

Contact plugs (region "f") are tungsten metal deposited using chemical vapor deposition (CVD-W). Since the contact plug is short, the higher resistance of the CVD-W is not a major concern ($1.7 * 10^{-8}$ $\Omega*$m for copper vs $5.6 * 10^{-8}$ $\Omega*$m for tungsten).

Engineering the resistance of CMOS transistors for high performance and low power is always a balancing act between low resistance, low capacitance, and low leakage current.

4.3.2   MOS Transistor Capacitance

Figure 4.7 is the electrical symbol for a capacitor.

Figure 4.8 is a copy of the NMOS transistor cartoon shown in Fig. 4.4 with the various capacitances in the NMOS transistor labeled a through i. Descriptions of each of these capacitances are listed in Tables 4.2 and 4.3.

There are two types of capacitors in the CMOS transistors: conventional capacitors with conductive capacitor plates (such as metal or doped silicon) separated by a

Fig. 4.7 Electrical symbol
for capacitance

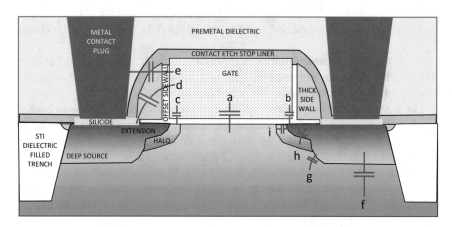

Fig. 4.8 Capacitances in an NMOS transistor

Table 4.2 Conventional capacitors in NMOS transistors

Conventional capacitor	First capacitor plate	Second capacitor plate	Capacitor dielectric
a	Gate	Substrate	Gate oxide
b	Gate	Extension	Gate oxide
c	Gate	Halo	Gate oxide
d	Gate	Source and drain	Thick sidewall and offset sidewall
e	Gate	Contact plug	Sidewall and contact liner

Table 4.3 Diode capacitors in NMOS transistors

Diode capacitor	First capacitor plate	Second capacitor plate	Capacitor dielectric
f—bottom wall	Drain	Substrate	Reverse bias depletion region
g—side wall	Drain	Substrate	Reverse bias depletion region
h—bottom wall	Extension	Halo	Reverse bias depletion region
i—side wall	Extension	Halo	Reverse bias depletion region

capacitor dielectric (Fig. 4.9) and reverse-biased diode capacitors where the depletion region functions as the capacitor dielectric (Fig. 4.10).

Figure 4.9 is a traditional capacitor connected to a battery. This capacitor has a dielectric such as silicon dioxide (SiO_2) between two metal plates. (Plates can be metal or heavily doped polysilicon for example.) The upper capacitor plate is charged positive by the battery and the lower capacitor plate is charged negative.

Fig. 4.9 Metal/dielectric/
metal capacitor

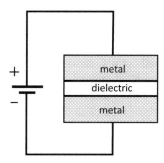

Fig. 4.10 Diode capacitor
doped silicon/depletion
region/doped silicon

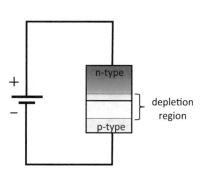

Figure 4.10 is a reverse-biased pn-diode capacitor (pn-junction capacitor). The
n-type side of the diode is connected to the positive terminal of the battery. The
p-type side of the diode is connected to the negative terminal of the battery. In this
example, the n-type silicon is more heavily doped than the p-type silicon. This
causes the depleted region in p-type silicon to be wider than the depleted region in
the n-type silicon. (The same number of electrons and holes are depleted from the
n-type and p-type depletion regions.) The n-type silicon that is not depleted is
charged to the positive battery voltage and the p-type silicon that is not depleted is
charged to the negative battery voltage. The charged (non depleted) n-type and
p-type silicon regions are the capacitor plates, and the depleted region is the capacitor dielectric.

4.3.3   MOS Transistor Conventional Capacitors

Conventional capacitors a–e in Fig. 4.8 are described in Table 4.2.
    In conventional capacitors, the dielectric constant and the thickness of the capac-
itor dielectric determine the magnitude of the capacitance. (It takes more electrons
to charge a capacitor that has a same thickness dielectric but with a higher dielectric
constant.) To minimize capacitance, the dielectric constant of the dielectric should
be a low as possible and the dielectric should be as thick as possible. Engineering
compromises for gate dielectric are required between thick dielectric for low capacitance, low gate leakage current, and good reliability and thin dielectric for good
control of the transistor channel.

After transistor gate length, the gate to extension capacitance in region b (extension-to-gate capacitance) is most critical for transistor performance. If this capacitance is too small, the resistance between the extension and the channel is suboptimal causing poor transistor performance. If this capacitance is too large, transistor switching speed is degraded because of the additional capacitance that needs to be charged when the transistor turns ON and discharged when the transistor turns OFF. The offset sidewall deposition thickness uniformity and etched thickness uniformity for dense vs isolated transistor gates across the wafer must be tightly controlled in manufacturing to produce tight transistor drive current distributions.

Not much can be done to reduce gate to halo capacitance in region c. Halo doping is dictated by the doping level needed to reduce V_T roll-off. Pocket area is small, so gate to halo capacitance is small.

Gate to source/drain capacitance (capacitor d) through the offset sidewall plus the thick sidewall is enough to degrade transistor performance. The thick sidewalls are usually silicon nitride (SiN) to take advantage of plasma etches that have been developed to etch SiN and stop on underlying silicon dioxide (SiO_2) (Fig. 4.11). In some high-performance MOS transistor manufacturing flows, transistor engineers replace part of the SiN (dielectric constant ~7) with SiO_2 (dielectric constant ~4.0) to reduce gate to source/drain capacitance (Fig. 4.12).

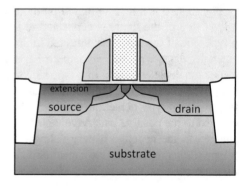

Fig. 4.11 NMOS transistor with silicon nitride thick sidewalls

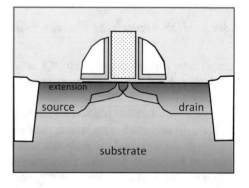

Fig. 4.12 NMOS transistor with a portion of the thick silicon nitride sidewall replaced by silicon dioxide to reduce capacitance and improve performance

The contact to gate capacitance (capacitor e) is determined by the spacing of the contact to the gate and primarily by the dielectric constants of the premetal dielectric (PMD) and the thick sidewall dielectric. In some high-performance transistor flows, the PMD dielectric is chosen to have the lowest dielectric constant compatible with structural integrity and void-free gap fill requirements. In some high-performance integrated circuit designs, circuit modeling programs are used to identify transistors in the speed path whose performance is degraded by contact to gate capacitance. Design for manufacturing design rules (DFM) are then added to increase the spacing between the gate and the contact plug when space allows for only these transistors.

4.3.4   MOS Transistor Diode Capacitors

Diode capacitors f–i in Fig. 4.8 are described in Table 4.3.

In diode capacitors, the width of the depletion region determines the magnitude of the capacitance. Lightly doped PN junctions have lower capacitance (wider depletion region), and they also have higher resistance. The engineering trade-off is between heavy doping for low resistance and light doping for low capacitance.

Bottom wall capacitance of the deep source and drain diodes (capacitance f) is reduced by grading the doping in the deep source/drains. The surface of the diode is heavily doped to provide a low resistance path to the transistor channel. A series of dopant implants with lower and lower doping at higher and higher energies are performed to lower the doping of the diode as it goes deeper into the substrate. The lower doping produces a wider depletion region with lower capacitance.

Engineering the capacitance of the deep source and drain diode (capacitor g) can be a bit tricky. As is explained in Chap. 5, a punch through implant is added to increase the doping under the channel between the deep source and drain to reduce leakage current between the source and drain under the channel when the transistor is turned OFF. This higher doping punch through region reduces the width of the depletion region. The reduced depletion width increases the capacitance of capacitor g and reduces the breakdown voltage of the deep drain diode. The doping levels of the punch through implant and doping level of the drain diode at this depth must be co-optimized for high speed (low capacitance), and low power (low OFF leakage).

Bottom wall capacitance of the source/drain extensions (capacitor h) is pretty much determined by the doping of the source/drain extensions and of the halos. The extensions need to be as low resistance as possible (high doping) and as shallow as possible (minimize leakage between extensions under the channel). The halo doping concentration is determined by what is needed to counter V_{TN} roll-off. Transistor engineers make source/drain extension lengths as short as possible for best transistor performance (low series resistance). The thickness of the thick sidewalls and the thermal drive at source and drain anneal are co-optimized. The goal is to have as short as possible source/drain extension linking the deep source/drains to the transistor channel while keeping leakage current between the deep source/drain diodes in specification.

Major roadblocks were encountered and surmounted as the area of the integrated circuits was cut in half every 2–3 years. A few of the major innovations and new equipment that enabled a new technology with smaller, higher speed, lower power transistors to be introduced into manufacturing every 2–3 years are described in greater detail in Chaps. 9–12.

4.4   MOS Transistor OFF Leakage Current

When MOS transistors are engineered to pass as high current as possible when switched ON (High I_{ON} for fast switching speed and high performance), it is simultaneously equally important they are engineered to leak as little current as possible when switched OFF (Low I_{OFF} for long battery life in portable devices). Three major contributions to I_{OFF} include: (1) leakage current under the transistor channel from source to drain, (2) leakage current from the reverse-biased drain diode to the substrate, and (3) leakage current from the transistor gate to the substrate through the gate dielectric (I_G).

4.4.1   Test Structures to Characterize MOS Transistor OFF Leakage Current

At the beginning of MOS transistor manufacturing flow development, test chips with design of experiment (DOE) test structures are processed through the manufacturing flow to provide data so the transistor engineers can select the best processing conditions. Transistor I_{OFF} data is generated by running process condition DOEs in conjunction with the test structure DOEs.

To select the processing conditions that deliver the highest MOS transistor I_{ON}/I_{OFF} ratio, processing condition DOE's are run using DOE test structures with transistors that are identical except for small changes in gate length. Transistors are drawn with 1 nm or less incremental gate lengths starting several increments above the target gate length to several increments below the target gate length.

To select the processing conditions that deliver the lowest reverse bias diode leakage, processing condition DOE's are run using DOE diode test structures that identify and quantify the various sources of diode leakage in a transistor. Diodes with various areas surrounded by shallow trench isolation characterize leakage to the substrate through the bottom of the diode (diode area leakage) plus leakage at the diode/STI interface (diode STI edge leakage). Diodes of various sizes surrounded by a transistor gate characterize leakage to the substrate through the bottom of the diode (diode area leakage) plus leakage at the diode/gate edge (gated diode leakage). Diodes with the same area but with different number of diode/STI corners characterize diode area leakage plus diode STI edge leakage plus diode STI corner leakage. Using data from these diode test structures a series of equations are solved to determine diode leakage per square nanometer of diode area, diode leakage per nanometer of STI edge, diode leakage per nanometer of transistor gate/drain diode, and diode leakage per STI/drain diode corner.

DOE gate dielectric capacitor test structures are designed to identify and quantify the various sources of gate dielectric leakage. Gate capacitors with various areas surrounded by shallow trench isolation characterize leakage through the gate dielectric (gate dielectric area leakage) to the underlying substrate plus leakage through the gate dielectric covering the STI edge (gate dielectric STI edge leakage). Gate dielectric capacitors of various sizes surrounded by a drain diode characterize leakage through the gate dielectric to the underlying substrate (gate dielectric area leakage) plus leakage through polyox surrounding the edge of the gate (gate to drain dielectric edge leakage) (Sect. 6.4.6, Fig. 6.24). Gate dielectric capacitors with the same area but with different gate/STI edge and with different gate/drain edge are used to quantify gate dielectric area leakage vs gate dielectric/STI edge and gate dielectric/drain edge leakage. Using data from these gate dielectric test structures a series of equations are solved to determine gate dielectric leakage per square nanometer of gate dielectric area, gate dielectric leakage per nanometer of gate/STI edge, and gate dielectric leakage per nanometer of transistor gate/drain edge.

4.4.2   Source to Drain Leakage Current

When capacitive coupling between the transistor gate and the substrate is insufficient to completely turn the channel of the MOS transistor OFF, subthreshold current can flow from the source to drain reducing battery life. Transistor engineers have three primary knobs they can turn to lower source to drain I_{OFF}: (1) They can make the gate length of the transistor longer since I_{OFF} increases exponentially with decreasing gate length, (2) They can make the gate dielectric thinner to increase the capacitive coupling between the gate and the substrate giving the transistor gate better control of the transistor channel, and (3) They can raise the turn ON voltage (V_T) of the transistor. If engineers make the gate length longer, I_{ON} is reduced and transistor performance takes a hit. If engineers decrease the gate dielectric thickness, leakage current (I_G) through the gate dielectric increases. If engineers raise the V_T of the transistor, again transistor performance takes a hit. The primary adjustment transistor engineers usually make to the process flow to address this source of I_{OFF} is to raise the doping (punch through implant) under the channel as much as possible without raising the doping at the surface of the substrate. This keeps the turn ON voltage (V_T) of the transistor the same while reducing source to drain leakage current immediately below the channel. In addition, this punch through implant raises the doping of the base of the parasitic bipolar transistor that underlies each MOS transistor. The increased base doping reduces the contribution of bipolar leakage current I_{OFF} (Sect. 5.4.1).

4.4.3   Area Diode Leakage Current

The drain diode of the MOS transistor is reverse biased with respect to the substrate. Reverse-biased diode leakage current flows through the diode depletion region from the drain (V_{DD}) to the substrate (V_{SS} or ground). See Sect. 3.6.2 and Fig. 3.12. The magnitude of the reverse-biased leakage current depends upon the width of the depletion region between the drain and the substrate. As explained in Sect. 4.3.4, reverse-biased diodes are capacitors in integrated circuits where the depletion region is the dielectric of the diode capacitor. The thickness of the diode capacitor dielectric (width of the depletion region) decreases as the p and n doping across the pn-junction diode increases. As the reverse-biased diode capacitor dielectric gets thinner, the reverse-biased diode current increases. Transistor engineers dope the surface of the drain as heavily as possible to reduce resistance between the drain contact and the transistor channel. They then grade the doping lighter and lighter below the surface to increase the width of the depletion region. Increasing the width of the depletion region (thickness of the diode capacitor dielectric) reduces reverse-biased diode leakage and therefore reduces I_{OFF}.

4.4.4   Gated Diode Leakage Current

Reverse-biased gated diode leakage is usually much higher than reverse-biased area diode leakage. A gated diode is formed when capacitive coupling from the transistor gate pins the surface potential near the drain diode junction (see Fig. 4.4). Away from the transistor gate, carriers are free to move and the depletion region is free to achieve the width needed to drop the voltage between the drain and the substrate. Capacitive coupling from the transistor gate next to the drain junction immobilizes (pins) the carriers at the surface preventing them from moving. Because the carriers under the gate (gated diode) near the drain diode are not free to move, the depletion region width is reduced and the reverse-biased diode leakage is increased. Careful engineering of the doping immediately under the edge of the transistor gate next to the drain and of the thickness of the gate dielectric at the edge of the gate over the drain helps reduce transistor I_{OFF} due to gated diode leakage (Sect. 6.4.6, Fig. 6.24).

4.4.5   Band-to-Band Tunneling Leakage Current

Engineering I_{OFF} leakage is particularly difficult between the end of the drain extension and the transistor channel. The drain extension is doped as high as possible to reduce resistance and the pocket implant surrounding the end of the drain extension under the transistor gate is also highly doped to reduce short channel effects (see Sect. 3.11.1). When the p and n doping across the pn-junction between the drain extension and the channel becomes too high, the capacitor dielectric becomes so thin (depletion region narrow) that carriers suddenly are able to penetrate the thin gate dielectric causing an exponential increase in I_{OFF} as drain voltage is raised. This tunneling current (band-to-band tunneling current) is a quantum mechanical phenomenon that enables carriers to freely pass through impenetrable barriers when the barrier is sufficiently thin. Quantum mechanical tunneling of electrons through the capacitor dielectric in conventional capacitors is used to program and erase non-volatile memory transistors such as EPROM and FLASH memories (see Sect. 2.2.1 and Fig. 2.3). Transistor engineers carefully engineer the drain extension doping, the pocket doping, and placement of the peak electric field in the drain/transistor channel junction to optimize transistor performance while at the same time avoiding band-to-band tunneling.

4.4.6 Gate Dielectric Leakage Current

When the MOS transistor is OFF, the voltage drop across the gate dielectric between the drain diode (V_{DD}) and the transistor gate (V_{SS} or ground) causes current to leak through the gate dielectric (I_G). I_G increases as the gate dielectric gets thinner and then increases exponentially when the gate dielectric gets so thin that quantum mechanical tunneling kicks in. For many technology nodes, transistor engineers kept the silicon dioxide gate dielectric sufficiently thick to avoid tunneling. When silicon dioxide gate dielectric scaled to where tunneling could not be avoided, transistor engineers replaced the silicon dioxide gate dielectric with nitrided silicon dioxide (SiON). SiON has a higher dielectric constant which enables a thicker gate dielectric (lower I_G) to be used while still providing sufficient capacitive coupling to the substrate to keep I_G low. As transistor scaling continued tunneling could no longer be avoided and the circuit designers had to account for the higher I_G in their designs. As scaling continues higher dielectric constant (high-k) gate dielectrics such as hafnium oxide ($k = 25$) are being used to improve capacitive control of the transistor channel while keeping I_G low.

4.4.7 Silicon on Insulator (SOI) Substrates for Low MOS Transistor OFF Current

To address the leakage from source to drain under the transistor channel plus the reverse-biased diode leakage from the drain to the substrate, silicon-on-insulator (SOI) substrates were introduced. Thin single crystal silicon is bonded to a thick dielectric replacing most of the single crystal silicon substrate. NMOS transistors are built in a thin layer of p-type silicon and PMOS transistors are built in a thin layer of n-type silicon. Since most of the substrate is replaced with dielectric, punch through implants under the channel are no longer needed. NWELLs and PWELLs also are no longer needed. Initially SOI wafers consisted of a few thousand nanometers of silicon bonded to an underlying dielectric. Partially depleted silicon-on-insulator (PDSOI) transistors are built in this single crystal silicon layer. With additional scaling the silicon layer was thinned to less than 1000 nm. Fully depleted silicon-on-insulator transistors (FDSOI) are built in this thinner single crystal silicon layer. Source to drain leakage under the channel is largely eliminated for transistors built on SOI substrates. Reverse-biased diode leakage from drain to substrate is also largely eliminated since the substrate is replaced by dielectric under the transistor drain. SOI ICs are often used in circuits that require exceptionally long battery life such as implanted ICs like heart pacemakers.

Reference

1. A. Chatterjee, M. Rodder, I. Chen, A transistor performance figure-of-merit including the effect of gate resistance and its application to scaling to sub-0.25-m CMOS logic technologies. IEEE Trans. Electron Dev. **45**(6), 1246–1252 (1998). https://ieeexplore.ieee.org/stamp/stamp.jsp?tp=&arnumber=678526

Chapter 5
Parasitic MOS and Bipolar Transistors in CMOS ICs

5.1 Introduction

The core NMOS and PMOS transistors in a CMOS integrated circuit are engineered to switch as fast as possible. They are engineered to provide as much current as possible when ON and leak as little current as possible when OFF. With each core transistor in an IC comes a parasitic bipolar transistor under the transistor channel. For example, under every NMOS transistor is a parasitic NPN bipolar transistor where the N+ source is the emitter, the p-substrate is the base, and the reverse-biased N+ drain is the collector. Under every PMOS transistor is a parasitic PNP bipolar transistor where the P+ source is the emitter, the NWELL is the base, and the reverse-biased P+ drain is the collector. If one of these parasitic NPN or PNP bipolar transistors turns ON during IC operation, usually the current flow is so high that heat generated by the high current melts the silicon and destroys the IC.

Likewise, when a conductive lead such as a polysilicon lead or a metal lead crosses over the shallow trench isolation (STI) dielectric, a parasitic MOS transistor is formed. For example when a polysilicon lead crosses the STI dielectric between two adjacent NMOS transistors, a parasitic MOS transistor is formed where the N+ diode from one NMOS transistor is the source, the polysilicon lead is the gate, the STI dielectric is the transistor gate dielectric, and the N+ diode from the adjacent NMOS transistor is the drain. If one of these parasitic transistors ever turns on during IC operation, the logic state from one of the NMOS transistors can leak through the parasitic MOS transistor to the adjacent NMOS transistor and cause the logic state to change. This can cause an IC calculation to fail or the IC to lock up.

As transistor engineers are engineering the core transistors to switch as fast as possible, they must also engineer the parasitic bipolar and parasitic MOS transistors to never turn ON.

© Springer Nature Switzerland AG 2020
H. Tigelaar, *How Transistor Area Shrank by 1 Million Fold*,
https://doi.org/10.1007/978-3-030-40021-7_5

5.2 Design Rules [1, 2]

Design rules (the rules the designers follow when they layout the transistors in an integrated circuit) are negotiated between the design engineers, the transistor engineers, and the process engineers in manufacturing. Design engineers want design rules to be a small as possible for high performance core transistors and as small as possible for more dies per wafer. Transistor engineers want transistor gate design rules to be sufficiently small that the drive current of core transistors meets transistor ON current specifications but sufficiently large, so the standby current of core transistors meets OFF current specifications. Process engineers from manufacturing want design rules to be large as possible to maximize yield, but sufficiently small for high profits.

Transistor engineers negotiate the minimum width of the STI trench to be sufficiently long, so the channel of the parasitic MOS transistor will not turn ON. Transistor engineers negotiate the minimum spacing between deep source/drains on core transistors and WELL boundaries to be sufficiently far apart that the parasitic bipolar transistors that cross well boundaries never turn ON.

Example design rules are illustrated in Fig. 5.1. For each design rule, there is a minimum allowed rule that the designers can use when laying out their circuits. For example, a minimum metal1 overlap of via1, a minimum transistor width, a minimum transistor gate length, a minimum contact width, and a minimum contact-to-gate space. There also are maximum design rules for the maximum allowed STI width, maximum allowed metal1 width, etc. To help manufacturing with chemical mechanical polishing, there are maximum allowed areas for STI, active, poly, and

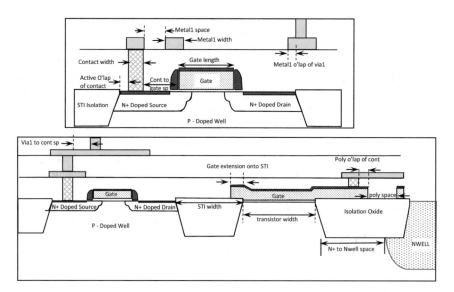

Fig. 5.1 Design rules for NMOS transistor. Top illustration is perpendicular through the transistor. Bottom illustration is parallel through the transistor

metal. To help manufacturing with etching uniformity, there are design rules for minimum and maximum allowed pattern densities. A design rule document can run hundreds of pages small print.

5.3 Parasitic MOS Transistors Between Core MOS Transistors

When building CMOS ICs with billions of core MOS transistors designed to turn ON and OFF as fast as possible, billions of parasitic MOS transistors are simultaneously built that must be designed to never turn ON. If one of these parasitic MOS transistors turns ON, at best a logic calculation fails and the blue screen of death results; at worst silicon melts and the IC is destroyed. Methods used by transistor engineers to keep these parasitic MOS transistors from turning on are to make the gate dielectric thick as possible, make the transistor channel (STI) as long as possible, and dope the channel under the STI as high as possible (punch through implant).

5.3.1 Parasitic MOS Transistors Between Adjacent Core MOS Transistors

In Fig. 5.2, NMOS 1 and PMOS 1 are parasitic MOS transistors with polysilicon gates over isolation dielectric (STI) between adjacent core MOS transistors acting as the gate dielectric. If these parasitic transistors turn ON, they connect a source or drain diffusion of one core MOS transistor to a source or drain of an adjacent core MOS transistor. Once one of these parasitic MOS transistors turns ON whatever logic state (voltage) was stored on the source or drain is lost and the computer crashes. The width of the isolation dielectric between the source and drain of these parasitic transistors must make the channel of the parasitic MOS transistor sufficiently long to prevent it from turning ON. In addition, transistor engineers heavily dope (channel stop implant) the substrate under the isolation dielectric (channel of the parasitic MOS transistor) to raise the turn ON voltage of the parasitic MOS transistor so high that it never turns ON.

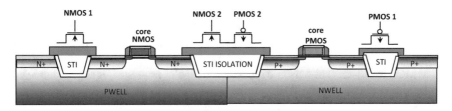

Fig. 5.2 Parasitic NMOS and PMOS transistors in an inverter integrated circuit

5.3.2   Parasitic MOS Transistors Between Core MOS Transistors and Adjacent WELLS

NMOS 2 and PMOS 2 in Fig. 5.2 are parasitic MOS transistors under the isolation dielectric across the NWELL/PWELL boundary. The isolation dielectric acts as the gate dielectric for these parasitic MOS transistors. The WELLs are the sources of these parasitic MOS transistors, and the source/drain junctions are the drains. If either of these parasitic transistors turns ON, the logic state (voltage) that was stored on the drain of the core MOS transistor is lost and the computer crashes. The width of the isolation dielectric between the drain and wells of these parasitic transistors is made sufficiently long to prevent these parasitic transistors from turning ON. See N+ to NWELL design rule in Fig. 5.1. In addition, the doping of the substrate under the isolation dielectric (channel stop implant) is increased to raise the turn ON voltage of these parasitic MOS transistors.

5.4   Parasitic Bipolar Transistors in Integrated Circuits

When building CMOS IC with billions of core MOS transistors designed to turn ON and OFF as fast as possible, billions of parasitic bipolar transistors are simultaneously built that must be engineered to never turn ON. When one of these parasitic transistors turns ON, at best a logic calculation fails and the blue screen of death results. Usually silicon melts and the IC is destroyed. Methods used by transistor engineers to keep these parasitic bipolar transistors from turning on are to make the bipolar base as wide as possible and dope the base as high as possible.

5.4.1   Parasitic Bipolar Transistors Under Core MOS Transistors

Under every core MOS transistor (designed to be as small and fast as possible) is a parasitic bipolar transistor that must never be allowed to turn ON. See NPN #1 and PNP #1 in Fig. 5.3. The emitter of the bipolar transistor is the source of the core MOS transistor. The base of the bipolar transistor is the WELL in which the core MOS source diode is formed. The collector of the bipolar transistor is the drain of the core MOS transistor. The drain diode of the core MOS transistor (collector/base diode of the bipolar transistor) is always reverse-biased. The source diode of the core MOS transistor (emitter/base diode of the bipolar transistor) is always intended to be unbiased (source and WELL at the same voltage) but during operation of the MOS tranistor can become forward biased turning the parasitic bipolar transistor ON.

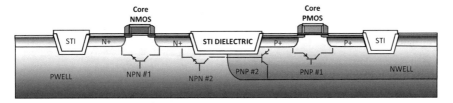

Fig. 5.3 Parasitic NPN and PNP bipolar transistors in silicon-based integrated circuits

Fig. 5.4 I/V curve of a
MOS transistor when the
underlying parasitic
bipolar transistor turns ON

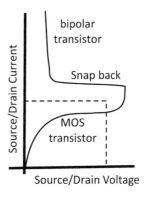

High-performance core CMOS digital transistors are turned ON as hard as possible (saturation) to provide as high a current as possible for fastest possible switching. Some of the carriers (electrons in NMOS and holes in PMOS) gain so much energy (channel hot carriers CHC) that they break bonds in the single crystal silicon substrate near where the transistor channel merges with the reverse-biased drain diode. These broken bonds create electron/hole pairs. In NMOS transistors electrons from the electron/hole pairs flow along with the electron channel current into the drain. Hole current flows through the substrate to the substrate ground contact. The substrate has a resistance. Hole current through the substrate resistance creates an IR voltage drop (voltage = current * resistance). When hole substrate current gets sufficiently high, the voltage drop generated can forward bias the source/substrate junction (emitter junction) and turn the parasitic NPN bipolar transistor ON. Similarly, sufficient hot holes can be generated in a PMOS transistor to forward bias the PMOS source/NWELL diode and turn the underlying parasitic PNP bipolar transistor ON.

Figure 5.4 shows the current/voltage (I/V) curve of a core MOS transistor where the drain voltage is increased to the point where channel hot carriers (CHC) generate sufficient electron/hole pairs and sufficient substrate current to turn ON (snapback) [3] the underlying parasitic bipolar transistor. The dashed box encloses the I/V curve of the MOS transistor (see also Fig. 3.35). The drain current of the core MOS transistor starts to rise as the source/substrate diode begins to forward bias. (emitter diode of the parasitic NPN bipolar transistor). Once the source/substrate junction becomes sufficiently forward-biased to turn the underlying parasitic bipolar transistor ON, the drain voltage on the core MOS transistor snaps back and the bipolar transistor current rapidly rises. The high current through the parasitic bipolar transistor can generate enough heat to melt silicon and destroy the core MOS transistor.

5.4.2 Parasitic Bipolar Transistors Between Core Transistors and Wells in Integrated Circuits

In Fig. 5.3, NPN #2 and PNP #2 are parasitic bipolar transistors across the NWELL/PWELL boundary. For parasitic bipolar transistor, NPN#2, the core NMOS transistor source diode is the emitter, the PWELL is the base, and the reverse-biased NWELL/p-type substrate is the collector. If the NMOS source diode becomes forward-biased, the parasitic NPN#2 bipolar transistor turns ON.

For parasitic bipolar transistor, PNP#2, the PMOS transistor source is the emitter, the NWELL is the base, and the p-type substrate is the base. If the PMOS source diode becomes forward-biased, the parasitic PNP#2 bipolar transistor turns ON.

Parasitic bipolar transistors NPN#2 and PNP#2 together form a silicon-controlled rectifier (SCR) circuit called a thyristor. The base of parasitic bipolar transistor PNP#2 is shorted to the collector of parasitic bipolar transistor NPN#2 and the base of NPN 2# is shorted to the collector of PNP#2. When this thyristor switch turns ON, it shorts the power supply to ground. The only way this thyristor can be turned OFF is to disconnect the battery. The condition when this thyristor turns ON is known as latch up. Latch up usually results in silicon melting and sparks flying.

To keep parasitic bipolar transistors from ever turning ON, the design rule for the base length between the emitter and collector must be sufficiently long and the doping of the base region must be sufficiently high. Transistor engineers determine the best compromise between base length and base doping to make the IC as small as possible while at the same time ensuring these parasitic bipolar transistors always stay turned OFF.

References

1. E. Sperling, Design rule complexity rising, Semiconductor Engineering (2018), https://semiengineering.com/design-rule-complexity-rising/
2. Design for Manufacturing (DFM): actions taken during the physical design stage of IC development to ensure that the design can be accurately manufactured, Semiconductor Engineering, https://semiengineering.com/knowledge_centers/eda-design/methodologies-and-flows/design-for-manufacturing-dfm/
3. Y. Zhou, D. Connerney, R. Carroll, T. Luk, Modeling MOS snapback for circuit-level ESD simulation using BSIM3 and VBIC models. Proceedings of the sixth international symposium on quality electronic design (ISQED'05), 21–23 March 2005, https://ieeexplore.ieee.org/stamp/stamp.jsp?arnumber=1410631

Chapter 6
CMOS Inverter Manufacturing Flow: Part 1
Wafer Start Through Transistors

6.1 Introduction to CMOS Inverter Manufacturing

A cross sectional view of the CMOS Inverter that we will be building is shown in Fig. 6.1.

Details of how this structure is actually built in manufacturing are explained in this and the next two chapters. The NMOS and PMOS transistors that are the inverter switches are built in a lightly doped p-type single crystal silicon substrate. The premetal dielectric layer (PMD) electrically isolates the transistor source, drain, and gate from the overlying first wiring layer (interconnect layer, M1) made of copper. Contact holes etched through the PMD layer are filled with tungsten metal to electrically connect the source, drains, and gates of the transistors to metal1. M1 wires are electrically isolated from each other with the first layer of intermetal-level dielectric (IMD1). M1 wires are electrically isolated from M2 wires by second intermetal-level dielectric layer IMD2. M2 wires are connected to M1 wires by via holes drilled through IMD2 and filled with copper metal2 (Via plug V1). After all the wiring layers are formed, a dielectric protective overcoat layer (PO) is deposited to protect the CMOS inverter IC from environmental hazards such as moisture and corrosive gasses. An aluminum bond pad is formed in an opening in the PO to provide outside electrical access to the IC. A bond wire (not shown) connects the bond pad to the power supply, battery, or to another integrated circuit.

Figure 6.2 is a cross-sectional cartoon of the CMOS inverter transistors indicating which of the transistor structures are connected to the battery terminals and which transistor structures are connected to input and output logic signals.

The NMOS transistor source diode and the PWELL substrate contact are connected to the negative or ground terminal of the battery. The PMOS transistor source diode and the NWELL substrate contact are connected to the positive terminal of the battery. The NMOS and PMOS transistor gates are shorted together and are connected to the input logic signal. The gates are shorted together, so both transistors switch simultaneously. The NMOS transistor switch turns ON when the PMOS

© Springer Nature Switzerland AG 2020
H. Tigelaar, *How Transistor Area Shrank by 1 Million Fold*,
https://doi.org/10.1007/978-3-030-40021-7_6

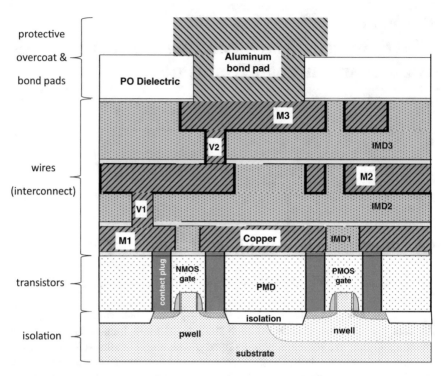

Fig. 6.1 Cross section of a CMOS inverter integrated circuit with three layers of copper metal interconnect plus an aluminum metal bond pad layer

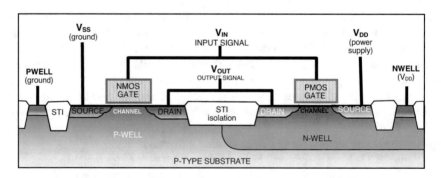

Fig. 6.2 Cross-sectional view of a CMOS inverter

transistor switch turns OFF and vice versa. The drain diodes of the NMOS and PMOS transistors are shorted together and provide the output logic signal from the CMOS inverter. Depending upon which of the NMOS and PMOS transistors is ON and which is OFF, the logic state stored on the shorted drain diodes is either V_{DD} (logic state 1) or ground (logic state 0). In the CMOS inverter, the logic state on the output is opposite the logic state on the input.

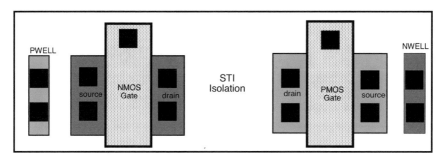

Fig. 6.3 Top-down view of the active, gate, and contact geometries of a CMOS inverter

Figure 6.3 is a top-down view of the active and gate patterns that are formed on the single crystal silicon p-type substrate (wafer) when building the CMOS inverter. PWELL and NWELL contact geometries are formed in the substrate to fix WELL potentials by connecting them to terminals of a power supply such as a battery. Transistor source, drain, and channel geometries are also formed in the substrate. The NMOS and PMOS transistor gates overlie the transistor channels and electrically isolate the transistor sources and drains. Gates are formed in a conductive polysilicon layer that is deposited on the substrate. The source and drains are formed by implanting dopants self-aligned to the gates. Source and drain diodes have doping opposite to the underlying substrate. The WELL contact and transistor source and drain area are called active areas because they are electrically active. These active areas are electrically isolated from each other by shallow dielectric filled trenches (STI).

6.2   Shallow Trench Isolation (STI)

6.2.1   Active Photoresist Pattern

The CMOS integrated circuit manufacturing process starts with a single crystal silicon wafer (p-type for this inverter). First, a silicon dioxide (SiO_2) layer (pad oxide) is thermally grown on the single crystal silicon surface (~20 nm). A silicon nitride (Si_3N_4) layer (~300 nm) is deposited on the pad oxide using low-pressure chemical vapor deposition (LPCVD). The SiO_2 layer prevents the Si_3N_4 layer from damaging the single crystal silicon when it is heated to high temperatures ($\geq$900 °C). The silicon nitride layer blocks oxidation (SiO_2 formation) of the silicon surface during subsequent high temperature oxidation steps. (Si_3N_4 is often abbreviated as SiN, but Si_3N_4 is the correct chemical formula for silicon nitride.)

Photoresist is coated on the silicon nitride layer and the active pattern is printed in the photoresist. Figure 6.4 is a top-down view of the active pattern as it appears on the photomask or reticle. Figure 6.5 is a cross-sectional view of the active photoresist pattern as it appears on the wafer. Active photoresist geometries cover regions

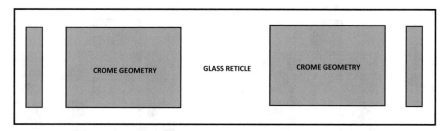

Fig. 6.4 Active pattern for CMOS inverter on reticle

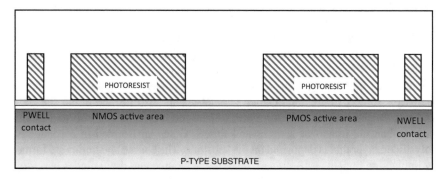

Fig. 6.5 Active pattern for CMOS inverter on p-type silicon substrate

of the wafer where transistors will be built (active areas). Active photoresist geometries also cover the substrate where electrical contacts to the wells are to be formed. The substrate is left uncovered where isolation trenches are to be etched into the substrate and refilled with dielectric to electrically isolate the active areas.

6.2.2   Shallow Trench Etch

Figure 6.6 shows the wafer after the silicon nitride layer and the pad oxide layer are etched from the unprotected regions and the active photoresist pattern is stripped. The active photoresist pattern blocks silicon nitride and silicon dioxide over the transistor active areas from being etched away during the plasma etch. The active photoresist pattern (also called a photoresist mask because it masks areas of the wafer from being etched) is removed after the nitride and oxide layers are etched.

The silicon nitride and silicon dioxide layers are etched using a plasma etch (see Sect. 2.1.2 in Chap. 2.)

The shallow trenches are then etched into the single crystal p-type substrate using the silicon nitride pattern as a hard mask (Fig. 6.7). The plasma etch erodes the photoresist pattern during the trench etch. Photoresist erosion introduces carbon into the etching plasma. More carbon is introduced where active pattern geometries

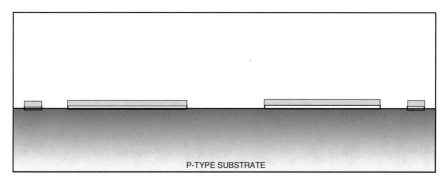

Fig. 6.6 Silicon nitride active hard mask for CMOS inverter

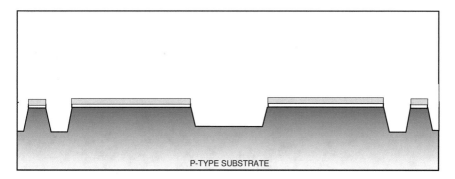

Fig. 6.7 STI trenches for CMOS inverter etched into substrate

are dense than where active pattern geometries are sparse. Carbon in the plasma deposits polymer on the sidewalls of the trenches that blocks the silicon from etching and causes sloped trench sidewalls. More carbon in the plasma forms more sloped sidewalls. Polymer also deposits on the bottoms of the trenches slowing the rate of silicon etching. Shallower trenches form where carbon concentration in the etching plasma is highest. Removing the active photoresist pattern before etching trenches into the silicon avoids etch loading effects caused by resist erosion.

6.2.3   Trench Dielectric Fill

A thin (~10 nm) layer of silicon dioxide (SiO_2) is thermally grown on the sidewalls and bottom of the trench (Chap. 2, Sect. 2.1.3). This ties up broken silicon bonds formed during the trench etch. The thin layer of thermal oxide suppresses leakage current.

A modified CVD (Chap. 2, Sect. 2.1.4), high aspect ratio process (HARP) using TEOS (tetraethyl-orthosilicate) plus ozone (O_3) is used to fill and over fill the STI

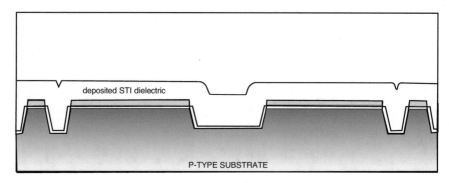

Fig. 6.8 STI trenches filled with silicon oxide dielectric

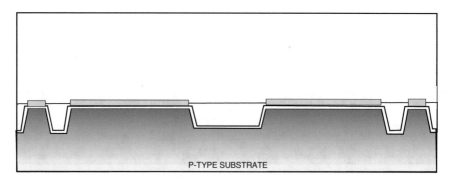

Fig. 6.9 STI trenches after dielectric planarization using CMP

trenches with a conformal layer of silicon dioxide (SiO_2) dielectric (Fig. 6.8). The next manufacturing step is to remove the SiO_2 overfill covering the active areas and to planarized the surface prior to the next patterning step.

6.2.4 Chemical Mechanical Polish Planarization

Chemical mechanical polish (CMP) is used to remove the SiO_2 overfill from the active areas. A CMP tool is essentially an abrasive wheel with a chemical slurry (abrasive particles suspended in a liquid) that polishes the surface of the wafer until the isolation oxide is removed from the active areas. The CMP polishing process stops on the silicon nitride layer (Fig. 6.9).

After CMP, the silicon nitride is removed by wet etching in a hot phosphoric acid bath (Fig. 6.10).

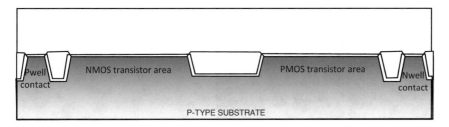

Fig. 6.10   STI trenches after CMP and silicon nitride strip

6.3   Transistor WELL and Transistor Turn ON Voltage Doping

6.3.1   NWELL and PWELL Formation

NWELL Formation

The next steps are to transform some regions of the wafer from p-type to n-type for PMOS transistors. PMOS transistors are built in n-type silicon (NWELLS) and NMOS transistors are built in p-type silicon (PWELLS). Shown in Fig. 6.11 is the NWELL photo pattern as it appears on the reticle (photomask). Figure 6.12 is a cross-sectional view of the NWELL photo pattern as it appears on the wafer. The NWELL photoresist pattern covers the NMOS transistor area blocking the n-type doping implant that forms NWELLs.

N-type dopant (phosphorus) is ion implanted into the PMOS transistor area to counter dope the substrate turning it from p-type to n-type silicon and form the NWELL. The NWELL implant is usually a series of chained phosphorus implants with different energies. The highest dose goes deepest to provide a low-resistance layer at the bottom of the NWELL under the PMOS transistors (retrograde NWELL). This helps to keep the bottom of the NWELL at the same potential (voltage). This prevents one transistor from affecting adjacent transistors when it turns ON. Midway, the NWELL is more lightly doped to reduce PMOS source/drain diode capacitance and to raise the breakdown voltage of PMOS source/drain pn-diodes. At the surface the NWELL is lightly doped. The surface doping sets the turn ON voltage (V_{TP}) of the PMOS transistor.

The NWELL pattern is stripped off the wafer using an oxygen plasma (called ashing). (Highly reactive oxygen atoms and molecules generated in the plasma react with the organic photoresist polymer which is composed primarily of hydrogen and carbon atoms to form carbon dioxide (CO_2) and water (H_2O) gasses which are pumped away.)

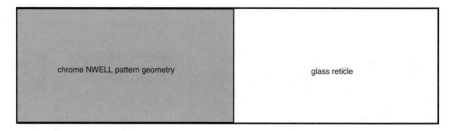

Fig. 6.11 NWELL and PMOS V$_{TP}$ photoresist pattern

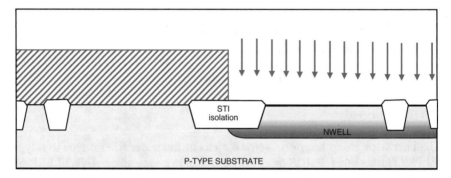

Fig. 6.12 NWELL ion implantation

PWELL Formation

The top-down view of the PWELL pattern as it appears on the reticle is shown in Fig. 6.13. A cross-sectional view of the PWELL photoresist pattern as it appears on the wafer is shown in Fig. 6.14. The PWELL pattern blocks the NWELL (PMOS transistor) area from the PWELL p-type implant (Fig. 6.14).

P-type dopant (boron) is ion implanted into the NMOS transistor area to form a PWELL. The PWELL implant is a series of chained boron implants with different energies. The highest dose goes deepest to provide a low-resistance layer under the NMOS transistors (retrograde PWELL). Mid depth, the PWELL is more lightly doped to reduce the capacitance and to raise the breakdown voltage of NMOS source/drain pn-diodes that will be formed later. At the surface of the PWELL light doping sets the turn ON voltage (V$_{TN}$) of the NMOS transistor.

After the PWELL is implanted, the photoresist is stripped and the wafer is annealed for a few seconds at a high temperature (900–1100 °C) to activate the dopant atoms (replace silicon atoms in the silicon lattice with phosphorus atoms in the NWELL and replace silicon atoms with boron atoms in the PWELL). The anneal time is adjusted to allow the dopant atoms sufficient time to diffuse through the single crystal silicon to the desired NWELL and PWELL depths (Fig. 6.15).

Silicon wafers containing integrated circuits are annealed in rapid thermal annealing (RTA) tools. These RTA tools take a silicon wafer from room temperature

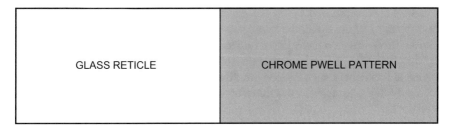

Fig. 6.13 PWELL pattern and NMOS V_{TN} pattern

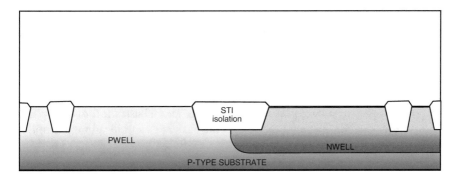

Fig. 6.14 PWELL implant

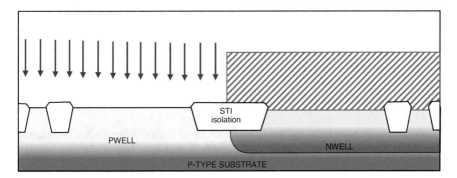

Fig. 6.15 NWELL and PWELL after WELL anneal

to 1000 °C or more and return it back to room temperature in a few seconds or even in a few milliseconds (1/1000 of a second). It took teams of scientists and engineers years to invent and make manufacturable RTA machines that could do this without shattering wafers due to thermal stresses. Flash anneal and laser anneal tools that raise the temperature of the surface of the wafer from room temperature to 1000+ °C and return the wafer back to room temperature in microseconds (millionths of a second) and even nanoseconds (billionths of a second) are running in production today.

Diffused and Implanted Wells

Prior to the 130 nm node, the nwell was formed by putting wafers with a silicon nitride (SiN) diffusion mask into a furnace with phosphine (PH_3) gas at a high temperature (~1000+ °C) to allow the phosphorus to diffuse into and dope the single crystal silicon where not covered with SiN. After doping, the wafers were annealed for many hours to activate the dopants and to diffuse the dopants to the desired well depth.

When transistors scaled below 130 nm, transistor structures could no longer tolerate the long diffusion times. In addition, the retrograde wells that were needed to keep scaling going could not be manufactured using diffusion doping.

PWELLs were formed by diffusion doping using diborane gas (B_2H_6).

6.3.2 Turn ON Voltage of NMOS and PMOS Transistors

NMOS Transistor Turn ON Voltage (V_{TN})

A photoresist pattern to adjust the turn ON voltage (V_{TN}) of the core NMOS transistors is now printed on the wafer. The V_{TN} implant pattern opens PWELL regions where core NMOS transistors are to be built and blocks NWELL regions where PMOS transistors are to be built. For the CMOS inverter, the pattern is identical to the PWELL pattern (Fig. 6.16). In most integrated circuits, multiple V_{TN} implant patterns plus multiple V_T adjust doping implants are performed to accommodate other transistors in addition to the core transistors on the IC. For example, typically there are separate, additional patterning and doping implant processing steps for SRAM memory cell NMOS transistors, input/output (high voltage) NMOS transistors, and analog NMOS transistors. The turn ON voltages for each of these NMOS transistors are usually different.

A p-type dopant is implanted into the channel of the NMOS transistor to increase the p-type doping near the surface of the PWELL. This adjusts the turn ON voltage (V_{TN}) of the core NMOS transistor to meet the designer's specification. In addition

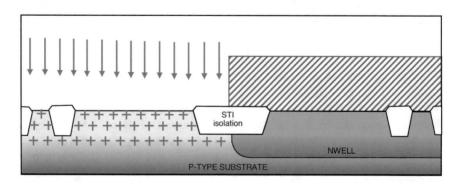

Fig. 6.16 NMOS V_{TN}, punch through, and isolation implants

to the V_{TN} implant, more p-type dopant (punch thru implant) is implanted at a slightly higher energy to raise the p-type doping under the transistor channel between the source and drain diffusions. This increases the base doping of the parasitic NPN bipolar transistors, so they do not turn ON. The doping of the punch through implant is a compromise between being high enough to keep the parasitic NPN bipolar transistor from turning ON, being low enough so the capacitance of the drain junction does not unacceptably degrade NMOS transistor performance (switching speed), and being low enough for the drain diode breakdown voltage to meet specifications. At an even higher energy, more p-type dopant is implanted under the isolation regions in the PWELL (channel stop implant). This raises the turn ON voltage of the parasitic NMOS transistors under the STI, so they always remain turned OFF.

PMOS Transistor Turn ON Voltage (V_{TP})

A photoresist pattern to adjust the turn ON voltage (V_{TP}) of the core PMOS transistors is now printed on the wafer. The V_{TP} implant pattern opens NWELL regions where core PMOS transistors are to be built and blocks PWELL regions where NMOS transistors are to be built. For the CMOS inverter, the pattern is identical to the PWELL pattern (Fig. 6.17). In most integrated circuits, multiple V_{TP} implant patterns and doping implants are performed to accommodate other types of PMOS transistors that are being built elsewhere in the IC.

An n-type dopant is implanted to adjust the n-type doping near the surface of the NWELL so that the turn ON voltage (V_{TP}) of the core PMOS transistor meets specification. In addition to the V_{TP} implant, more n-type dopant (punch thru implant) is implanted at a slightly higher energy to raise the n-type doping under the transistor channel between the source/drain diffusions. This increases the base doping of parasitic PNP bipolar transistors, so they do not turn ON. At an even higher energy, more n-type dopant is implanted under the STI isolation regions in the NWELL (channel stop implant). This raises the turn ON voltage of parasitic PMOS transistors, so they always remain OFF.

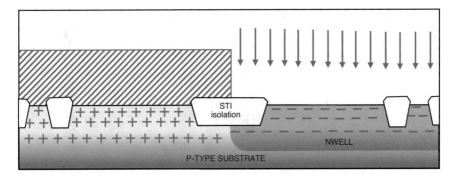

Fig. 6.17 NMOS V_{TP}, punch through, and isolation implants

6.4   NMOS and PMOS Transistor Gates

6.4.1   Introduction

After the transistor WELLS are formed and V_T dopants implanted, the transistor gates, sources, and drains are formed. Transistor dielectric (gate dielectric) is grown on the surface of the wafer, and the transistor gate polysilicon is deposited. These processes are now performed sequentially in one multimillion-dollar manufacturing tool with multiple processing chambers such as Applied Materials Centura© Gate Stack Tool (Fig. 6.18).

6.4.2   Transistor Gate Dielectric

At the 130 nm technology node, the SiO_2 gate dielectric had become so thin that excessive gate leakage current flowed through the thin SiO_2 when the transistor was turned OFF. In addition, on PMOS transistors, boron (B) from the p-type doped PMOS polysilicon gate diffused through the thin gate dielectric into the PMOS channel. This addional boron dopant unacceptably changes the PMOS transistor turn ON voltage. The solution is to nitride (add nitrogen to) the SiO_2 gate dielectric. A lot of research, invention, and development went into engineering the nitrogen profile in the gate dielectric. A little nitrogen at the gate dielectric/channel interface is good—it reduces channel hot carrier (CHC) degradation and improves negative-biased temperature instability (NBTI) of the PMOS transistor. Too much nitrogen at

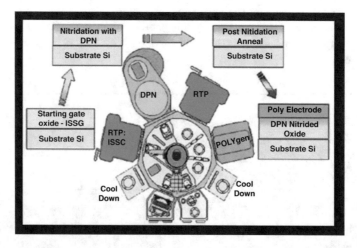

Fig. 6.18 Applied materials gate stack tool with multiple processing chambers. https://www.edn.com/its-all-in-the-plasma/ [1]

the gate dielectric/channel interface degrades carrier mobility (It takes more time for holes to travel from the source to the drain. Too little nitrogen at the gate dielectric/gate polysilicon interface allows boron to penetrate through the thin gate dielectric and additionally dope the PMOS channel. The preferred nitrogen profile has a high concentration of nitrogen at the polysilicon gate/gate dielectric interface to block boron penetration, just enough nitrogen at the gate/channel interface to reduce CHC degradation without significantly reducing hole mobility, and a high nitrogen concentration in the bulk of the gate dielectric to raise the dielectric constant. A higher dielectric constant enables a thicker gate dielectric to control the transistor channel. A thicker gate dielectric reduces the gate leakage contribution transistor OFF current.

Core Transistors and Input/Output Transistors

Core transistors are designed to operate with low voltages to minimize power and prolong battery life. Power is proportional to the voltage squared ($P = V^2/R$) where P is power, V is voltage, and R is resistance. (As voltage doubles from 1 to 2 V, the power consumed quadruples from 1 to 4 W.) Most electronic devices that the integrated circuits interact with operate at 2.5–5 V. In addition to the low voltage core transistors, higher voltage input/output (I/O) transistors are typically also built on the IC chip to enable the core logic transistors to communicate with external circuits. The higher voltage transistors require a thicker gate oxide. To simultaneously build low voltage core transistors and 2.5 V I/O transistors, the I/O transistor gate oxide is first partially grown on all exposed silicon surfaces on the wafer. An I/O photoresist pattern is then formed on the wafer to protect gate oxide where the I/O transistors are to be built. The partially grown gate oxide is etched from core transistor areas. The I/O transistor pattern is stripped, and the wafer is put back into the gate oxide furnace to grow the core gate oxide where core CMOS transistors are being built and to increase the I/O gate dielectric to its target thickness where I/O CMOS transistors are being built.

Figure 6.19 shows a cross section of the CMOS inverter after the gate dielectric is grown on the exposed silicon surfaces of the wafer and is nitrided. For the 45 nm technology node, the nitrided gate dielectric thickness is about 1 nm (about 5 times thicker than a silicon atom).

6.4.3 Transistor Polysilicon Gate

After the gate dielectric is grown and nitrided, a layer of polysilicon is deposited on the wafer. The polysilicon layer is deposited in a single wafer reactor using chemical vapor deposition (CVD). The silane (SiH4) and hydrogen (H_2) reaction occurs thermally at a temperature between about 580 and 650 °C and at a pressure between about 25 and 150 Torr (Fig. 6.20). Dopant gas can be introduced during the deposition to deposit doped polysilicon gates.

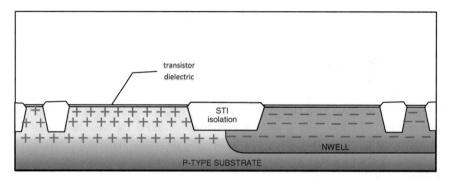

Fig. 6.19 Gate dielectric grown on the wafer surface

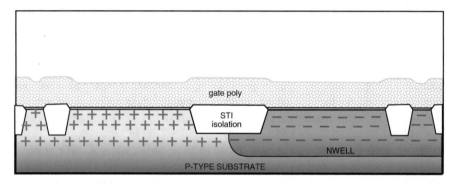

Fig. 6.20 Gate polysilicon deposited on the gate dielectric

6.4.4   Transistor Gate Patterning

A photoresist gate pattern is formed on the polysilicon. The transistor gate pattern is the most critical pattern on an integrated circuit. The most advanced photoresist chemistry, the most advanced optical proximity correction, and the most advanced photolithography tools (i.e., immersion scanners that cost over $100 million each!) are used to print the gate-level photoresist pattern.

Figure 6.21 is a top-down view of the CMOS inverter gate pattern as it appears on the gate reticle. Figure 6.22 is a cross-sectional view of the CMOS inverter gate photoresist pattern as it appears on the silicon wafer.

6.4.5   Transistor Gate Etching

An anisotropic (etches vertically but not horizontally) plasma etch removes the poly silicon not protected by the photoresist gate pattern. A cross-sectional view of the CMOS inverter after gate etch is presented in Fig. 6.23.

Fig. 6.21   Gate pattern on the gate reticle

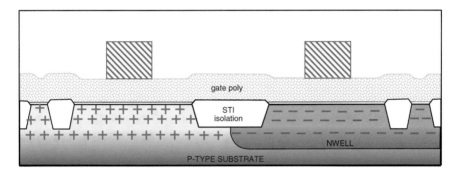

Fig. 6.22   Gate photoresist pattern

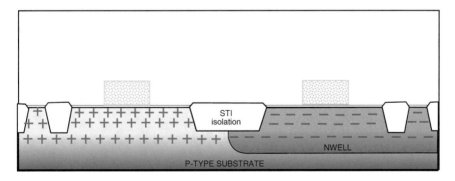

Fig. 6.23   Etched polysilicon gates

Gate etch is the most critical etching process in integrated circuit manufacture. The sides of the gates must be vertical, straight, and uniform. The polysilicon must be completely etched away from the substrate surface adjacent to the gate leaving no residues and not penetrating through the very thin gate dielectric (10s of nanometers) and etching trenches into the underlying semiconductor substrate. The semiconductor substrate is single crystal silicon that etches rapidly when the gate plasma etch penetrates the thin gate dielectric.

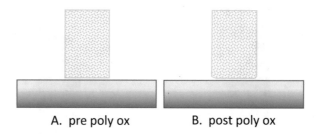

A. pre poly ox   B. post poly ox

Fig. 6.24 Poly oxidation post gate etch. (**a**) Pre poly ox. (**b**) Post poly ox

6.4.6 Gate Poly Oxidation

After the gates are etched, the wafers are loaded into a single wafer rapid thermal oxidation (RTO) tool to grow a thin oxide on the sidewalls and to oxidize the bottom corners of the transistor gates (Fig. 6.24a). Electric fields are enhanced at corners. The slight reduction in the electric field at the bottom corners of the transistor gates due to the slight corner rounding and slight thickening of the gate dielectric reduces reliability failure due to gate dielectric breakdown at the bottom gate edges (Fig. 6.24b). The reduced electric field at the at the bottom corners of the gate (Fig. 6.24b) also reduces channel hot carrier (CHC) degradation. CHC electrons with sufficient energy to penetrate the gate dielectric are generated in the channel when the transistor is operated full ON—in saturation. The high vertical field between the channel and the gate near the drain alters the path of some of these electrons toward the gate instead of toward the drain. Some of these electrons penetrate and get trapped in the gate dielectric. Over time this causes the turn ON voltage of the transistor to increase. This degrades transistor performance and can cause ICs to fail. The slightly thicker oxide at the bottom corner of the gate lowers the electric field and reduces the CHC degradation.

6.5 Transistor Source and Drain Extensions

After the transistor gates are formed, the transistor source/drain extensions are formed to the electrically connect the transistor channel with the deep source/drain diodes that will be formed later. Halo dopant is also implanted using the same photoresist pattern to counter turn ON voltage roll-off (short channel effects).

6.5.1 NMOS Extension and Halo Dopant Pattern and Implants

The wafer is patterned with NMOS transistor source/drain extension photoresist geometries (Figs. 6.25 and 6.26). Low-energy phosphorus and/or arsenic is implanted self-aligned to the transistor gate to dope n-type source/drain extensions. Source/drain extensions provide the electrical connection between the transistor channel and the deep source/drain pn-junction diodes. Metal contact plugs will electrically connect the deep source/drain diodes to overlying metal wires (Fig. 6.1).

Originally for NMOS transistors phosphorus was used as the dopant atom. As NMOS transistors scaled and extensions needed to be shallower and sharper, the diffusion of phosphorus became a liability. Arsenic being a larger atom with slower diffusion replaced phosphorus as the dopant of choice for NMOS extensions. To additionally, retard diffusion and to additionally increase the shallowness and sharpness of the NMOS extension, a carbon implant is added. Carbon being a smaller atom than silicon and gets trapped in the lattice spaces of the single crystal silicon (interstitial carbon). These trapped carbon atoms retard diffusion of extension dopant atoms resulting in sharper, shallower, NMOS extensions.

Extensions are implanted at zero degrees. Arsenic is a large atom and pulverizes the surface changing the single crystal silicon to amorphous silicon. The amorphous silicon prevents smaller atoms such as phosphorus or boron from channeling when they are implanted. (At zero degrees small atoms can pass through the spaces in the single crystal lattice without colliding with any silicon atoms. When this happens, they go deep into the silicon (channel) causing the pn-junction to be ill defined.) When arsenic is not used, silicon or germanium atoms may be implanted to pulverize the surface prior to implanting of phosphorus to prevent channeling.

6.5.2 NMOS Halo Dopant Implants

Using the same photoresist pattern p-type halos are implanted (Fig. 6.27) at an angle typically between about 20 and 40 ° to increase doping in the channel of the NMOS transistor around the ends of the extensions to counter short channel effects (V_T roll-off when the transistor gets shorter) (see Chap. 3, Sect. 3.11.1). The halo implant is split into four doses. A quarter dose is implanted at each of 0, 90, 180, and 270 ° to form pockets around the source and drain extensions on transistors laid out both vertically and horizontally.

Missed Patent Opportunity on Halo Implants
In the mid-1980s, I worked on electrical programmable read-only memory integrated circuits (EPROMS). An EPROM transistor is an NMOS transistor with an additional gate between the NMOS transistor gate and the transistor channel (see Fig. 2.3).

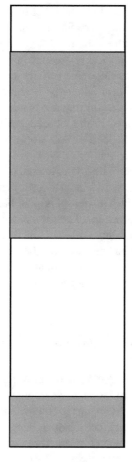

Fig. 6.25 NMOS source and drain extension pattern on reticle

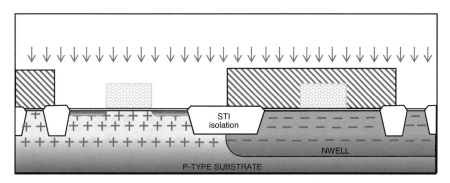

Fig. 6.26 Patterned NMOS source and drain extension photoresist pattern and implant on wafer

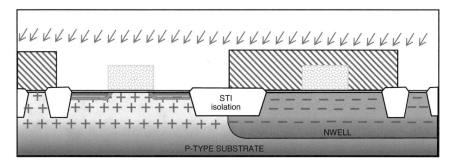

Fig. 6.27 NMOS source and drain halo implant

As I shrunk the EPROM for the next technology node, I also reduced the temperature and time of the anneal of the source/drain pn-junctions to reduce their size. The first time I tested my newly scaled EPROM transistor, I discovered it programmed well when I biased the source and drain in one direction but programmed poorly when I swapped voltages on the source and drain. The source/drain dopants were implanted with a seven-degree angle to avoid channeling.

The EPROM transistor gate is taller than the CMOS gate, so the seven-degree implant resulted in more shadowing (a greater distance between implanted dopant and base of the transistor gate) on the EPROM transistor gate than on the core CMOS gate (see Figs. 6.28 and 6.29). The reduced time and temperature of the source and drain anneal required to make the source/drain junctions smaller did not drive the dopants sufficiently to connect the shadowed pn-junction with the channel. Hence the asymmetry in programming.

I thought this was an interesting new phenomenon and submitted a patent suggestion to TI's patent committee. (A group of experts who kept Texas Instruments from spending money to patent useless ideas.) My idea was to implant dopants at an angle to engineer source/drain junctions. The patent committee decided this idea was not manufacturable because halting implanting and rotating the wafers before resuming implanting was not manufacturable. Less than 10 years later, halo or

Fig. 6.28 NMOS
transistor with angled
source/drain implant

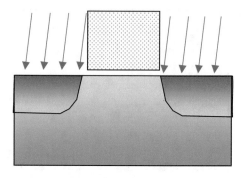

Fig. 6.29 EPROM
transistor with angled
source/drain implant

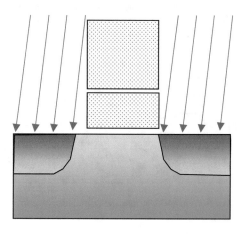

pocket implants were invented, and implant equipment manufacturers were selling ion implanters that implanted dopants at an angle and automatically rotated the wafers. Was not such a bad idea after all.

PMOS Source and Drain Extensions are formed by implanting p-type dopant that form the conductive extensions plus n-type halo dopant to counter PMOS turn on voltage (V_{TP}) roll-off.

6.5.3   PMOS Extension Pattern and Implant

The wafer is patterned with PMOS transistor source/drain extension photoresist geometries (Fig. 6.30). Low-energy boron (B) or boron difluoride (BF_2) is implanted self-aligned to the transistor gate to form p-type source/drain extensions (Fig. 6.31).

Originally boron (B) was implanted to form PMOS source and drain extensions. As transistors scaled requiring shallower source/drain extensions, BF_2 replaced B. The BF_2 molecule is heavier than the B atom so is not implanted as deeply. Prior to the boron or BF_2 implant, a heavy atom such as germanium or indium is implanted

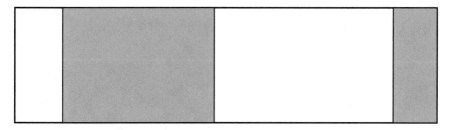

Fig. 6.30 PMOS source and drain extension pattern on the reticle

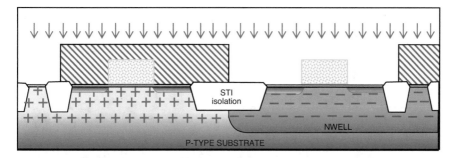

Fig. 6.31 PMOS source and drain extension photoresist pattern and dopant implant

first to amorphize the silicon surface to prevent channeling. As PMOS transistors scaled further and extensions needed to be even shallower and sharper, carbon atoms are also implanted into the PMOS transistor extensions to for interstitial carbon that retards diffusion of the boron atoms.

6.5.4 PMOS Halo Pattern and Implant

Using the same photoresist pattern, n-type halos are implanted at an angle typically between about 20 and 40 ° to increase doping in the channel around the ends of the PMOS source/drain extensions to counter short channel effects (V_T roll-off when the transistor gets shorter) (Fig. 6.32; see Chap. 3, Sect. 3.11.1). The halo implant is split into four doses. A quarter dose is implanted at 0, 90, 180, and 270 ° to form pockets on transistors with gate orientations in both x and y directions. Arsenic is now used as the n-type halo dopant because it is a larger and heavier atom than phosphorus and diffuses less during thermal processing steps.

Following source/drain extension and halo implants, an anneal of about 900 °C or more is performed to repair the damage done to the silicon crystal lattice by the amorphizing implant (regrow amorphous silicon at the surface back into single crystal silicon) and to activate the dopant (replace silicon atoms in the silicon crystal

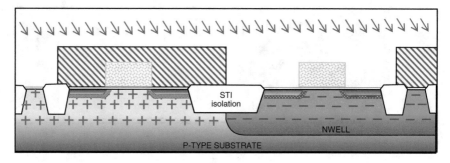

Fig. 6.32 Patterned PMOS source and drain halo implant

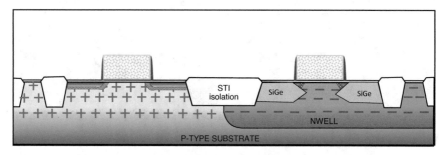

Fig. 6.33 PMOS with performance enhancement using compressive stress applied by replacing silicon crystal with silicon germanium (SiGe) crystal

lattice with dopant atoms). This anneal is performed in a rapid thermal annealing (RTA) tool for a millisecond or less at a temperature of 950 °C or more.

High-Performance PMOS Transistors

The mobility of holes in PMOS transistors is about three times slower than the mobility of electrons in NMOS transistors. To compensate, the width of PMOS transistors in an CMOS inverter is approximately three times wider than NMOS transistors. For high-performance CMOS inverters, compressive stress applied to the channel of the PMOS transistors improves hole mobility. To apply compressive stress to the PMOS transistor channel, trenches are etched into the source/drain areas of PMOS transistors and refilled with epitaxially grown, single crystal silicon germanium (SiGe) (see Figs. 6.33 and 6.34). The Ge atom is larger than the Si atom, so single crystal SiGe is larger than single crystal Si. The larger SiGe crystalline material in the PMOS source/drains pushes against the PMOS channel applying compressive stress which boosts the speed of the hole carriers.

Fig. 6.34 Intel 45 nm PMOS transistor with silicon germanium source and drain. https://en.wikichip.org/wiki/File:intel_45nm_gate.png [2]

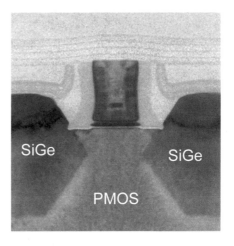

6.6   Transistor Deep Source and Drain Diodes

6.6.1   Transistor Dielectric Sidewalls

After the source/drain extensions are formed, dielectric sidewalls are formed on the vertical sides of the transistor gates to define the spacing between the deep source and drain diodes. The deep source/drain dopants are implanted self-aligned to the sidewalls. The thickness of the sidewalls is very tightly controlled. Tight control of the distance between the deep source/drains and the transistor gate provides narrow transistor drive current distributions.

A layer of silicon nitride (Si_3N_4) is conformally deposited over the transistor gates and over the surface of the wafer using plasma enhanced chemical vapor deposition (PECVD) (see Fig. 6.35). This nitride layer is then plasma etched anisotropically (etched vertically but not horizontally) to form sidewalls on the gates of the CMOS inverter transistors as shown in Fig. 6.36.

Originally the sidewalls on MOS transistors were made of silicon dioxide and called sidewall oxide. As MOS transistors scaled, the silicon oxide was replaced with silicon nitride to take advantage of the high selectivity of the silicon nitride etch to underlying silicon dioxide. One drawback of using silicon nitride is that the dielectric constant of the silicon nitride is larger (7.5) than the dielectric constant of silicon dioxide (3.9). This means that gate to drain capacitance (see capacitance d in Fig. 4.6) is larger with silicon nitride sidewalls. For this reason, in some CMOS process flows, a thick layer of silicon dioxide is deposited onto a thin layer of silicon nitride to form composite SiO_2/SiN sidewalls. (Compare sidewalls in Figs. 6.37 and 6.38 to Figs. 6.39 and 6.40.)

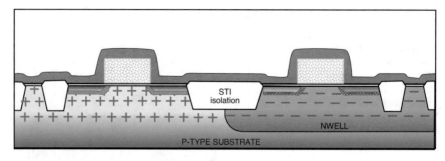

Fig. 6.35 Conformally deposited sidewall nitride

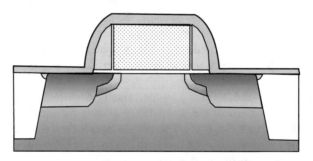

Fig. 6.36 Sidewalls post side wall etch

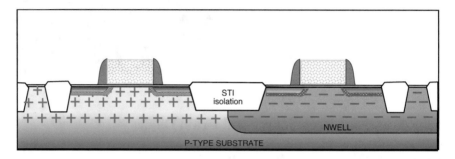

Fig. 6.37 CMOS transistor with composite silicon nitride/silicon dioxide sidewall

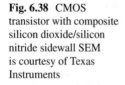

Fig. 6.38 CMOS transistor with composite silicon dioxide/silicon nitride sidewall SEM is courtesy of Texas Instruments

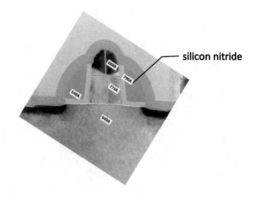

silicon nitride

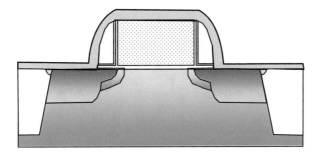

Fig. 6.39 CMOS transistor with silicon nitride sidewall

Fig. 6.40 CMOS
transistor with silicon
nitride sidewall. SEM is
courtesy of Texas
Instruments

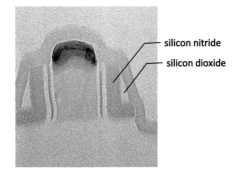

silicon nitride

silicon dioxide

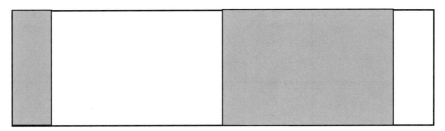

Fig. 6.41 NMOS deep source and drain pattern

6.6.2   NMOS Deep Source/Drain Dopant Pattern and Implant

NMOS deep source/drain diodes are formed by implanting n-type dopants with
series of energies and concentrations to engineer compromises between requirements for low resistance, low capacitance, and high breakdown voltage.

The NMOS deep source/drain (NSD) pattern (Fig. 6.41) is formed with photoresist on the surface of the wafer (Fig. 6.42). NSD photoresist pattern geometries
cover the PMOS transistor regions and cover the PWELL contacts. N-type dopants
(phosphorus and arsenic) are implanted self-aligned to the sidewalls on the NMOS
transistor gate to form the deep NSDs (Fig. 6.42). The NSDs are doped heavily with
As at the surface to provide a low-resistance path between the metal-filled contact

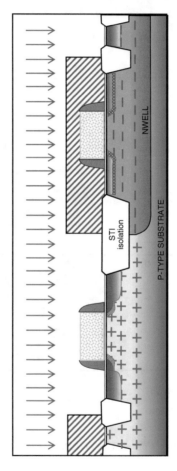

Fig. 6.42 Patterned NMOS deep source and drain implant

plug and the NMOS transistor channel. Lighter doses of As and P are implanted deeper to grade the junction. The lighter doping below the highly doped surface provides a wider depletion region between the bottom of n-type NSD and the p-type substrate. The wider depletion region gives lower capacitance, higher diode break-down voltage, and reduced reverse-biased diode leakage. The depletion region must be kept sufficiently narrow, however, to prevent the source and drain depletion regions from approaching each other under the channel allowing leakage current to flow when the NMOS transistor is OFF and causing the parasitic NPN transistor to turn ON when the NMOS transistor is turned ON. The NSD dopants are also implanted into the NWELL contact to lower the resistance of the NWELL contact.

6.6.3   PMOS Transistor Deep Source/Drain Dopant Pattern & Implant

A PMOS deep source/drain (PSD) pattern (Fig. 6.43) is formed on the surface of the wafer. PSD photoresist pattern geometries cover the NMOS transistor region and cover the NWELL contact (Fig. 6.44). P-type dopant, boron (B), is implanted self-aligned to the silicon nitride sidewalls on the PMOS transistor gate to form the deep

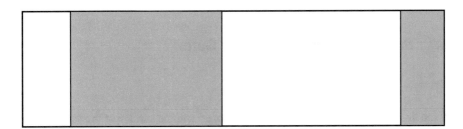

Fig. 6.43 PMOS deep source and drain implant pattern on reticle

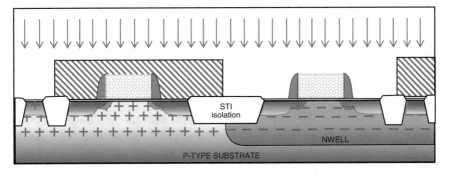

Fig. 6.44 Patterned PMOS deep source and drain photoresist pattern and dopant implant on silicon wafer

PSDs (Fig. 6.44). The PSDs are doped heavily with B at the surface to provide the PMOS transistor with low series resistance. Lighter doses of B are implanted deeper to grade the junction and provide a wider depletion region between the p-type PSD and the n-type NWELL. The deep, lighter doping gives lower capacitance, higher diode breakdown voltage, and reduced reverse-biased diode leakage. The depletion region must be kept sufficiently narrow, however, to keep the source and drain depletion regions sufficiently apart to prevent leakage current under the channel when the PMOS transistor is turned OFF and to prevent the parasitic NPN bipolar transistor from turning ON when the PMOS transistor is turned ON. The PSD dopant is also implanted into the PWELL contact to lower the resistance of the PWELL contact.

6.6.4 Source/Drain Extension and Deep Source/Drain Diode Anneal

The NMOS and PMOS deep source/drain diode and source/drain extension implants are then annealed at a temperature of about 900 °C or more to activate the dopants (replace silicon atoms in the single crystal silicon lattice with dopant atoms). The anneal is performed for a second or less in rapid thermal anneal (RTA) equipment.

6.7 Silicided Source, Drains, and Gates

6.7.1 Introduction to Silicides

Even though polysilicon gates are heavily doped and the surface of the source/drain pn-junctions are heavily doped, the resistance of the gates and the source/drains still limits IC performance. For example, the resistance of a heavily doped polysilicon gate word line delays an electrical signal when it crosses a memory array causing the circuit to wait for the signal to arrive (Electrical signals travel close to the speed of light and the distance across a memory array is less than a quarter inch!) To reduce the resistance of the heavily doped polysilicon and the heavily doped source/drains, nickel is deposited on the surface of the wafer and reacted with the single crystal silicon that is exposed on the surface of the source/drains and gates to form nickel silicide. The resistance of the nickel silicide is orders of magnitude lower than the resistance of heavily doped silicon.

The resistivity of heavily doped silicon is on the order of 15 ohm-cm (depends upon the doping level). The resistivity of nickel silicide is on the order of 15 $\mu\Omega$-cm ($\mu = 10^{-6}$)—a million times less! Electrical signal traveling near the speed of light can now cross the memory array in an IC chip on your cell phone with time to spare.

6.7.2 Nickel Silicide Formation

Metallic nickel (sometimes with platinum and titanium nitride) is deposited (Fig. 6.45) onto the wafer and reacted for several seconds using rapid thermal anneal (RTA) at a temperature of about 300 °C. Where silicon is exposed on the polysilicon gates and on the source/drains, the nickel metal reacts with the exposed silicon to form a mixture of NiSi and $Ni_{31}Si_{12}$ crystalline phases of nickel silicide. Over dielectric regions, where silicon is not exposed, the nickel does not react with SiO_2 and remains nickel metal.

6.7.3 Nickel Metal Strip

After the nickel silicide is formed, the wafers are immersed in a liquid etching bath that etches metallic nickel without etching nickel silicide. The metallic nickel that otherwise would short the gates and source/drains together is etched away from the dielectric regions (Fig. 6.46).

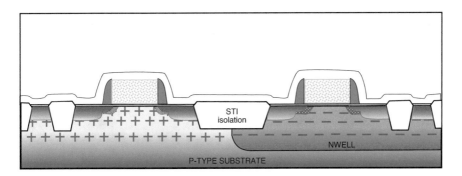

Fig. 6.45 Nickel deposited for nickel silicide formation

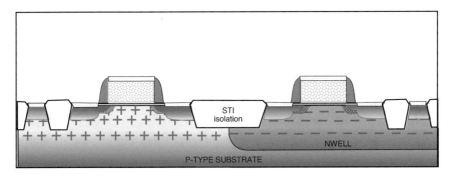

Fig. 6.46 CMOS inverter with silicided gates, sources, and drains

6.7.4   Nickel Silicide Resistance Lowering Anneal

After the unreacted nickel metal strip, the wafers are annealed for a few microseconds using a spike (about a thousandth of a second) anneal at a temperature of about 500 °C to covert the mixed NiSi and $Ni_{31}Si_{12}$ crystalline phases that are formed during the 300 °C nickel silicide reaction anneal into a single, low-resistance phase of NiSi.

The nickel silicide 300 °C formation anneal temperature needs to be high enough to form nickel silicide over exposed silicon source/drain and gate regions but not so high that it forms nickel silicide over silicon dioxide isolation dielectric regions. Resistance must be low over the source/drain and gate regions, but infinite over the dielectric regions. If a formation temperature of 500 °C is used, nickel silicides also forms on silicon dioxide dielectric and cannot be etched away by the nickel metal etching solution. Nickel silicide on the dielectric regions shorts the source, drains, and gates together.

References

1. J. Chappell, It's all in the plasma, EDN (2001), https://www.edn.com/its-all-in-the-plasma/
2. Intel 45 nm Metal Replacement Gate with SiGe Replacement Source and Drains, WikiChip, https://en.wikichip.org/wiki/File:intel_45nm_gate.png. Accessed 4 Mar 2020

Chapter 7
CMOS Inverter Manufacturing Flow: Part 2 Transistors Through Single-Level Metal

7.1 Introduction

A cross-sectional view of the CMOS inverter [1] with a first level of metal wiring (metal1) is shown in Fig. 7.1. A multilayered premetal dielectric (PMD) electrically isolates the metal1 wiring layer from the underlying transistors. Contact holes are etched through the PMD and filled with metal plugs to electrically connect the metal1 wires to the transistor source/drain and gate terminals.

The cross section is taken through the signal in node (shorted gates) of the CMOS inverter. The contact holes to the WELLs (V_{SS} and V_{DD}) are much deeper than the contact holes to the NMOS and PMOS transistor gates.

After silicide is formed on the surface of the MOS transistor source, drain, and gate terminals, a composite dielectric layer is formed to prevent metal wires in overlying interconnect layers from shorting to the transistors. The composite dielectric layer typically is a thin contact etch-stop layer (CESL) dielectric layer, a gap fill dielectric layer (PMD1) and dielectric capping layer (PMD2). The contact etch-stop layer enables shallow contacts to the top of the transistor gates to be etched at the same time as the deep contacts to the wells and to the transistor source and drains. The gap fill layer is deposited using a modified CVD process that fills the narrow gaps between minimum spaced transistor gates without forming voids. After planarization using CMP, a premetal dielectric capping layer (PMD2) is deposited to a final PMD thickness specification.

© Springer Nature Switzerland AG 2020   103
H. Tigelaar, *How Transistor Area Shrank by 1 Million Fold*,
https://doi.org/10.1007/978-3-030-40021-7_7

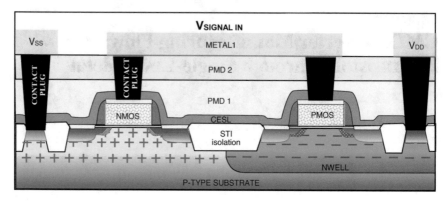

Fig. 7.1 CMOS inverter with first level of metal wiring

7.2   Premetal Dielectric Layers (PMD)

7.2.1   Contact Etch-Stop Layer (CESL)

The contact etch-stop layer (CESL) is a layer of silicon nitride (SiN) with a thickness in the range of 150–350 nm. SiN is deposited by low-pressure chemical vapor deposition (LPCVD) using silane SiH_4 and ammonia NH_3 plus Argon (Ar) carrier gas.

As is illustrated in Fig. 7.1, contact holes that are etched through the PMD layers to the top of the transistor gates are much shallower than contact holes etched through the PMD layers to the transistor source and drains and to the well contacts. The contact plasma etch opens the contact holes to the top of the gates long before the contact holes to the substrate open. Severe damage to the transistor gate and to the transistor gate dielectric occurs if the long plasma over etch is allowed to etch the exposed transistor gate.

The contact etching equipment and process engineers developed contact etch equipment and processes specifically for this problem. They developed a silicon dioxide (SiO_2) dielectric etch that does not etch silicon nitride (SiN) dielectric and developed a SiN etch that does not etch silicide. They also developed a plasma etching tool that can perform both etches sequentially in the same etching chamber.

The contact plasma etch drills contact holes through the PMD silicon dioxide and stops when it hits the CESL layer. The CESL layer protects the transistor gates while the contact holes over the source/drains continue to etch until they are open. After the contact holes to the top of the gates and to the source/drains are open down to the CESL layer, the contact plasma etch chemistry is changed to etch through the CESL layer and stop when it hits the underlying nickel silicide on the transistor source/drains and gates (Fig. 7.2).

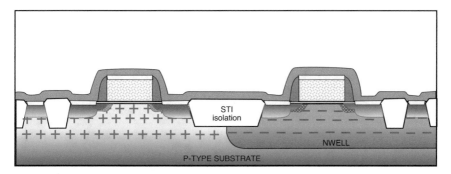

Fig. 7.2 Silicon nitride contact etch-stop liner (CESL)

Fig. 7.3 Void in PMD1
gap fill dielectric between
minimum spaced transistor
gates. SEM is courtesy of
Texas Instruments

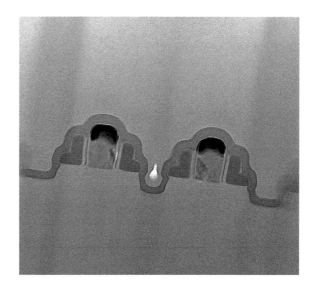

7.2.2   Gap Fill Dielectric: PMD1

A first layer of premetal dielectric (PMD1) that can fill the narrow gaps between transistor gates that are spaced minimum design rule apart is deposited on the CESL layer. Figure 7.3 is a SEM cross section showing where the PMD1 gap fill was inadequate and left a void. Voids in PMD1 can later fill with the chemical-vapor-deposited tungsten (CVD-W), while the CVD-W is filling the contact plugs. This shorts adjacent transistors together causing ICs to fail Fig. 11.4.

PMD1 is deposited using a highly modified SACVD process called eHARP™ (enhanced high aspect ratio process). eHARP™ SiO_2 is deposited using TEOS, ozone (O_3), and water (H_2O) at a temperature of about 450 °C and pressure of about 600 Torr. The eHARP™ process fills narrow gaps void-free.

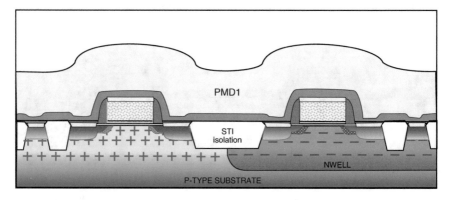

Fig. 7.4 Premetal dielectric layer (PMD1)

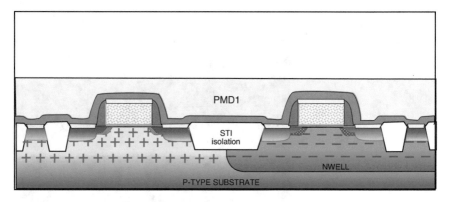

Fig. 7.5 Chemical mechanical polish (CMP) planarized PMD1 layer

The PMD1 layer is deposited to a thickness sufficient to completely fill the gaps between the transistor gates with a thickness at least as high as the transistor gates Fig. 7.4.

7.2.3 PMD1 Planarization

After PMD1 is deposited, it is planarized as shown in Fig. 7.5 using chemical mechanical polish (CMP). A planar PMD surface is critical to provide adequate depth of focus for patterning small contact holes.

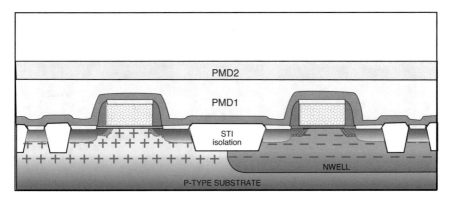

Fig. 7.6 USG or PSG capping dielectric layer (PMD2)

7.2.4   PMD2

A second pre-metal dielectric layer (PMD2 or capping layer) is deposited on the planarized PMD1 dielectric layer (Fig. 7.6). The PMD2 layer can be undoped silicate glass (USG) or phosphorus doped silicate glass (PSG). The PMD2 layer is deposited using a plasma-enhanced chemical vapor deposition (PECVD). Deposition gases are tetraethyl ortho silicate (TEOS, $Si(OC_2H_5)_4$) in Argon (Ar) carrier gas for USG or TEOS plus PH_3 in Ar for PSG.

7.3   Contacts

A contact photoresist pattern is formed on the PMD layer and contact holes are etched through the underlying PMD dielectric stack exposing the nickel silicide on the source/drain diodes, gates, and WELL contacts. Contact holes also expose the nickel silicide on the tops of the doped polysilicon transistor gates. These contact holes are filled with metal to form electrical connection between the transistor terminals and the overlying wiring layer.

7.3.1   Contact Pattern

A top-down view of the CMOS inverter contact pattern as it appears on the contact reticle is shown in Fig. 7.7. A first cross-sectional view of the contact photoresist pattern through dashed line B-B as it appears on a silicon wafer is shown in Fig. 7.8. The first cross-sectional view in Fig. 7.8 is through the signal output node of the CMOS inverter. A second cross-sectional view of the contact photoresist pattern through dashed line C-C as it appears on a silicon wafer is shown in Fig. 7.9. The

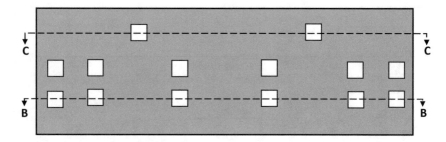

Fig. 7.7 Top-down view of contact pattern on reticle

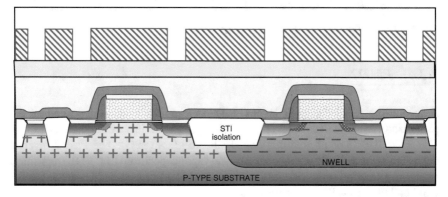

Fig. 7.8 Cross-sectional view of the contact photoresist pattern through the output node of the CMOS inverter

second cross-sectional view in Fig. 7.9 is through the signal input node of the CMOS inverter.

7.3.2   Contact Etch

Figure 7.10 is a cross section view through the source/drain contact openings etched through the PMD and the CESL layers. As discussed previously, an anisotropic (etches vertically but not horizontally) plasma oxide etch with high selectivity to silicon nitride (etches oxide but does not etch silicon nitride) is used to etch contact openings through PMD layer stopping on the silicon nitride CESL layer. The plasma etch chemistry is then changed to a silicon nitride etch with high selectivity to the underlying nickel silicide.

The contact openings to the source/drains (Fig. 7.10) are much deeper than the contact openings to the transistor gates (Figs. 7.11 and 7.1).

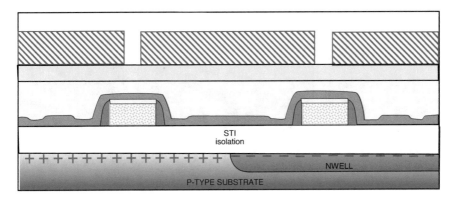

Fig. 7.9 Cross-sectional view of the contact photoresist pattern through the input node of the CMOS inverter

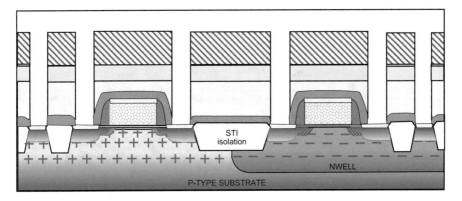

Fig. 7.10 Cross-sectional view through substrate contacts post etch

7.3.3 Contact Barrier Layer

The contact holes are filled with chemical-vapor-deposited tungsten metal (CVD-W). CVD-W is deposited using the thermal decomposition of tungsten hexafluoride (WF_6). When a WF_6 molecule hits the hot surface on the bottom and walls of the contact hole, it decomposes depositing tungsten atoms which fill the narrow, deep contact holes with tungsten metal. Fluorine gas (F_2) produced during the decomposition gets pumped away. WF_6 molecules are small and readily diffuse into the very small, deep contact holes filling them without forming voids.

If not protected by a barrier layer, the nickel silicide at the bottom of the contact openings and the silicon dioxide sidewalls of the contact holes would be attacked and etched by the fluorine gas generated during the decompositionof the WF_6. To protect the nickel silicide and the PMD, a thin titanium (Ti) adhesion layer followed

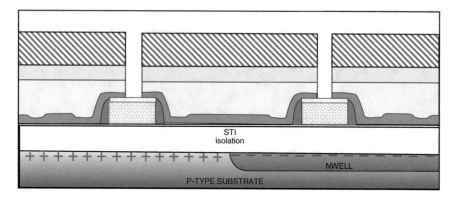

Fig. 7.11 Cross-sectional view through gate contacts post etch

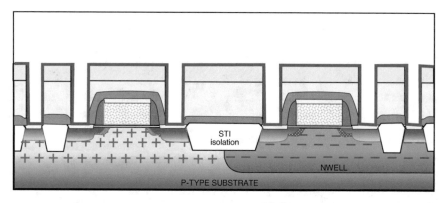

Fig. 7.12 Titanium adhesion layer plus a titanium nitride barrier contact liner protects the nickel silicide and PMD from attack by fluorine gas

by a thin titanium nitride (TiN) barrier layer are deposited on the bottom and sidewalls of the contact holes (Fig. 7.12). The Ti lowers contact resistance by reacting with any residual oxide on the surface of the nickel silicide turning it into titanium oxides (TiO$_X$) which are conductive. The titanium nitride barrier layer blocks F$_2$ from etching the underlying nickel silicide and from etching the PMD. Titanium is deposited using ion metal plasma (IMP) deposition. The titanium nitride barrier layer is deposited using a chemical vapor deposition (CVD) process with titanium tetrachloride (TiCl$_4$) gas as the titanium source plus ammonia (NH$_3$) as the nitrogen source. The titanium IMP process was invented and developed when conventional sputtered titanium could no longer deposit sufficient titanium in the bottom of high aspect ratio (narrow and deep) contact holes. The CVD TiN process was invented and developed when reactive sputtered TiN no longer formed a sufficiently uniform layer on the sidewalls of the high aspect ratio contact holes.

Ion Metal Plasma (IMP) Sputter Deposition

The IMP equipment and process were invented to resolve the problem of coating the bottoms and sidewalls of high aspect ratio (narrow and deep) contact openings and metal trenches.

In the ion metal plasma process (IMP), positively charged Argon atoms are used to sputtered metal atoms and metal clusters from a metal target. The sputtered metal atoms and clusters are then ionized in an intense plasma and are accelerated with an electric field toward the surface of the wafer and into the contact holes. The high directionality of IMP titanium atoms enables them to reach the bottom of very deep and narrow contact holes. When titanium is being IMP sputtered, nitrogen (N_2) can be added to the argon plasma to produce highly reactive nitrogen atoms. The IMP sputtered titanium atoms readily react (reactive sputtering) with the nitrogen ions to form titanium nitride (TiN).

7.3.4 Contact Hole Metal Fill

Chemical-vapor-deposited tungsten (CVD-W) deposits conformally onto the surfaces of the wafer filling and overfilling the contact holes (Fig. 7.13).

7.3.5 Contact CMP

Chemical mechanical polishing (CMP) is used to remove the CVD-W overfill and to remove the Ti/TiN liner from the surface of the PMD (Fig. 7.14).

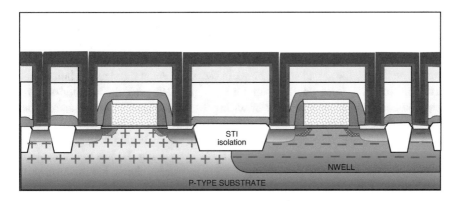

Fig. 7.13 CVD tungsten metal filling the contact holes

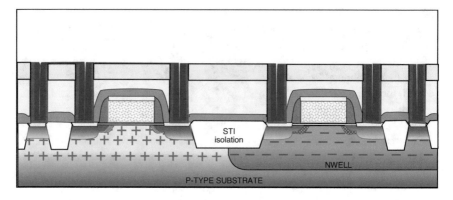

Fig. 7.14 CVD tungsten metal contact plugs post CMP

7.4   Metal1 Wiring Layer

7.4.1   Overview

Metal1 copper wires are formed using a damascene process [2]. A single damascene process is used to fill single level metal1 (SLM) trenches. A dual damascene process is used to sumultaneously fill vias and metal trenches in additional layers of multi-level metal (MLM). The damascene name comes from the process for making damascene jewelry. Damascene jewelry is made by etching trenches into a piece of metal jewelry and then refilling the trenches with a metal of a different color (inlaying). In the metal1 copper damascene process, metal1 trenches are etched into a dielectric layer and then refilled with electroplated copper. Copper chemical mechanical polish removes excess copper leaving copper wires filling the metal1 trenches.

Some integrated circuits have a dozen or more layers of metal wiring. As integrated circuits shrank and the transistor count on integrated circuit chips rose into the billions (i.e., 32, 64, and 128 Gbit memory chips. 1 Gbit = 1 billion), more and more wiring layers were required to route the signals between the billions of transistors. Even though the signal wires shrank in size from about the size of a human hair (100 µm) to 1000 times smaller than a human hair (100 nm), the number of wires required could not fit on one layer of wiring. Multiple layers of wires stacked one on top of the other were engineered to provide the needed wiring resources.

A cross-sectional view of the CMOS inverter showing wiring connections to the power supply and to the input/output signals is shown in Fig. 7.15. A top-down view showing the metal1 wiring layer on the CMOS inverter is shown in Fig. 7.16.

A cross-sectional view of the CMOS inverter showing the transistors and the first wiring layer is shown in Fig. 7.17 (taken along dashed line B-B in Fig. 7.16).

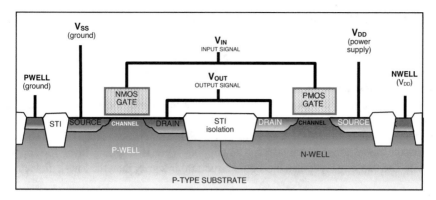

Fig. 7.15 CMOS inverter showing signal and power connections OS

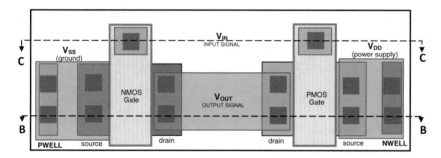

Fig. 7.16 Top-down view of CMOS inverter metal1 layer on a wafer

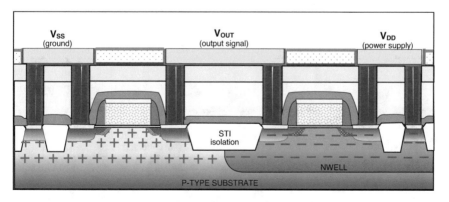

Fig. 7.17 Cross-sectional view of CMOS inverter taken along dashed line B-B in Fig. 7.16 through the inverter output node

7.4.2 First Intermetal Dielectric (IMD1)

The metal1 wiring layer is formed by etching metal1 trenches into the IMD1 dielectric layer and then refilling the trenches with copper metal. The IMD1 dielectric stack has three dielectric layers: a thin metal1 trench etch-stop layer (M1TESL), a low dielectric constant (low k) layer [3], and a thin dielectric capping layer. The metal1 trench etch stops on the M1TESL stopping layer. The low-k dielectric layer forms the bulk of the IMD1 layer. The low-k dielectric reduces cross talk of signals between the adjacent metal wires. The capping layer is stopping layer for copper CMP. The capping layer also has better structural integrity than the low-k dielectric.

IMD1 Metal1 Trench Etch-Stop Layer

One M1TESL material is silicon carbon nitride (SiCN) with k ~4.8. It can be deposited at about 300–400 °C using PECVD with trimethyl silane $((CH_3)_3SiH)$ plus ammonia (NH_3). The SiCN M1TESL is deposited on the PMD layer and over the contact plugs (Fig. 7.18). This metal1 trench etch-stop layer (M1TESL) functions for the metal1 etch much the same as the contact etch-stop layer (CESL) functions for the contact etch.

IMD1 Low-k Dielectric

A low-k dielectric constant (less than about 3.5) film is deposited on the M1TESL (Fig. 7.19). There are a number of low-k dielectric materials used in semiconductor manufacturing (Table 12.2). Black Diamond II™ and Coral™ are two examples of organosilicate glass (OSG) low-k dielectrics deposited using chemical vapor

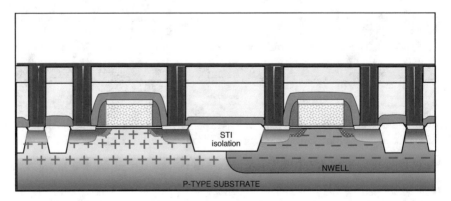

Fig. 7.18 CMOS inverter with metal1 trench etch-stop layer (M1TESL) deposited on the PMD and contact plugs

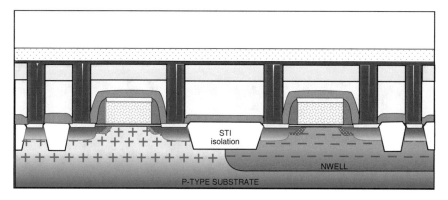

Fig. 7.19 CMOS inverter with low-k dielectric deposited on the M1ESL

deposition (CVD) processes. The dielectric constant of these OSGs is in the range of about 2.7–3.0.

Why Low k Is Important
An electrical pulse traveling down one wire can induce a mirrored pulse on an adjacent wire via capacitive coupling. This coupling is stronger when the dielectric constant is high (hi-k). This mirrored pulse slows and degrades the electric signal and can cause logic errors. When the dielectric constant of the dielectric between the wires in an IC is lower, electrical signals travel faster and are degraded less.

IMD1 Capping Layer

A capping layer such as a PETEOS or silicon nitride (SiN) is deposited on the low-k IMD1 layer Fig. 7.20. The capping layer performs dual functions. It is a chemical mechanical polish stopping layer for copper CMP. It also caps the low-k dielectric with a more structurally robust dielectric. When the capping layer is silicon nitride it also functions as an antireflective coating for the metal1 trench pattern.

7.4.3   Metal1 Trench Pattern and Etch

Metal1 Trench Pattern

A top-down view of the metal1 trench pattern as it appears on the photomask is shown in Fig. 7.21.

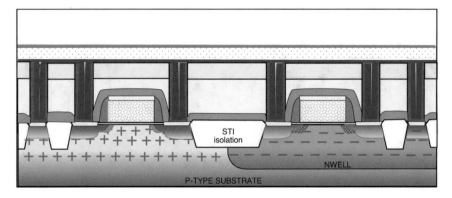

Fig. 7.20 CMOS inverter IMD1 layer comprised of a metal1 etch-stop layer, a low-k dielectric layer, and a capping layer

The metal1 trench photoresist pattern as it appears in cross section through the output node of the CMOS inverter (line B-B in Fig. 7.16) on the IMD1 dielectric layer on the wafer is shown in Fig. 7.22.

Metal1 Trench Etch

Using the metal1 trench pattern, metal1 damascene trenches are etched through the IMD1 stack exposing the tops of the contact plugs (Fig. 7.23).

7.4.4   Metal1 Trench Copper Fill

Before the metal1 trenches are filled with copper, a tantalum barrier layer is deposited to prevent copper diffusion into the surrounding IMD1 dielectric. Copper is a fast diffuser and would diffuse to and adversely affect the MOS transistors. A thin copper seed layer is then deposited on the Ta barrier layer to provide a low resistance path for the copper electroplating current. Electroplated copper metal fills and overfills the metal1 trenches. Copper overburden is removed using CMP.

Metal1 Trench Barrier Layer

After the metal1 damascene trenches are etched, the photoresist pattern is removed and a tantalum nitride/tantalum adhesion/barrier layer is deposited. Tantalum nitride (TaN) adheres to silicon dioxide and is a good barrier to copper metal diffusion. Tantalum metal is also a good barrier to copper diffusion and enhances adhesion

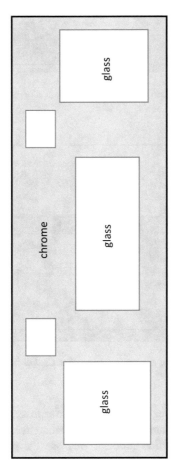

Fig. 7.21 Top-down view of CMOS inverter metal1 pattern on a reticle

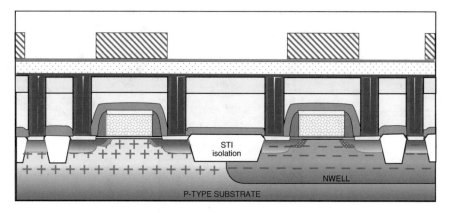

Fig. 7.22 Metal1 photoresist pattern on IMD1 layer taken along through the CMOS inverter output node

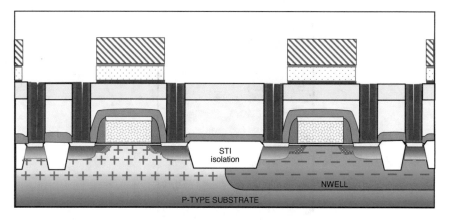

Fig. 7.23 Cross-sectional view of the output node of a CMOS inverter with metal1 trenches etched through IMD1

between the copper seed layer and the TaN barrier layer. The tantalum nitride is deposited using CVD. The tantalum layer is deposited using an enhanced ion metal plasma (IMP) sputter deposition process. Because the resistance of the TaN/Ta (~500 μΩ-cm) is much higher than copper (~0.02 μΩ∗cm), the layers of TaN/Ta are deposited as thin as possible while still maintaining barrier integrity.

Metal1 Trench Copper Seed Layer

A thin copper seed layer which is required to provide sufficient electrical current for copper electroplating is deposited on the TaN/Ta barrier layer using a self-ionized plasma (SIP) process (Fig. 7.22).

Metal1 Trench Copper Fill

The wafers are then immersed into a copper plating solution. Electroplating current flows through the thin copper seed layer causing copper ions from the solution to deposit on the copper seed layer as metallic copper. The negative voltage attracts positive copper ions (Cu^{++}) to the surface of the copper seed layer. The copper ions pick up electrons from the copper seed layer. The electrons neutralize the positive charged copper ions and covert them to copper metal ($Cu^{++} \rightarrow Cu^0$). Copper metal deposits on the copper seed layer filling and overfilling the metal1 damascene trenches with copper (Figs. 7.24 and 7.25).

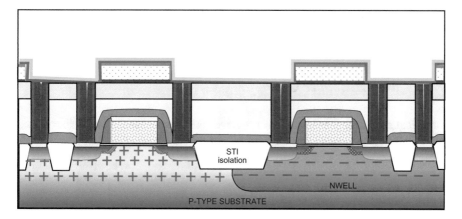

Fig. 7.24 Cross section of CMOS inverter through output node. Copper seed and TaN barrier layers cover trench surfaces

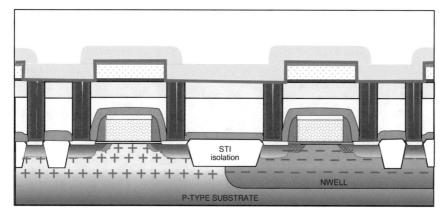

Fig. 7.25 Cross section of CMOS inverter through output node. Metal1 trenches are filled with electroplated copper

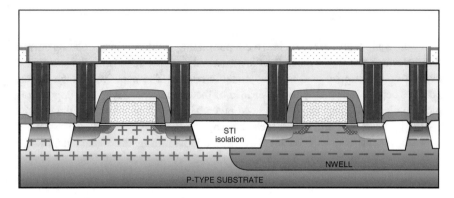

Fig. 7.26 Cross section through output node of CMOS inverter after metal1 copper CMP

7.4.5   Metal1 Copper CMP

Chemical mechanical polish (CMP) removes the copper overfill and also removes the TaN/Ta barrier layer from the surface of the IMD1 layer (Fig. 7.26).

7.5   Protective Overcoat and Bond Pads

Single-level metal (SLM) IC chip wiring processing stops with metal1. In SLM chips a protective overcoat (PO) dielectric layer such as silicon nitride (SiN), silicon oxynitride (SiO_XN_Y), or polyimide covers the IC to protect it from environmental gases such as moisture or corrosive gases. Aluminum bond pads are formed on the PO layer and shorted to metal1 through openings in the PO (see Fig. 7.27). These processes are described in more detail in Sect. 8.4.2 of Chap. 8.

Later when the IC chip is packaged, bond wires are connected to the bond pads to provide electrical connections to power supplies and to enable communication with other integrated circuits.

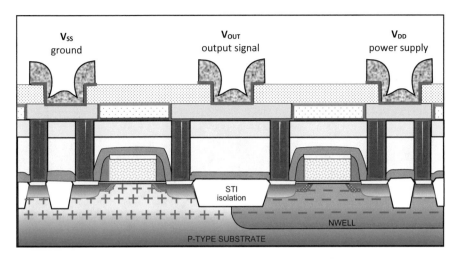

Fig. 7.27 Cross section through the output node of a SLM CMOS inverter after protective over-coat (PO) and aluminum bondpad formation

References

1. A. Mason, Chapter 7 CMOS Inverter: DC Analysis, ECE 410 Lecture Notes, Michigan State University, https://www.egr.msu.edu/classes/ece410/mason/files/Ch7.pdf. Accessed 4 Mar 2020
2. Damascene Jewelry, Wikipedia, the free encyclopedia, https://en.wikipedia.org/wiki/Damascening. Accessed 4 Mar 2020
3. K. Saraswat, Low-k Dielectrics, EE311 Lecture Notes, Stanford University, https://web.stanford.edu/class/ee311/NOTES/Interconnect%20Lowk.pdf. Assessed 4 Mar 2020

Chapter 8
CMOS Inverter Manufacturing Flow: Part 3 Additional Levels of Metal Through PO

8.1 Overview

Additional wiring layers are added on top of the metal1 wiring layer using a dual damascene process. The process for building the metal2 wiring (interconnect) layer is similar to the process for building the metal1 interconnect layer with additional steps to form damascene copper-filled via1 holes that electrically connect the metal2 copper wires to the metal1 copper wires. The dual damascene process simultaneously fills the metal trench and the vias that connect the metal trench to underlying metal leads with copper.

A cross-sectional view of the CMOS inverter in Fig. 8.1 shows the terminal connections to the power supply, ground, input and output signals.

A top-down view of the via1 hole and the metal2 wiring layers on the CMOS inverter is shown in Fig. 8.2. The metal2 layer shorts the NMOS and PMOS transistor gates together through via1 holes (Fig. 8.3). The input signal to the CMOS inverter is applied to this metal2 lead. The via1 holes on the V_{SS}, V_{DD}, and V_{OUT} terminals provide electrical connection between the bottom of metal2 leads and the top surface of the underlying metal1 interconnect (Fig. 8.4).

8.2 Metal2 Wiring Layer

The metal2 interconnect process is complicated by the addition of via1 holes that connect the bottom of the metal2 damascene trenches to the top of the metal1 leads. Because both the via1 holes and the metal2 trenches are simultaneously filled with electroplated copper, this process is called dual damascene. The dual damascene process requires two photomask patterning and etching steps: A first pattern and etch for the via1 holes and a second pattern and etch for the metal2 trenches.

© Springer Nature Switzerland AG 2020
H. Tigelaar, *How Transistor Area Shrank by 1 Million Fold*,
https://doi.org/10.1007/978-3-030-40021-7_8

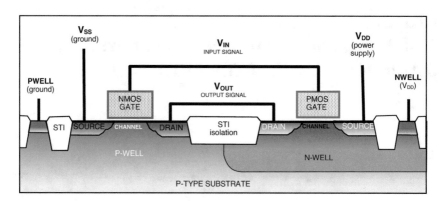

Fig. 8.1 Cross section of a CMOS inverter showing terminal connections to power supply, ground, input and output signals

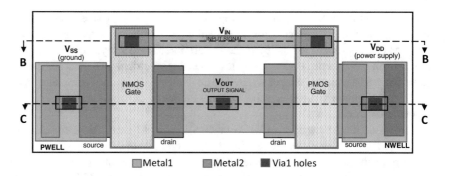

Fig. 8.2 Top-down view of metal2 on CMOS inverter.

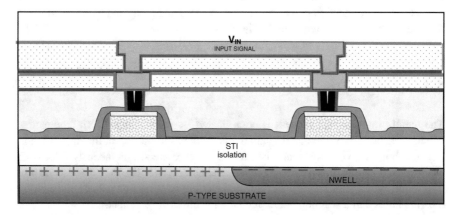

Fig. 8.3 Cross-sectional view of CMOS inverter through gate contacts and through the CMOS inverter metal2 input signal lead. (Through dashed line B-B in Fig. 8.2)

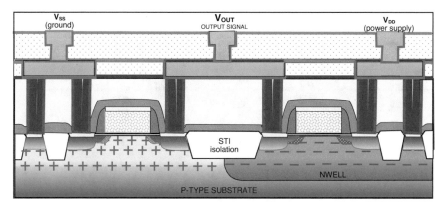

Fig. 8.4 Cross-sectional view of CMOS inverter through substrate contacts and through the CMOS inverter metal1 output signal lead. (Through dashed line C-C in Fig. 8.2)

8.2.1 Metal2/Metal1 Intermetal Dielectric: IMD2

The IMD2 dielectric stack is deposited on the IMD1 dielectric stack and on the metal1 interconnect layer. The IMD2 dielectric stack is SiCN1/OSG1/SiCN2/OSG2/SiN. See Table 12.2 for dielectric compositions. SiCN1 is an etch-stop layer for Via1 plasma etch (V1ESL). SiCN2 is an etch-stop layer for metal2 trench plasma etch (M2TESL). The IMD2 stack materials are the same as IMD1 stack materials except the capping layer is silicon nitride (SiN) instead of PETEOS. Even though SiN has a higher dielectric constant (k ~7) than PETEOS (k ~4) it is used because it provides the dual function of being an excellent antireflection layer for metal2 trench and via1 patterning and a good stopping layer for copper CMP. Most of the high dielectric constant SiN layer is removed with copper CMP over polish (Fig. 8.5).

8.2.2 Via1 Pattern and Etch

The via1 pattern is shown as it appears on the reticle in Fig. 8.6. A cross section through the via1 photoresist pattern on the CMOS inverter is shown in Fig. 8.7. This cross section is along dashed line B-B in Fig. 8.6 (Through the input node of the CMOS inverter where the NMOS and PMOS transistor gates are shorted together).

Figure 8.7 is after via1 etch. The via1 plasma etch sequentially etches through the SiN dielectric capping layer, the OSG2 low-k dielectric layer, the SCN2 M2ESL, and part of the way through the OSG1 low-k layer. This is performed in situ in the same plasma etching chamber by changing etching gases and processing conditions.

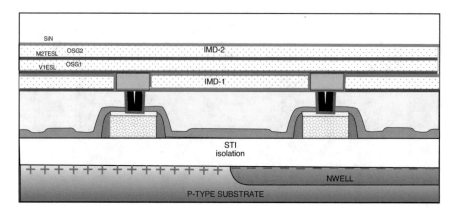

Fig. 8.5 Cross-sectional view of CMOS inverter with the IMD2 layer stack deposited on the IMD1 layer and on the metal1 interconnect

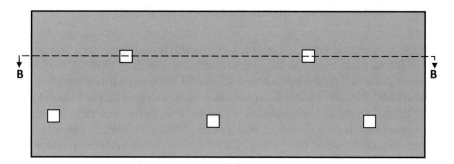

Fig. 8.6 Dual damascene via1 pattern on the reticle

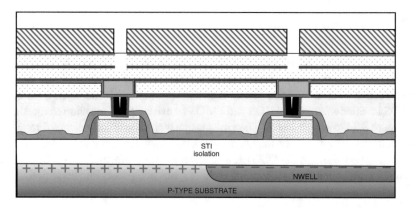

Fig. 8.7 Dual damascene via1 photoresist pattern on the CMOS inverter. Cross section along dashed line B-B through input node

8.2.3   Metal2 Trench Pattern and Etch

After via1 etch, the via1 photopattern is stripped off and the metal2 trench pattern is printed in photoresist on the IMD2 layer. The metal2 trench pattern as it appears on the reticle is shown in Fig. 8.8. The metal2 trench photoresist pattern as it appears on the input signal node (through cut line B - B) of the CMOS inverter is shown in Fig. 8.9.

The metal2 plasma trench etch removes the SiN capping layer and etches the OSG2 layer stopping on the SiCN M2TESL. The metal2 plasma etch also etches the OSG1 layer in the via1 holes stopping on the SiCN V1ESL. The plasma chemistry is then changed to remove the V1ESL from the bottom of the vial1 holes and remove the M2TESL from the bottom of the metal2 trenches.

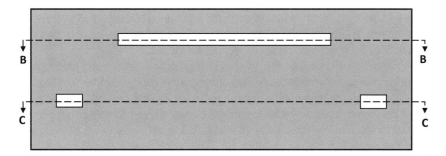

Fig. 8.8 Dual Damascene metal2 trench pattern on the reticle

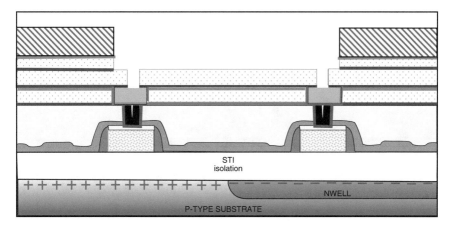

Fig. 8.9 Cross section through input node of CMOS inverter after metal2 etch

8.2.4 Via1 and Metal2 Trench Fill

Via1 and Metal2 Barrier and Copper Seed Layers

Depositing uniformly thick barrier and copper seed layers in dual damascene metal2 is much more difficult than in single damascene metal1. Getting uniform TaN/Ta barrier plus Cu seed layers onto the bottom and sidewalls of the small via1 openings at the same time as on the bottom and sidewalls of the metal2 trenches is extremely challenging. These challenges drove virtually constant upgrades to the deposition processes and the invention of a number of new deposition processes and new deposition tools as integrated circuit geometries scaled (Fig. 8.10).

Via1 and Metal2 Trench Copper Fill

Figure 8.11 shows a cross section through the input node of the CMOS inverter after metal2 copper is electroplated filling and overfilling the via1 holes and the metal2 trenches. Via1 holes connect metal2 wires to metal1 wires.

The chemistry and engineering of the electroplating solutions and electroplating equipment used in semiconductor manufacturing is extremely complex. Uniform current is applied at the edges of the wafer and must be uniform across the wafer to deposit the copper uniformly and deposit the copper without trapping voids. Special additives to the electroplating solution enable the small vias to fill without voids before the overlying metal2 trenches fill. This complicated chemistry plus very specialized sequencing of electroplating voltages and current fills both the via1 holes

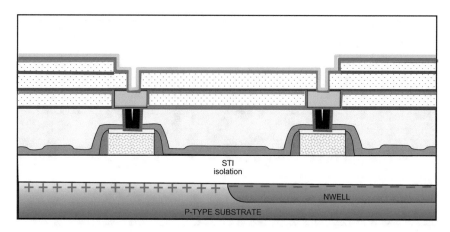

Fig. 8.10 Cross section through input node of CMOS inverter after via1 holes and metal2 trenches are coated with Ta/TaN barrier plus copper seed layers

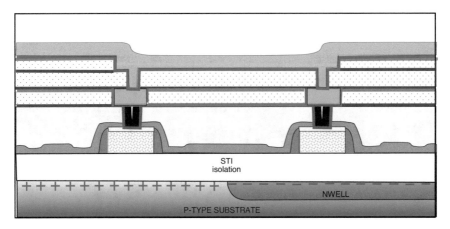

Fig. 8.11 Cross section through input node of CMOS inverter after via1 holes and metal2 trenches are filled and overfilled with electroplated copper metal

and the metal2 trenches without forming voids. Metallization engineers work with designers, photolithographers, barrier and seed deposition engineers, and copper plating engineers to co-optimize the layout, the optical proximity corrections, the depth and size of the via1 openings and the metal2 trenches, the thickness of the barrier and seed layers, and copper plating process. Many compromises between these process variables and tweaks to these processes are made before a manufacturable a dual damascene copper process for ICs is realized. Electroplating Copper Interconnects [1].

8.2.5   Metal2 Copper CMP

Figure 8.12 shows a cross section through the input node of the CMOS inverter after the copper overfill is removed using chemical mechanical polish (CMP). CMP also removes the TaN/Ta barrier layer from the surface of the IMD2. The barrier layers are conductive and would cause shorts between the metal2 leads if not removed. The metal2 lead shorts the NMOS and PMOS gates together to form the input signal node of the CMOS inverter.

A metal1 lead shorts the drains of the NMOS and PMOS transistors together to form the output node of the CMOS inverter. A first metal1 lead shorts the NMOS source to the PWELL contact. A second metal1 lead shorts the PMOS source to the NWELL contact (Fig. 8.13). A first metal2 copper lead and copper via connects the NMOS source to ground. A second metal2 copper lead and copper via connects the PMOS drain to the power supply voltage (V_{DD}).

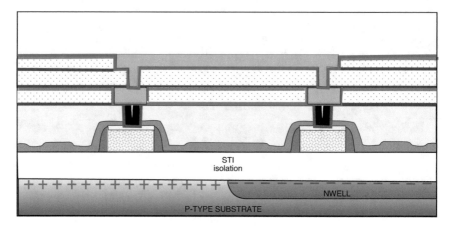

Fig. 8.12 Cross section through input node of CMOS inverter after copper overfill in via1 holes and metal2 trenches is removed using CMP (through dashed line B-B in Fig. 8.8)

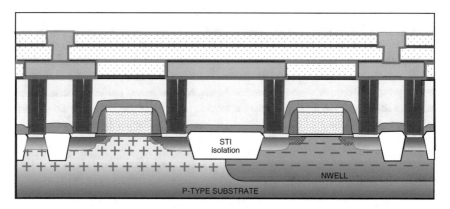

Fig. 8.13 Cross section through output node of CMOS inverter after copper overfill in via1 holes and metal2 trenches is removed using CMP (through dashed line C-C in Fig. 8.8)

8.3   Additional Metal Wiring Levels

Complex integrated circuits can have 15 levels of interconnect or more—15+ layers of intermetal dielectric (IMD) plus 15+ layers of metal. All additional copper metal wiring layers are manufactured by repeating the processes described for via1 and metal2 over and over. The top one or two wiring layers of an integrated circuit are usually aluminum alloy. The aluminum wires are used primarily to distribute the power supply voltage and ground voltages to subcircuits but can also be large power busses for carrying high current. Design rules for these upper aluminum alloy layers are much more relaxed than the copper metal layers.

In the following descriptive manufacturing flow, intermediate levels of copper and aluminum alloy interconnect are omitted. Manufacturing steps for a protective overcoat and an aluminum alloy bond pad are described. The process is illustrated with relaxed design rules (large geometries) and with sputtered aluminum alloy for the bond pad filling a large via hole through the PO.

8.4 Protective Overcoat (PO) and Bond Pad

After the additional wiring layers are completed, a protective overcoat dielectric stack (PO) is deposited on the upper most metal layer to protect it from environmental hazards such as humidity and other corrosive gasses. Holes are then etched through the PO, and bond pads are formed to enable the IC to be powered up by a battery or power supply and to send and receive signals to other ICs and other electrical devices.

8.4.1 PO Layer Deposition, Pattern, and Etch

In Fig. 8.14, an PO via etch-stop layer of silicon nitride (~50 nm) is deposited on the IMD2 and the metal2 interconnect. A protective overcoat (PO) layer of silicon nitride (SiN) or silicon oxynitride (SiO_XN_Y) or polyimide is deposited (~400–600 nm) on the PO via etch-stop layer to protect the integrated circuit from moisture, mobile ions, and other gases and in the environment that would be harmful to the integrated circuit.

Figure 8.15 shows a cross section of the CMOS inverter through the input signal node after PO via photoresist patterning and etching.

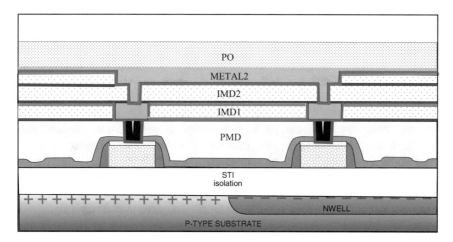

Fig. 8.14 Cross section through input node of CMOS inverter with protective overcoat (PO) etch-stop layer and PO layer

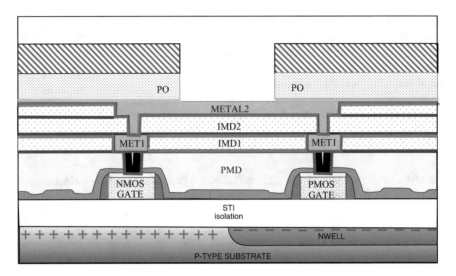

Fig. 8.15 Cross section through input node of CMOS inverter with PO via photoresist pattern

8.4.2 Bond Pad Deposition, Pattern, and Etch

Figure 8.16 shows the CMOS inverter the PO via photoresist pattern is removed, and a tantalum nitride (TaN) adhesion/barrier layer is deposited on the PO layer and into the PO via opening. A titanium/tungsten (Ti/W) barrier layer or a titanium nitride (TiN) are also sometimes used instead of the TaN under aluminum/copper alloy.

Bond pad metal, an alloy of aluminum (Al) and copper (Cu) (~98% Al: 2% Cu), is sputter deposited onto the TaN adhesion/barrier layer. Aluminum alloy thickness is dictated by the aluminum lead resistance requirements. Thickness can range from 400 to 1200 nm (Fig. 8.17). Sputter deposited aluminum fills the PO via hole nonuniformly as shown in Fig. 8.17.

A bond pad photoresist pattern is then formed on the AlCu bond pad metal (Fig. 8.17).

Aluminum alloy plasma etch using chlorine (Cl_2) plus boron trichloride (BCl_3) etching gases. Reactive chlorine atoms and ions produced in the plasma react with the exposed aluminum alloy surface to form volatile aluminum chloride gas ($AlCl_3$) which is pumped away. Argon ions also produced in the etching plasma are accelerated toward the surface of the aluminum by applying a bias to the wafer chuck holding the wafer. The argon atoms impinge on the etching aluminum surface accelerating the vertical etching while not accelerating the horizontal etching (anisotropic aluminum sputter-etch). This produces straight, vertical sidewalls on the etched aluminum bond pad.

A cross section through the bond pad with electrical connection to the input signal node of the CMOS inverter is shown in Fig. 8.18. A cross section through the bond pads with electrical connection to the output signal node of the CMOS inverter and to the voltage terminals of the CMOS inverter is shown in Fig. 8.19.

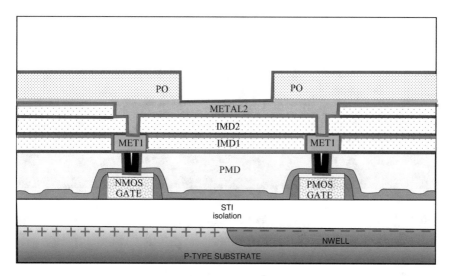

Fig. 8.16 Cross section through input node of CMOS inverter after PO via pattern photoresist is removed and the adhesion/barrier layer is deposited

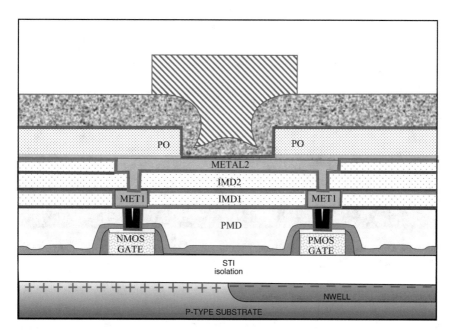

Fig. 8.17 Cross section through input node of CMOS inverter after PO via pattern photoresist is removed and the adhesion/barrier layer is deposited

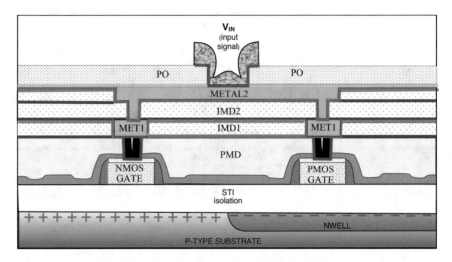

Fig. 8.18 Cross section through the bond pad on the input node of CMOS inverter (taken along dashed line B-B in Fig. 8.2)

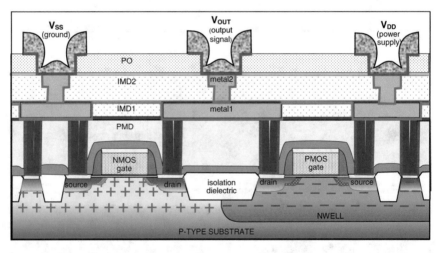

Fig. 8.19 Cross section through the bond pads on the output node of the CMOS inverter (taken along dashed line C-C in Fig. 8.2)

Because of the difficulty in forming reliable wire bonds to the uneven bond pad surfaces such as are in Figs. 8.18 and 8.19, most IC manufacturers add the extra manufacturing steps of filling the PO via with CVD tungsten and then using CMP to planarize the tungsten PO via plug. They then sputter the aluminum alloy on the flat surface and form aluminum bond pads with flat topography.

8.5   Forming Gas Anneal (Sintering)

The final step in an integrated circuit manufacturing flow is a forming gas (95% nitrogen plus 5% hydrogen) or hydrogen anneal at a temperature in the range of about 400–450 °C. This floods the gate dielectric/single crystal silicon interface with hydrogen and converts any unpaired silicon atom electrons to silicon-hydrogen bonds. This stabilizes and narrows the transistor turn-on voltage distribution.

This completes the manufacturing steps required to build a CMOS inverter. Additional processing steps to manufacture I/O transistors, precision resistors, capacitors, inductors, varactors, and other electronic devices required in the integrated circuit are omitted. Many additional processing steps are required to add a memory array to an integrated circuit—memory arrays such as dynamic random-access memory (DRAM), static random-access memory (SRAM), electrically programmable read only memory (EPROM), and FLASH, to name a few. The evolution of the designs and manufacturing flows for each of these types of memory cells as they scaled are interesting stories on their own.

8.6   Example Packaged ICs

The completed integrated circuit wafer now leaves the semiconductor fab and is sent to a packaging house where the scribe lanes between the dies are cut through with a diamond saw or laser into individual dies (die singulation process). The singulated dies are then packaged by mounting them on a lead frame and encasing them in molding compound. [2] See Section 2.3.2 and Figs. 2.2 through 2.8. Example packaged integrated circuits are shown in Fig. 8.20. The leads on the IC packages are electrically connected to bond pads on the IC dies with wire bonds prior to encapsulation with the black plastic molding compound (usually a filled epoxy resin).

The packaged ICs are typically mounted on circuit boards by soldering the leads on the IC packages to leads on the circuit board.

Fig. 8.20 [2] Packaged integrated circuits. Dual Inline Package (DIP); Small Outline Integrated Circuit (SOIC); Quad. Flat Package (QFP); Quad. Flat No lead package (QFN). Chip Scale Package (CSP)/ Wafer Level Package (WLP). https://sst. semiconductor-digest. com/2005/08/ materials-and-methods-for-ic-package-assemblies/

References

1. R. Carpio, A. Jaworski, Review—management of copper damascene plating. J. Electrochem. Soc. **166**(1), D3072–D3096 (2019). https://iopscience.iop.org/article/10.1149/2.0101901jes/pdf
2. J. Fjelsted, Materials and Methods for IC Package Assemblies, Packaging Design Review, Solid State Technology, Semiconductor Digest (2005), https://sst.semiconductor-digest.com/2005/08/materials-and-methods-for-ic-package-assemblies/

Chapter 9
The Incredible Shrinking IC: Part 1
Enabling Technologies

9.1 Introduction

Over a period of about 45 years, from about 1970–2015, transistor geometries scaled about 1000 times (10^3 times) smaller. Transistor geometries scaled from about 10 μm (ten one millionths of a meter—about 1/10th the size of a human hair) in 1980 to about 10 nm (ten one billionths of a meter—10,000 times smaller than a human hair) in 2020. The area of integrated circuits decreased in size by about one million. The size of a computer chip stayed about the same but became enormously more complex and more powerful. The incredible increase in the computation power of Intel microprocessor chips is summarized in Table 9.1. [2] Transistor count ballooned from about 2000 in 1970 to about two billion in 2015. At the same time, transistor switching speed increased from about 200,000 cycles per second to about 4.5 billion cycles per second. This simultaneous increase in transistor count and transistor switching speed enabled far more computing power to be packed into the cell phone we stick in our back pockets today than could be packed into the computer that filled a very large room in 1970. Computers that required tons of air conditioning to remove the heat being generated.

I was fortunate to begin work (1982) at Texas Instruments at the beginning of CMOS IC technology. By luck I secured a job where I was able to take part in and watch it all happen. It was an incredibly exciting place and time to work. New science and technologies were being invented and innovated on an almost daily basis. For a tech nerd like me, it was heaven. For the almost 30 years I worked at TI, a new technology node (geometries 25% smaller; integrated circuit area 50% smaller) was introduced every 2–3 years. Engineers at TI and other semiconductor companies worked miracles to make Moore's Law a self-fulfilling prophesy. (In a 1965 paper, Gordon Moore, cofounder of Intel, expounded the principle that the speed and capability of computers could be expected to double about every 2 years.) Each new technology node pushed the manufacturing equipment and manufacturing processes to their limits. Each new technology node pushed the computers used to design the

© Springer Nature Switzerland AG 2020   137
H. Tigelaar, *How Transistor Area Shrank by 1 Million Fold*,
https://doi.org/10.1007/978-3-030-40021-7_9

Table 9.1 Evolution of intel computer chips from 1971 to 2014

Node	Approx. year	Power supply	Levels of metal	Intel CPU	Microprocessor clock speed cycles/s	Approx. # transistors in CPU
10,000	1971	5.0	1	4004	200,000	2300
6000	1976	5.0	1	8080	2,000,000	6000
3000	1979	5.0	1	8088	8,000,000	29,000
1500	1982	5.0	1	80286	12,000,000	134,000
1000	1987	5.0	2	386SX	33,000,000	275,000
800	1989	4.0	3	486DX	33,000,000	1.2 million
500	1993	3.3	4	IntelDX4	100,000,000	1.6 million
320	1995	2.5	4	Pentium	133,000,000	3.3 million
250	1998	1.8	5	Pentium2	300,000,000	7.5 million
180	1999	1.6	6	Pentium3	700,000,000	28 million
130	2001	1.4	6	Celeron	1,330,000,000	44 million
90	2003	1.2	7	Pentium4	3,060,000,000	125 million
65	2005	1.0	8	Pentium4	3,800,000,000	184 million
45	2007	1.0	9	Xenon	3,000,000,000	410 million
32	2009	0.75	10	Core i7 G	3,460,000,000	1.17 billion
22	2011	0.75	11	Core i7 IB	3,900,000,000	1.4 billion
14	2014	0.70	12	Core i7 BU	4,500,000,000	1.9 billion

https://en.wikichip.org/wiki/intel/process [1]
https://www.intel.com/pressroom/kits/quickrefyr.htm [2]

integrated circuits and run simulated transistor manufacturing flows and integrated circuit designs to their limit. At each new technology node, most of the existing equipment and/or processes had to be reengineered to accommodate the smaller geometries, tighter spaces, and denser circuitry. A 10% variation on 100 μm geometry is 10 μm. A 10% variation on a 100 nm geometry is 10 nm - one thousands time smaller. Equipment that was pushed to the limit to print and etch a 100-μm line needed improvement to print and etch the 75-μm line in the next generation technology. Equipment and processes that could fill a 100-μm trench without forming a void had to be improved to fill a 75-μm trench without a void. The variation in thickness of thin films across wafer after deposition and after planarization had to be improved to meet specifications for the new technology node. Equipment and processes were pushed to meet tighter and tighter tolerances until the equipment and/or process capability was exceeded. When this occurred, scientists and engineers invented and developed new equipment and new processes to keep integrated circuit scaling going. The new equipment and processes had to be manufacturable. If not manufacturable, they were abandoned. Invention and innovation did not just focus on just one manufacturing step or just one manufacturing process. At each node, it was pretty much a full court press for multiple tools and multiple processes across the entire manufacturing flow. Thousands of scientists and engineers from industry and academia collaborated to attack and overcome thousands of technical challenges at each new technology node. Multiple equipment, process, and material approaches were pursued to address each issue. Most approaches were abandoned. Only a few made it into production. Thousands of inventions and innovations succeeded in producing manufacturable solutions that kept IC scaling going. The advancements in communication (radio, television, cell phones) and productivity that this amazing collaboration produced raised the standard of living worldwide. I was very fortunate to be a member of one of these development teams.

9.2 Clean Room and Wafers

9.2.1 Introduction

Ambient outdoor air in a typical city contains 35,000,000 particles that are 0.5 μm or larger in each cubic meter of air. A class 1 cleanroom in which ICs are manufactured has no particles 0.5 μm or larger (a human hair is about 100 μm) and 12 particles 0.3 μm or smaller in each cubic meter of air (https://en.wikipedia.org/wiki/Cleanroom [3]). This extreme degree of cleanliness is mandatory for high IC yield. As IC geometry size decreased, clean rooms had to get cleaner and cleaner to maintain acceptable yields.

Clean room technology was in a state of constant revolution as IC geometries scaled smaller and smaller. As IC geometries scaled, the size of particles that caused an IC to fail got smaller and smaller. When IC geometries were large, clean rooms were crowded with operators, technicians, equipment engineers, and process engineers. They would don a hair net, lab coat, safety glasses, and lab gloves before entering the clean room. As ICs scaled smaller, people in the clean room became a major source of IC yield killing particles. Today few people work in clean rooms. Today wafers are moved through the fab and are processed by robots. When someone does enter the clean room, they are covered from head to toe to prevent shedding skin cells from killing ICs (Fig. 9.1).

9.2.2 Clean Room Attire

When I started working in a wafer fab, I put on a hairnet, lab coat, gloves, and safety goggles before entering the clean room. A few years later, booties and a surgical face mask were added. Chewing gum was banned. Only drinking from water fountains in the fab was allowed—sodas and coffee were banned. Then a hood that covered my head except for my face, a bunny suit that covered my clothes, and lab gloves that covered my hands and wrists and were tucked into the sleeves of my bunny suit were added. Special booties with grounding straps that tucked into my shoes were added to prevent electrostatic sparks from killing ICs. All makeup was banned. (Not a problem for me but a huge problem for some of my female friends. Some went out and got makeup tatoos). Next came a hood with full facial coverage except for my eyes and clean room boots covering my shoes and my leggings up to

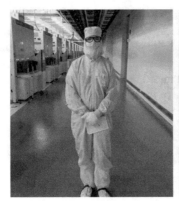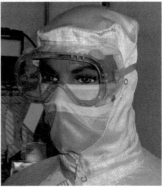

Fig. 9.1 Wafer fab person in bunny suit. Light is yellow in lithography areas to avoid photoresist exposure.   https://techreport.com/review/33337/globalfoundries-gears-up-for-the-next-genera-tions-of-chip-manufacturing/   [4];   https://en.wikipedia.org/wiki/Cleanroom#/media/File:Cleanroom_Garment2.JPG [5]

my knees (Fig. 9.1). Finally, I was kicked out of the fab and was replaced with robots because I still was bringing too many particles into the fab.

All personnel entering the fab are trained to dress top down (from head to foot) to prevent particles from above contaminating the lower fab garb. Undressing is the opposite (from foot to head) for the same reason. After dressing, each person enter-ing the fab first steps on a special machine to verify they are properly grounded and will not build up electrostatic charge that could cause an electrical spark and destroy an IC. Each person then steps onto a sticky mat to remove debris from the bottom of their booties and passes through and air shower that blows jets of air from above and all sides for a half minute or so to remove particles before entering the fab.

Safety glasses are manditory in the fab at all times. Special safety glasses are required to protect the eyes of lab personnel entering areas where lasers are used.

9.2.3   Evolution of Clean Rooms

Clean room technology underwent constant innovation and improvement to keep up with the yield demands of the smaller IC geometries. The increasingly onerous gowning procedures became mandatory when failure analysis engineers discovered more failing IC dies due to particles that were shed from fab personnel. Air filtration underwent constant improvement. A curtain of air constantly flows down through high-efficiency particulate air (HEPA) filters that cover the ceilings in the fab rooms, down through perforated floor tiles, under the raised floors and up through hollow walls to be recirculated through the HEPA filters. The filtration improved year by year to keep pace with the rapidly decreasing size of killer particles. The fab was segmented into rooms to accommodate the particle demands and safety demands of

Fig. 9.2 Wafer fab run by humans outside and inside the fab. https://www.criticalenvironmentso-lutions.co.uk/what-is-a-cleanroom/ [6]

various processes. Most rooms in the fab are positive pressure, so no particles are pulled into the fab from the outside. A few rooms where dangerous gasses (phosphine, silane, hydrogen cyanide, etc.) are used have negative pressure, so no hazardous gas escapes should a leak occur.

As IC geometries scaled, the demands on fab air quality became more stringent. A yield correlation with time of the year lead to the discovery that humidity changes in the fab produced yield changes. Tightly controlled low humidity improved yield but also produced an uptick in IC fails caused by sparking (electrostatic discharge or ESD). Specialized air ionizers were developed and installed in the fab to inject equal numbers of positive and negative ions into the fab air. Negative ions neutralize positively charged surfaces. Positive ions neutralize negatively charged surfaces. The uptick in ESD fails went away.

Early on all foods and drinking cups were banned from the fab. Soon chewing gum was banned. All packaging materials had to be removed before new equipment or chemicals were brought into the fab. All fab personnel were trained on procedures to wipe down new equipment and other materials before bringing them into the fab. Pencils and all non-clean-room-approved pens and wipes were banned. Finally, only fab-approved materials could be taken into the fab. Now everything including all tools, cell phones, bottles of chemicals, and lab notebooks must be wiped down according to protocol before being taken into the fab.

Until about the 90 nm technology node and the introduction of 300 mm wafers, operators, technicians, and engineers ran the manufacturing equipment and transported wafers through the fab (Fig. 9.2). Failure analysis engineers determined that humans moving through in the fab were one of the primary sources of particles that caused IC fails. This and the difficulty in handling and transporting 300 mm wafers (a box with twenty five 300 mm wafers weighs about 20 pounds) instigated the replacement of wafer fab humans with wafer fab robots (Fig. 9.3).

Fig. 9.3 Wafer fab run by humans outside the fab and by robots inside the fab. https://www.forbes.com/sites/patrickmoorhead/2018/02/04/the-u-s-already-has-bleeding-edge-technology-manufac-turing-with-globalfoundries-fab-8-in-malta-ny/#1ecc2fff23af [7]

Since humans have been for the most part removed from the fabs and replaced with robots, extremely low particulate counts in fab air are no longer required. Wafers no longer are exposed to fab air. Wafers are enclosed in wafer carriers called FOUPs (front opening unified pods) when they are not being processed inside a machine. After processing, wafers are loaded back into the FOUP and the FOUP is closed. A robot transports the FOUP to the next manufacturing tool or to a wafer storage area. Humans now only enter the fab to install equipment, repair equipment, and clean. Other than those activities, all wafer processing in modern fabs is performed almost entirely by robots.

Prior to the introduction of FOUPs, yield once suddenly started to decline. With passing days, the yield drop increased. The failures correlated to wafers that were processed in a certain area of the fab beginning on a certain day. Failure analysis engineers identified the cause to be microscopic bits of polymer fiber from HEPA filters. Turns out when the wind was in the East, fumes from an acid hood that exited a vent on the roof of the fab, were sucked into an air intake for makeup air for the area of the fab correlated to the failures. Acid fumes degraded the material in the HEPA filters causing microscopic particulates to rain down on the wafers. All the HEPA filters in the fab were replaced and the air intake was relocated at great cost.

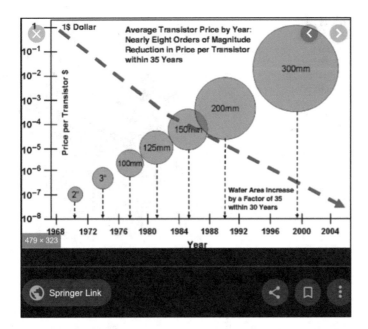

Fig. 9.4   Wafer size versus year

9.2.4   Increasing Wafer Size

As IC geometries scaled smaller, wafer size increased Fig. 9.4. Wafer fabs could print more IC dies on the larger wafers. Processing time for the larger wafers is about the same as for the smaller wafers. Larger wafers significantly reduce the manufacturing cost of each IC die. To accommodate the smaller IC geometries, wafer specifications became tighter as the wafers got larger.

Smaller IC geometries demanded that the larger wafers have less defects and better planarity than the previously used smaller wafers. Crystal defects such as small surface pits and crystal dislocations that were not a problem with larger geometries on previous technology nodes became killer defects for smaller geometries on the next generation technologies. Trace impurities that were not a problem with larger geometries had to be removed or gettered (rendered immobile) as transistors got smaller. The surface of the larger wafers had to become increasingly flat to accommodate the smaller transistors.

Wafer size in the early 1970s was 3 in. Wafer size was limited by the size of the largest crystal of silicon that could be grown at the time (Fig. 9.5). Wafer size increased from 4 in. (100 mm) in the late 1970s, to 6 in. (150 mm) in the mid-1980s, to 8 in. (200 mm) in the early 1990s, and to 12 in. (300 mm) in 2000.

Each time the wafers became larger, new wafer handling techniques were introduced to handle the larger wafers without introducing defects or breakage. Each time the wafers became larger, every piece of manufacturing equipment in

Fig. 9.5 Single crystal silicon ingots or boules from which starting material wafers are cut. https://www.chipsetc.com/silicon-wafers.html [8]

the fab had to be adapted to accommodate the larger wafer size or had to be replaced with new equipment.

The new manufacturing equipment had to be capable of meeting the tighter deposited film thickness specs, tighter photo printed geometry specs, and tighter post etch geometry specs. In addition, the new equipment had to meet more stringent across wafer process uniformity specifications for the larger wafers.

Most high-volume IC fabs today process 300 mm (12 in.) wafers. These wafers are cut from 300 mm diameter silicon single crystals. Most wafers used as starting material in IC manufacturing now have a customized layer of doped single crystal silicon epitaxially grown (epi) on the wafer. The large variety of custom ICs has different starting material requirements. A unique epi layer can be engineered for the transistors in each custom IC. Silicon epi is most often grown on the single crystal silicon wafer by exposing the wafer to silane dichloride (SiH_2Cl_2) and hydrogen gas (H_2) or silane trichloride ($SiHCl_3$) and hydrogen (H_2) gas at a temperature in the range of 1050–1150 °C. The concentrations of dopant gases such as phosphine (PH_3) or diborane (B_2H_6) can be adjusted to provide a variety of uniquely doped epi n-type or p-type starting material wafers.

Each time the wafers became larger, every operator, technician, and engineer in the fab underwent extensive training to properly handle and transport the larger wafers without generating particles. Boats containing wafers are to be carried level and held with two hands. When transporting a wafer boat, walk slowly so wafers do not rattle. When setting the wafer boat on a table, set it down gently—no sound from the wafers. If you can hear the wafers in the boat, you are generating particles and are killing yield.

Wafers kept getting larger as the transistors kept getting smaller. As wafers got larger and transistors smaller, across wafer uniformity requirements for deposited thin films and for etched geometries kept getting tighter. Across wafer thin film thickness uniformity that was acceptable for 6-in. wafers was unacceptable for 8-in. wafers. Post etch geometry variation what was acceptable for 8-in. wafers was unacceptable for 12-in. wafers. Acceptable variation for a half micron gate (500 nm) on

Fig. 9.6 80 mm (3 in.) wafers in wafer carrier. https://www.mtixtl.com/onesetof3diameter-25groupwaferscarrierboxsp5-3-25.aspx [9]

a 6-in. wafer is about 10% or 50 nm. 50 nm variation is larger than the entire gate length of 45 nm node transistors on 12-in. wafers. Not only does the new equipment have to control gate length variation an order of magnitude better, it also must control this variation across wafers that are two times bigger!

9.2.5   Wafer Carriers

A second motivation for implementing robots in 300 mm (12 in.) wafer fabs is a fully loaded wafer carrier (FOUP) weighs about 9 kg (~20 lbs.). Moving these carriers without jostling the wafers inside and generating particles is difficult for humans.

ICs are typically manufactured in lots of 12 or 24 wafers. Wafers usually come from the wafer supplier in wafer boxes containing 24 wafers (Figs. 9.6 and 9.7). When wafers were smaller and transistor geometries larger, the wafers were transported from one manufacturing tool to the next in these boxes. As transistor geometries scaled smaller, it was discovered the friction between the lid and base of these boxes generated particles that killed yield. Wafer transport boxes were redesigned and engineered to minimize particle generating friction between the plastic lid and base. The boxes open and close with hinges and are held closed by clamps instead of friction. The hinges and clamps are designed to minimize particle generation during opening and closing (Figs. 9.8 and 9.9).

300 mm (12 in.) wafer fabs are for the most part run by robots. 300 mm wafers are transported in Front Opening Universal Pods (FOUPS) (Fig. 9.10). The manufacturing tools are designed to interface with FOUPs. When processing is completed in one tool, the wafers are loaded into the FOUP and the FOUP cover is put

Fig. 9.7 125 mm (5 in.)
wafers in wafer carrier.
https://hncrystal.en.
made-in-china.com/
product/pXBxfRNGsuYh/
China-N-Type-5-Inch-
Polished-Monocrystalline-
Silicon-Wafer-for-
Semiconductor.html [10]

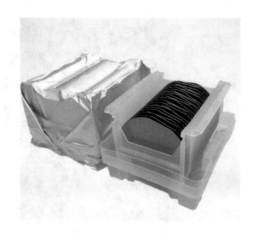

Fig. 9.8 150 mm (6 in.)
wafers in wafer transport
box. http://www.
spisemicon.com/category/
products/wafer-containers/
[11]

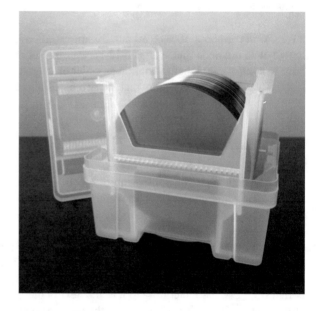

on sealing the wafers with an inert (nitrogen) atmosphere. A robot then transports
the FOUP to the next machine, aligns the door of the FOUP with the wafer input
port of the next machine, and couples the FOUP to the machine. The surface of the
FOUP seals against the input port of the machine. The input port removes the FOUP
cover and transfers the wafers from the FOUP into the machine.

After processing the machine returns the wafers into the FOUP and puts the
cover back on the FOUP sealing the wafers inside with nitrogen.

Fig. 9.9   200 mm (8 in.) wafers with wafer transport box. https://www.brooks.com/products/semi-conductor-automation/semiconductor-contamination-control/wafer-carriers-boxes/wafer-trans-port-boxes [12]

Fig. 9.10   300 mm (12 in.) wafers in a front opening universal pod (FOUP). https://industryre-ports24.com/338393/300mm-wafer-carrier-boxes-market-2019-increasing-demand-growth-anal-ysis-and-future-outlook-entegris-miraial-co-ltd-shin-etsu-polymer/ [13]

9.2.6   Wafer Handling Tools

Before wafers were handled by robots, fab personnel primarily used wafer tweezers such as are shown in Figs. 9.11, 9.12, and 9.13. Because the metal tweezers contacted the wafers and generated particles, they were to be used to handle wafers only when absolutely necessary. Vacuum wands such as shown in Figs. 9.14 and 9.15 were positioned by most machines and inspection stations. Vacuum wands have the advantage of being soft plastic instead of hard metal, of contacting only the back-side of the wafer instead of topside edges, and of using vacuum pressure instead of

Fig. 9.11 100 mm (4 in.) wafer tweezers. https://www.microtonano.com/EM-Tec-precision-wafer-handling-tweezers.php [13]

Fig. 9.12 150 mm (6 in.) wafer tweezers. https://www.microtonano.com/EM-Tec-precision-wafer-handling-tweezers.php [13]

Fig. 9.13 200 mm (8 in.)
wafer tweezers. https://
www.microtonano.com/
EM-Tec-precision-wafer-
handling-tweezers.php [13]

tweezer leverage to secure the wafers. Vacuum wands generate far fewer particles than metal tweezers.

300 mm wafers are handled almost exclusively by robots. Outside the fab when an engineer wants to inspect a 300-mm wafer, he/she loads the boat with the 300 mm wafers onto the inspection tool. A robot takes the wafer from the wafer boat and loads it onto the stage of the inspection tool. After inspection, the robot loads the wafer back into the wafer boat.

I think everyone who worked in the fab including me hated giving up their beloved wafer tweezers in favor of those damn vacuum wands. Before we got the bugs worked out, vacuum hoses kinked, vacuum leaks developed, and simultaneous use of several vacuum wands in close proximity caused the vacuum to fall below a critical level, so a wafer you had been processing for weeks dropped on the floor and shattered. I did become proficient with vacuum wands, but still kept my wafer tweezers in my drawer just in case.

Fig. 9.14 Vacuum wand.
https://www.microtonano.
com/EM-Tec-precision-
wafer-handling-tweezers.
php [14]

Shown with PELCO® Vacuum Pick-Up
System (Product No. 520), not included
with ExP Vacuum Wand Kit.

Fig. 9.15 Vacuum wand.
https://www.wandshop.
com/ [15]

9.3 MMST Program: IC Chips Manufacture in 3 Days!

In 1988, the U.S. Air Force and the Advanced Research Projects Agency funded a forward-looking development program at Texas Instruments that revolutionized IC manufacturing. The objective of the MMST (Microelectronics Manufacturing Science and Technology) [16] program was to invent and develop new manufacturing technologies to: (a) reduce cycle time, (b) reduce manufacturing costs, and (c) build manufacturing tools that would be required to provide the across wafer deposition and etch uniformity when the large 12 inch (300 mm) wafers became available.

The MMST team was managed by Dr. Robert Doering. On June 30, 1993, the MMST team [17] announced that they had manufactured fully functional double-level metal IC wafers in less than 3 days! To this day, this amazing feat has not been repeated.

The MMST program was far ahead of its time. The manufacturing line they developed in 1993 is amazingly similar to manufacturing lines in modern-day factories. Much of the single-wafer equipment, technology, and processes they invented and developed for their MMST manufacturing line forms the basis for the single-wafer manufacturing tools in today's IC manufacturing fabs (Table 9.2). In the MMST manufacturing line, wafers were transported from one single-wafer processing tool to the next in vacuum carriers. Today wafers are transported from one single-wafer processing tool to the next in FOUPS. The MMST manufacturing line was computer controlled. Today's manufacturing lines are computer controlled. MMST tools and processes had sensors with computerized feedback control. The state of tools and processes in today's fabs are monitored with sensors with feedback control.

Without the single-wafer processing equipment that was invented and perfected in the MMST program, the tight wafer-to-wafer process control and the tight across wafer uniformity demanded by IC scaling on 300 mm wafers would not be possible. Without these single-wafer tools, 300-mm manufacturing would not be possible.

Texas Instruments built many of the MMST single-wafer plasma etch, plasma clean, and rapid thermal processing tools in-house. Texas Instruments partnered

Table 9.2 Comparison of production lines in Modern Waferfabs with 1993 Waferfab production lines and with the 1993 MMST production line at Texas Instruments

Parameter	1993 Factory	Modern factory	MMST factory
Equipment	Mostly batch	Mostly single wafer	All single wafer
Wafer carriers	Open boats	FOUPs	Vacuum carriers
Wafer transport	Hand transport in open boats	Robot transport in FOUP	Hand transport in Vac. carrier
Thermal processing	Batches of wafers in furnace	RTA, flash, spike, laser	RTP
Oxidation	Furnace	RTO	RTO
Sputter deposition	Multiple wafers in batch tool	Single-wafer tools	Single-wafer tools
CVD deposition	Multiple wafers in batch tool	Single-wafer tools	Single-wafer tools
Factory control	Operators hand enter lot data into computers	100% computer control	100% computer control
Tool control	Engineers enter tool data on control charts	Sensors with feedback control	Sensors with feedback control
Process control	Engineers enter post processing metrology data on control charts	Robots enter post processing metrology data on control charts	In situ metrology sensors with real-time feedback

with companies such as Applied Materials to build single-wafer thin film deposition tools and Gasonics to build single-wafer high-pressure oxidation tools. At the conclusion of the MMST program, Texas Instruments sold many of the single-wafer equipment patents to Applied Materials.

The MMST manufacturing technology was decades ahead of its time. Most of the MMST innovations are now incorporated in modern-day semiconductor manufacturing. One area of MMST innovation that has not yet been implemented is in situ process sensors with feedback for real-time process control and endpoint detection. When implemented these process sensors could adjust across wafer deposition and etching uniformity in real time and also trigger endpoints. In situ metrology could eliminate the need for separate metrology processing steps, could eliminate the need for separate metrology equipment, and significantly reduce cycle time.

9.4   Lithography

9.4.1   Introduction

Lithography which is the printing of billions of miniscule transistor geometries on an IC chip in the size of your thumb nail is the most challenging process in IC manufacturing. My common sense says it should not be possible. My common sense says it should not be possible to print clearly defined geometries with dimensions smaller than the wavelength of light used to print them. Yet this is done millions of times every day in modern wafer fabs.

9.4.2   Printing IC Dies on a Wafer

A semiconductor wafer with integrated circuit chips (dies) printed on it is shown in Fig. 9.16. Each small square is an integrated circuit die. Each green die is a good IC. Each colored die is a failed IC. The different colors correspond to different failure modes. Different failure modes include the following: the die is completely dead, metal opens cause die to fail, metal shorts cause die to fail, the IC is too slow, the IC off current is too high, a subcircuit is not working, a memory cell fails to program or erase, plus a host of others. The horizontal and vertical lines between the integrated circuits are scribe lanes. Test structures in the scribe lanes are electrically tested at parametric probe to determine if transistors, diodes, capacitors, and resistors meet specifications. After parametric probe the wafers are sent to multiprobe where each IC die is tested to determine if the IC passes or fails. After the integrated circuits complete multiprobe, the wafers are sent to a packaging house (factory) where individual dies are singulated by cutting through the scribe lanes and the individual dies are packaged.

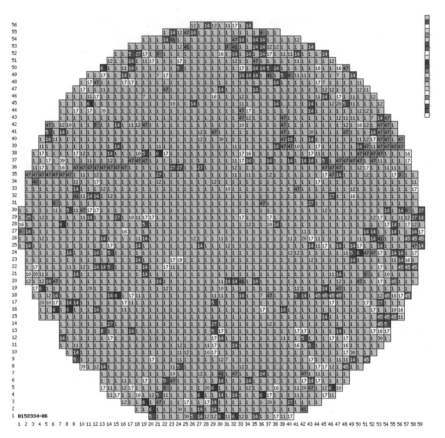

Fig. 9.16 Integrated circuits printed on a silicon wafer. Green dies are good die. Failed dies are color coded to indicate which test they failed

Scribe Lanes or Scribe Streets

Early on, when dies were large and wafers were small, a diamond-tipped scribing tool was used to make a small scratch in the scribe lane. A small amount of pressure caused a crack to initiate at the scratch and propagate along a crystal plane across the single crystal silicon wafer. This method was used to cut the wafer up into individual dies (singulation). The scribe lane name stuck. Scribing and breaking are no longer used to singulate dies. Diamond-tipped saws or lasers are now used to cut through scribe streets.

Pattern Alignment

During the manufacture of an integrated circuit, many patterns are stacked one on top of another. Each new pattern is aligned to an underlying pattern. For example, the metal1 interconnect pattern is aligned to the contact pattern to ensure metal1 completely covers the contact plug. As contacts got smaller and smaller, alignment requirements got tighter and tighter. A 0.2-μm misalignment to a 2-μm contact plug is not a problem, but a 0.2-μm misalignment to a 0.4-μm contact plug is a huge problem. The alignment mechanisms and alignment software programs in lithography printers have undergone enormous changes to keep up with scaling. Teams of engineers with masters and Ph.D. degrees make this problem their life's work.

As geometries get smaller and closer together, printing them gets increasingly difficult. The hole patterns have gotten so close on the contact pattern that at the most advanced technology nodes they can no longer be printed using one reticle. A solution is to put half the contacts on a first reticle and the other half of the contacts on a second reticle. Contacts on each of the two separate reticles are not as close to each other as when they are on the same reticle. After both contact patterns are printed on the wafer the contacts have the desired closeness. A problem arises when the first contact pattern is misaligned in one direction and the second contact pattern is misaligned in the opposite direction. The width of the metal1 lead must be made wider so that both sets of contact plugs will be fully covered. Making the metal1 lead wider makes the integrated circuit area larger. Lithography equipment engineers have done an amazing job of improving alignment capability so that double and even triple patterning is now being considered in high-volume integrated circuit manufacturing for the most advanced technology nodes.

Vibrations in Printing Small Geometries
In the mid-1980s, I was talking with a photolithography engineer in TI's South Building fab. She said vibrations from semi-trucks rolling past on Highway 635 were causing her printed images to blur. Her printer was sitting on a huge granite slab that was sitting on top of a reinforced column of concrete that reached down to bedrock. Still her printed geometries were blurred by vibrations from passing trucks. Images printed on third shift when truck traffic on 635 was minimal were better than images printed on first and second shifts. To me this looked to be a showstopper. How could we manufacture integrated circuits with geometries less than a half micron if the gates were blurry? To solve this problem, scientists and engineers invented and developed anti-vibration technology that acts much like the noise cancelation headphones used to block noise on airplanes. This anti-vibration technology detects a vibration and then generates a canceling vibration so fast that the lithography printer never sees it. With the anti-vibration technology, lithography printers no longer need to sit on reinforced concrete pedestals down to bedrock.

9.4.3 Photolithography Printers: Introduction

Large teams of scientists and engineers, from industry and academia, collaborated to revolutionize lithography multiple times. Each time IC geometries approached the wavelength of the light in the lithography printer, an entirely new photolithography system had to be developed to keep IC scaling going. Materials scientists developed new lens materials for the optical columns needed to focus the smaller wavelength. Scientists and engineers developed new light sources with shorter wavelengths and designed the new lithographic printers—printers with improved alignment and registration capability, with improved stage flatness and registration, and with improved depth of focus. Polymer scientists and chemists developed new photopolymers and new photochemistries for the photoresists using shorter wavelength light. Materials scientists and engineers developed new materials and methods to make photomasks compatible with the shorter wavelength light. Optical physicists developed software programs that performed increasingly complex calculations to compensate for light interference effects when building photomasks. Close collaboration between all these groups was mandatory to deliver each new photolithographic system capable of running in production day in and day out. These teams did not perform this miraculous feat just once. These teams performed this miraculous feat multiple times.

Except for transistor sidewalls which are formed by depositing and etching a thin dielectric film, photoresist patterns are used to define every structure in the integrated circuit manufacturing flow. Photoresist patterns mask etching steps (the photoresist pattern blocks etching) and mask implant doping steps (the photoresist pattern blocks ion implants). A light-activated polymer layer (photoresist) is coated on the surface of the wafer. Light is projected through a glass plate with a chrome geometry pattern (reticle or photomask) exposing the photoresist. The light triggers a chemical reaction in the photoresist changing the properties of the photoresist rendering it soluble in developer. The photoresist pattern is then developed by immersing it in the photoresist solvent (developer). Exposed photoresist is washed away. The unexposed photoresist pattern remains.

As transistor geometries scaled smaller, the dimensions approached the wavelength of the light being used to print them. As the size of the geometries approaches the wavelength of light, light interference creates distortions rendering the printed patterns unrecognizable. Each time scaled geometries approached the wavelength of the lithography printer, to keep scaling going, scientists and engineers invented and developed light sources and photolithography systems that utilized a shorter wavelength. Materials scientists and optical engineers developed new lens materials and new lithography projection printers. Chemists and polymer scientists developed new polymers (photoresists) compatible with the shorter wavelength light. At least 4 times, IC scaling would have stopped dead in the water, had not a completely new photolithography system been invented and made manufacturable.

Both negative and positive photoresists are used. A negative photoresist becomes insoluble in developer when exposed to light. A positive photoresist becomes soluble in developer when exposed to light.

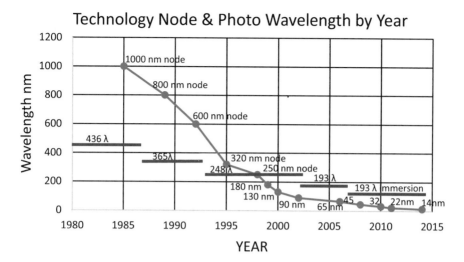

Fig. 9.17 Lithography wavelength by year. Technology node by year. Beginning in the mid-1990s, the length of the transistor gate is smaller than the wavelength of the light used to print it

9.4.4 Printers: Light Wavelength as ICs Scaled

The wavelength of light used in the lithography printers versus year and versus technology node is summarized in Fig. 9.17.

In the early 1980s transistor geometries were printed with G-line light from a mercury vapor lamp (436 nm wavelength). At that time the width of transistor gates was about 1000 nm. As the width of the transistor gate geometry approached the wavelength of the light, a new light source with a shorter wavelength and a new photoresist compatible with the new wavelength was developed. At the 0.8-μm technology node, when transistor gates had scaled to 800 nm, I-line (365 nm wavelength) was introduced. At the 0.32 μm technology node, when transistor gates had scaled to 320 nm, lithography printers with fluoride excimer laser light source (248 nm wavelength) were introduced.

Transistor geometries scaled faster than new shorter wavelength photolithographic systems could be developed. At the 0.25 μm technology node, the transistor gate length (250 nm) was about equal to the 248 nm wavelength of light being used to print it. In Fig. 9.17, the blue transistor gate length line crosses the magenta lithography printer wavelength line in about 1997. To keep IC scaling on track, the lithography team developed incredible new photo reisists and incredible new methods to enable them to print geometries smaller than the light used to print them. The next two technology nodes, 180 nm and 130 nm transistor gate lengths, were manufactured using 248 nm wavelength lithography printers. A new lithography system with shorter 193 nm wavelength light from an argon fluoride excimer laser did not become available until the 90 nm technology node in about 2002. Lithography printers with 193 nm wavelength light were also used to manufacture the next technology node with 65 nm transistor gate lengths. Enormous effort went into the development of a manufacturable 157 nm lithography system. Unfortunately, these efforts were unsuccessful. To keep IC scaling going, the lithography teams developed an absolutely amazing work around. They pulled a rabbit out of the hat. They developed a 193-nm immersion system that uses 193 nm wavelength light but gives resolution equivalent to a lithography printer with a 135-nm wavelength light source. This technology is incredible. I am blown away they pulled it off. Just one of these 193 nm immersion lithography printers costs in excess of $120 million today! These printers are being used for technology nodes with 32, 22, and 14 nm transistor gate lengths.

9.4.5   Printers: Intel's 2002 Lithography Roadmap

Intel's 2002 Technology Roadmap [18] is included to give an appreciation for just how dynamic and exciting photolithography technologies were during this time of rapid scaling (Table 9.3). Intel's technology roadmap was Intel's best projection of what lithography technology they would be using in manufacturing by year. Intel's lithography roadmap plan remained true for about 2 years. Intel's roadmap antici-

Table 9.3 Comparison of Intel's 2002 Lithography Roadmap to what actually happened

Year	Roadmap node (nm)	Roadmap litho.	Actual litho.
1981	**2000**	**i/g-line steppers**	i/g-line steppers
1984	**1500**	**i/g-line steppers**	i/g-line steppers
1987	**1000**	**i/g-line steppers**	i/g-line steppers
1990	**800**	**i/g-line steppers**	i/g-line steppers
1993	**500**	**i/g-line steppers**	i/g-line steppers
1995	**350**	**i-line plus DUV introduction**	248 nm
1997	**250**	**DUV**	248 nm
1999	**180**	**DUV**	248 nm
2001	**130**	**DUV**	248 nm
2003	**90**	**193 nm**	193 nm
2005	**65**	**193 nm plus 157 nm introduction**	193 nm
2007	**45**	**157 nm plus EUV introduction**	193 immersion
2009	**32**	**EUV = ~13.5 nm**	193 immersion
2010	22		193 immersion
2014	14		193 immersion
2016	10		193 immersion
2018	7		193 immersion

Bold data are from page 57 of Intel Technology Journal [18]

pated Intel would implement 157 nm lithography in 2005 and implement Extreme UV (EUV) lithography in 2009. Neither of these lithography technologies was ready for manufacturing by these dates. A revolutionary new lithography technology that was not even on Intel's 2002 lithography roadmap that was implemented in 2007 kept IC scaling going. The expense of this new 193 nm immersion lithography technology may be the reason it was not on Intel's lithography roadmap (One stepper costs over 120 million dollars). I am certain Intel was following 193 nm immersion technology very closely.

As with lithography other technologies such as dielectric, metal, and epi thin film deposition; thin film etch; dielectric and metal CMP, wafer cleaning, starting material crystal growth, and others were just as dynamic. In all cases, various technological approaches were pursued in parallel at multiple companies and universities. Equipment, process, and transistor engineers who developed the IC manufacturing flows followed these multiple approaches with great interest. During the process of generating the IC manufacturing flow for the next-generation technology, they usually would select a couple of the most promising approaches and bring them in to evaluate side-by-side in-house. Eventually one of the approaches was selected to be transferred into manufacturing.

Although hundreds of millions of research and development dollars have been invested in 157 nm and EUV lithography technologies for decades, neither has yet passed the manufacturability test. Many millions of dollars are still being invested in these lithography technologies. It is anticipated that EUV will ramp in production at Samsung in 2020 (https://optics.org/news/10/4/31) [20] and at Intel in 2021 (Intel EUV Roadmap) [19, 21]. Whether this actually happens remains to be seen. To continue scaling in spite of the failure to introduce manufacturable 157 nm or manufacturable EUV (~13.5 nm) lithography processes, scientists and engineers developed work around solutions with 193 nm immersion lithography such as double, triple, and quadruple patterning plus phase shift masks.

9.4.6 Printers: Steppers and Scanners

The first photolithography tools called steppers would project light through the reticle and print the pattern on one entire die at a time. The steppers would expose one die and then step and expose the next die until the entire surface of the wafer was patterned. At the 130 nm technology node, lenses for the 248 nm steppers could not be manufactured with the uniformity required to print gate geometries across the whole die in one shot. Engineers were able to build lenses with the capability to print a narrow stripe of gate pattern uniformly across a die. This printer, called a scanner, exposes a narrow strip of the gate pattern across the bottom of the die and scans across the pattern and die until the whole pattern is printed on the die from bottom to top. The scanner then steps to the next die and scans across it. This process, called step and scan, is repeated until all dies on the wafer are exposed.

9.4.7   Printers: 193 nm Immersion Scanner

To keep transistor scaling going without a 157-nm lithography system, the photolithography engineering team had to invent/develop a means of focusing light to a smaller point. The lithography team invented and developed 193 nm immersion lithography. The 193 immersion scanner uses the higher index of refraction of water to bend the light more and focus it to a sharper point. A 193-nm immersion scanner is equivalent to a 135-nm scanner without immersion. In 2004 the first immersion scanner was delivered by Dutch company ASML to Albany Nanotech. The immersion scanner interposes a layer of water between the final lens of the 193 nm scanner and the photoresist on the integrated circuit wafer. The immersion scanner dispenses a thin film of water onto the photoresist ahead of the lens as it scans across the die and then sucks it back off the photoresist after the lens passes Fig. 9.18. Any micro air bubble or dust particle in the water ruins the image. In production, a 193-nm immersion scanner prints the pattern on 350 dies in less than 1 min! https://www.nikon.com/products/semi/lineup/pdf/NSR-S635E_e.pdf [22].

Think of all the engineering that went into making the water dust and bubble-free and at a precisely controlled temperature to keep the index of refraction constant. Think of the engineering that went into completely filling the space between the lens and the photoresist with a droplet of water and then completely vacuuming it up over 350 times per minute. Think of the engineering that went into the nozzles that dispense and then retrieve the water. Think of the research and development that went into developing and applying hydrophobic coatings to the lens and other parts of the machine the water touches, so the water does not wet those surfaces. Think about the amount of research and development that went into development of multilayer photoresist stack to provide the photo properties required for 193 nm light exposure and to provide the surface properties required for immersion lithography. Their challenges were enormous. The 193 nm immersion lithography team met and overcame every one of them. I am still amazed they were able to do it!

Water has a higher index of refraction than air (1.436 vs 1.0). Water bends light more than air, so it can focus light to a sharper point. The 193 nm light source on the immersion scanner is converted to the equivalent of a 135-nm light source (193 nm/1.436 = 134.7 nm) by inserting the water layer between the lens and the photoresist on the wafer. The immersion lithography team that developed this amazing technology was composed of many engineers across many disciplines. (Optical physicists and engineers, mechanical engineers, fluid mechanical engineers, microfluidic engineers, photoresist chemists, OPC engineers, simulation engineers, robotics engineers, etc.)

In 2004 I was on assignment at IMEC in Belgium working on hi-k, metal gate transistor development. IMEC had a prototype immersion scanner from ASML. A colleague of mine from Texas Instruments was also working at IMEC on the 193-nm immersion lithography team. At that time, their challenges were so great and their results so poor that I doubted they would ever get immersion lithography to work consistently enough for manufacturing. Within a few years they overcame these enormous obstacles and transferred 193-nm immersion photolithography into production. Amazing!

9.4.8 Printers: The 157-nm Photolithography System Saga

Many hundreds of millions of dollars plus an enormous number of R&D hours went into attempts to develop a manufacturable 157 nm (argon fluoride excimer laser) photolithography process. The target date for introduction of 157 nm lithography was first in the late 1990s and then because of difficulties was pushed back to the early 2000s. So far, the enormous effort and investment has not produced a manufacturable 157 nm photolithography process.

Fig. 9.18 In immersion scanners, a film of purified water between the lens and the wafer bends light more than air does. https://www.slideshare.net/anandhus/immersion-lithography [23]

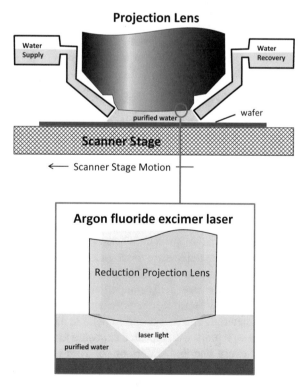

157 nm lithography presented a whole host of technical challenges that teams of optical physicists, chemists and equipment engineers, materials scientists, and many other scientific disciplines attacked and conquered 99+ %. Since the lens material used for previous 248 nm and 193 nm light sources strongly adsorbs 157 nm light, processes to grow large defect free crystals of calcium fluoride (CaF_2) for the lens were developed. Complex catadioptric (employing both lens and mirrors to correct for birefringence inherent in CaF_2 crystals) optical systems were designed and developed by optical physicists and engineers. Because trace contaminants such as oxygen, carbon dioxide, and water vapor strongly absorb 157 nm light and adsorb on the surfaces of CaF_2 lenses and mirrors degrading 157 nm light transmission, engineers constantly bathe the 157 nm optical column in high-purity nitrogen (expensive). By early 2003, all the major lithography tool manufacturers (ASML, Canon, Nikon) announced they would have 157 nm steppers ready for sale within a year—and they did. http://spie. org/news/progress-report-157-nm-lithography-prepares-to-graduate?SSO=1 [21]

Because fused silica reticle materials used for reticle blanks at 248 nm and 193 nm strongly adsorb 157 nm light, and because 6 in. × 6 in. reticle blanks could not be made of CaF_2, material scientists and engineers developed an acceptable fused silica reticle blank material by reducing the hydroxyl (OH) content of the fused silica glass and by incorporating fluorine (F) into the SiO_2 network of the fused silica glass.

Imagine the profound disappointment that the thousands of engineers who worked for many years on the 157 nm photolithography source, lens system, scanner, photoresists, and reticles felt when their work failed to bear fruit. They met all their goals. One small item outside their control sunk all their efforts. My group worked on a project for 3 years and met all objectives. Just when we were ready to transfer our process into production, management decided to sell the technology to another company. Our job then became teaching the other company how to manufacture the technology we developed. Very disappointing to say the least.

The showstopper for 157 nm lithography seems to be the failure of chemists and materials scientists to invent and produce a material for 157 nm pellicles. A pellicle is a thin layer of transparent material mounted on the reticle just above the pattern. The pellicle keeps particles out of the focal plane. When a particle is not in the focal plane, the particle image is defocused and is not printed in the photoresist. Without a pellicle, pattern defects are too high and yield is too low for manufacturing. There has been an enormous investment in time and money, and it continues to be expended in attempts to develop a 157 nm pellicle material. Chemists and material scientists synthesized a plethora of new polymers in attempts to find an acceptable 157 nm pellicle material. The best thin film pellicle material they have come up with so far degrades so rapidly with repeated exposure to 157 nm light that the pellicle must be replaced every hundred wafers or so. Thin pellicles can be made of fused silica reticle material, but they become part of the optical system and require the optical column be recalibrated each time a new reticle is put into the printer.

9.4.9 Photoresist: Introduction

Photoresists are organic polymers that contain molecules with chemical bonds that break apart in response to light. Light is projected through a photomask or reticle containing chrome IC geometries drawn by the IC designers. The chrome geometries on the reticle block light from exposing the photoresist. Light is transmitted through the glass spaces between the chrome geometries exposing the photoresist. In positive photoresist, light breaks the chemical bond releasing a capping molecule attached to the end of a photoresist polymer chain. The capping molecule protects the photoresist polymer chain from depolymerization. With the capping molecule gone, the photoresist depolymerizes and becomes soluble in developer. In negative resist, the light breaks photoactive chemical bonds in the resist forming highly reactive molecules. These highly reactive molecules then react with and form bonds to other polymer chains in the photoresist. (Called cross linking). These additional bonds render the photoresist insoluble in developer solution.

For each new wavelength, the materials scientists, chemists, and photo engineers had to come up with new photoactive molecules, had to develop new polymers for the photoresists, and had to develop new developer chemistry for the photolithography.

The photolithography team also developed antireflective coatings to reduce unwanted exposure caused by light from reflective surfaces and planarizing coatings to increase depth of focus over surfaces with topography.

9.4.10 Photoresist: Coating Wafers

Photoresist (and other liquid thin films) are deposited on the wafer using a spin coater. The wafer is held firmly in place on a vacuum chuck. A nozzle deposits a precise volume of photoresist in a puddle in the center of the wafer. The chuck then slowly ramps up to a first spinning speed to spread the photoresist uniformly across the wafer. Next the wafer is ramped up to a second much faster spinning speed to remove excess photoresist off the edges of the wafer and to give the photoresist a very uniform thickness across the wafer. After coating, the wafer is transferred to a heated chuck to bake the resist driving off excess solvent from the photoresist.

9.4.11 Photoresist: Kodak Thin Film Photoresist

The first photoresist used in semiconductor manufacture was a Kodak thin film resist (KTFR) discovered by Kodak chemists Martin Hepher and Hans Wagner. It was a negative photoresist. KTFR exposed to light underwent chemical reactions that rendered it insoluble. Areas of the KTFR that were blocked from light exposure by the pattern on the reticle dissolved away when immersed in developer (KTFR solvent). This photoresist was used for transistor gates down to about 2 microns. Around 2 μm loss of resolution due to swelling of the KTFR pattern during development ended the use of KTFR.

9.4.12 Photoresist: Novolak Positive Photoresist

German chemist Oskar Sues of Kalle Company in Wiesbaden is credited with inventing DNQ-novolak positive photoresists. These photoresists became the workhorse of the semiconductor manufacturing industry for many years. Chemists made many modifications to the novolak resists over the years to improve resolution and to adapt them to different wavelengths. Modified novolak positive resists are still being used in high-volume semiconductor fabs today for patterns with large geometries.

DNQ-novolak photoresist consists of a photoactive compound 2,1-diazonaphtha quinone-5-sulfonic acid esterified with 2,3,4-trihydroxy benzophenone (DNQ) in novolak resins. Novolak resins are polymers of cresols and other phenolic compounds with formaldehyde (Novo means new and lak means lacquer). When DNQ-novolak photoresist is exposed, the DNQ forms a ketene which reacts with ambient water to form indene carboxylic acid. The indene carboxylic acid is soluble in an aqueous basic solution. The most commonly used developer for DNQ-novolak photoresist is a dilute solution of tetramethylammonium hydroxide (TMAH).

9.4.13 Photoresist: Chemically Amplified Photoresist

For transistor scaling to continue, light sources with shorter and shorter wavelengths were required to print the smaller and smaller geometries. When lithography transitioned from 365 nm wavelength light (I-line) to 248 nm light (KF laser), a new type of photoresist was needed. The 248 nm light source is about 30 times dimmer than the 365 nm light source—not bright enough to expose novolak resist. A new resist with increased light sensitivity was needed. Photoresist polymer chemists C. Grant Wilison, Jean Frechet, and Horishi Ito at IBM invented and developed chemically amplified (CA) resists. CA resists have over a hundred times more sensitivity to light than the novolak resins. The invention and development of production worthy CA resists made transistor scaling to smaller sizes possible.

In 1979, C. Grant Willson, a chemist at IBM, and Jean Fréchet, a French chemist on sabbatical at IBM, came up with the idea of a single photo event triggering a chemical chain reaction in the photoresist. Exposure by a single photon would be chemically amplified in the photoresist by the chain reaction. While trying to invent a CA resist with a 30-fold increase in sensitivity to light to compensate for the 30-fold dimmer 248 nm light sources, Willson, Frechet, and Hiroshi Ito (a student of Frechet) invented a chemically amplified resist with over a 100-fold increase in light sensitivity.

Their idea was to construct a polymer chain that could easily be depolymerized and cap it with a chemical group that would prevent the depolymerization. Their idea was to attach the cap to the polymer using a chemical bond that could be cleaved by acid. They proposed combining the capped polymer chain with a known photo acid generator (a light-sensitive chemical that changes into an acid when exposed to light). The photo generated acid would cleave the cap off the polymer chain allowing the polymer chain to depolymerize. The depolymerized resist would then be soluble in developer. They started working on this approach in 1979. Four years later in 1983, they introduced their super light-sensitive photoresist to the new technology development team at IBM.

The polymer chain they came up with was poly(p-hydroxy styrene), or PHOST. The chemical group they capped the polymer chain with was tertiary butoxy carbonyl, or tBOC. The polymer chain with the cap is poly(p-t-butyloxycarbonyloxystyrene), or PBOCST. At General Electric, chemist James Crivello invented a photo acid generator (PAG) triphenyl sulfonium hexafluoro antimonate (TPSHFA). He synthesized this chemical to cure epoxy resins using UV light. Wilison, Frechet, and Ito mixed TPSHFA with their PBOCST capped polymer chain to form their chemically amplified photoresist. After exposure to light, the photoresist is heated in a post-exposure bake. The acid generated by the TPSHFA catalyzes the cleavage of the tBOC groups. The resulting fragments then generate additional acid, catalyzing additional tBOC cleavages and resulting in a cascade of de-protection and depolymerization. Their new chemically amplified resists are 100–200 times more sensitive to light than the previous generation novolak photoresists.

9.4.14 Photoresist: Bottom Antireflective Coating

Light can reflect from the surface under the photoresist and expose adjacent photoresist that should not be exposed. This is especially true of shiny surfaces such as metal layers. Optical physicists, materials scientists, and polymer chemists developed special coatings that are engineered to block reflections.

When you immerse a stick in water, the stick looks bent where it enters the water. Because the index of refraction (ability to bend light) is different in water than it is in air, light reflected from the stick under the water is bent at a different angle than light reflected from the stick above the water.

At night when you look out a window from inside a room, you see light reflected from objects outside the window and you also see images of objects that are inside the room with light reflected from the surface of the window.

When light hits the flat surface of the window at just the right angle (critical angle), 100% of that light is reflected from the glass. Light that does not hit the window at the critical angle is transmitted through the window.

Optical physicists, materials scientists, and polymer chemists use the principles of light reflection and transmission to engineer bottom antireflective coatings (BARC) that stop light reflection from an underlying surface. The BARC layer is deposited on the wafer before the photoresist is deposited (Fig. 9.19). A sputter-deposited aluminum film on a wafer is a high-quality mirror. A bottom anti-refection coating deposited on the sputter-deposited aluminum film looks black.

The index of refraction, thickness, and light absorption properties of BARC is chosen to bend the light to the critical angle, so it 100% reflects from the BARC/metal and photoresist/BARC interfaces. The BARC is engineered so that the light gets trapped in the BARC layer due to internal reflections.

9.4.15   Photoresist: Top Antireflective Coating

Top Antireflective Coatings (TARC) are deposited on top of the photoresist to reduce reflected light (Fig. 9.20). TARC takes advantage of light interference to reduce reflections. When two identical light waves with the same phase (Fig. 9.21a)

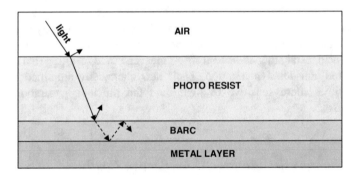

Fig. 9.19 Bottom antireflection coating (BARC). BARC is engineered to trap any and all light reflected from the underlying layer

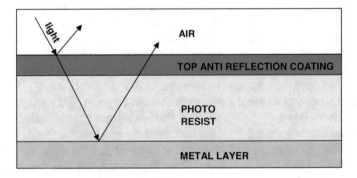

Fig. 9.20 Top antireflective coating designed to reduce the intensity of reflected light that would expose resist and create unwanted geometries

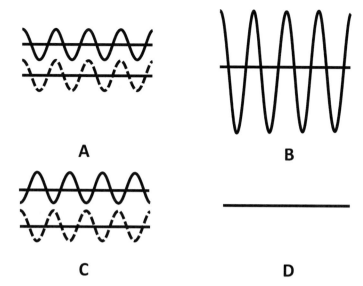

Fig. 9.21 (**a**) Light waves in phase. (**b**) Constructive interference of two identical light waves results in a fourfold increase in light intensity. (**c**) Light waves 180° out of phase. (**d**) Destructive interference of two identical waves results in zero light intensity (total darkness)

are added together, the light is four times brighter (Fig. 9.21b). When two identical light waves with opposite phase (Fig. 9.21c) are added together, they cancel each other out resulting in total darkness (Fig. 9.21d).

Optical physicists, materials scientists, and polymer chemists engineer the photoresist/TARC stack to reduce reflections between the interfaces, and to ensure that light reflected from the air/TARC and the TARC/photoresist interfaces is out of phase and cancels the light reflected from the photoresist/metal interface.

9.4.16   Photoresist: Bilayer Resists

Planarizing bilayer resists have been developed to improve imaging (depth-of-focus) on uneven surfaces. A first layer of a thick planarizing polymer is spun on the wafer over the topography to provide a planar surface over the hills and valleys. A second thin layer of photoresist is then spun on the planarized surface. This method provides improved depth of focus. Plasma etch first transfers the photoresist geometry patterns through the planarizing polymer layer down to the surface of the thin film to be etched. The plasma etch is then changed to etch the underlying thin film.

9.4.17   Ultraviolet Hardened Resists

Some semiconductor manufacturing processes such as sputtering and etching can raise the temperature of the patterned wafer above the softening point of the photoresist polymer (~115 to ~130 °C). The photoresist polymer can then flow into small spaces between photoresist geometries resulting in scumming and electrical shorts.

Special photoresists with two different photoactive molecules were developed as a solution to this problem. The first photoactive molecule is activated by the lithography printer to form the IC geometries. After the pattern is developed, the second photoactive molecule is activated by exposing the wafer to UV light (UV hardening). The reactive molecules generated during the UV exposure form chemical bonds between the photoresist polymer chains (cross linking). This significantly raises the softening temperature of the photoresist. UV hardened resist geometries remain unchanged when used in processes with higher temperatures.

9.4.18   Photomasks (Reticles)

Reticles used in semiconductor manufacturing are glass plates with chrome metal geometries that the circuit designers lay out. To make a reticle, the chrome on the glass plate is coated with a photoresist polymer that chemically changes when exposed by an electron beam (e-beam). A computer-driven e-beam directly writes (exposes) the desired pattern in the e-beam photoresist. For complex patterns this can take days. The pattern is then developed to dissolve away the unexposed photoresist. This pattern is etched into the chrome layer using anisotropic plasma etching. One reticle with a complicated photoresist pattern such as the gate level or the metal1 level can cost several hundred thousand dollars.

An unusual across wafer correlation was discovered. Transistor drive current increased monotonically across the wafer from top to bottom. After considerable effort, photo engineers finally determined that this pattern correlated perfectly with the order in which the dies were printed on the wafer. The photolithography manager told me her team found the root cause to be the difference in time between when the first die was printed and developed versus the time between when the last die was printed and developed. The dies are printed sequentially but are developed simultaneously. The acid molecules generated when the chemically amplified resist is exposed have more time to diffuse for the first printed die than for the last printed die. This difference in diffusion time results in a slightly shorter gate length (higher drive current) for the first printed die than for the last printed die. Her team was able to eliminate this correlation by slightly increasing the exposure time as dies are printed across the wafer. The slightly increased exposure time slightly increases the number of photons exposing the die. The slightly increased number of photons slightly increases the number of acid molecules that are generated. The slightly increased number of acid molecules slightly increases the acid molecule diffusion (diffusion is a function of concentration). They compensated for difference in diffusion time by producing a difference in acid molecule concentration. Clever!

9.4.19 Photomasks: Phase Shift Masks

For a few years, considerable effort went into the development and manufacture of phase shifting reticles (masks). By shifting the phase of some of the light going through the reticle, light waves at the edges of geometries being printed can be canceled out using light of the opposite phase. This gives the images sharper edges (better resolution) (see Fig. 9.20c, d).

Two types of phase shifting masks were developed: alternating phase shifting masks and attenuated phase shifting masks.

9.4.20 Photomasks: Alternating Phase Shift Masks

On alternating phase shifting masks, designers separate geometries onto two phase shifting patterns. The glass on the reticle for one of the patterns is etched thinner than for the other pattern. The light traveling through the thin glass regions of the reticle is shifted out of phase with respect to the light traveling through thicker glass regions. Design and layout of patterns for alternating phase shift masks are extremely complex. Complicated design rules to prevent phase shift conflicts make layout extremely challenging. Double patterning is often required when phase shift conflicts cannot be resolved on one reticle.

9.4.21 Photomasks: Attenuated Phase Shift Masks

In attenuated phase shifting masks, the chrome pattern geometries are thinned to permit a small amount of light to be transmitted through (typically a few percent). This light is too dim to expose the photoresist but is strong enough to interfere with the light coming through the transparent parts of the attenuated phase shifting mask. This interference improves the resolution of the image.

As transistors scaled, alternating phase shift technology usage decreased because of the difficulty in designing and laying out alternating phase shift patterns. Attenuated phase shifting was utilized more because this technology is transparent to designers.

With the introduction of 193 nm lithography and 193 nm immersion lithography, phase shift technology decreased but is now coming back as geometries scale smaller. At the most advanced technology nodes today, phase shifting technology with double, triple, and quadrupole exposures is being developed to print sub 20 nm patterns with 193 immersion lithography.

Engineers are working hard (as they have been for decades) to develop manufacturable electron beam lithography (ebeam lithography) but so far, the process is not cost-effective.

9.4.22　Photomasks: Compensating for Light Interference and Etch Loading

By the mid-1990s, scaling drove transistor gate lengths to be the same size as the 248 nm light being used to print them. Light interference effects became a show-stopper. The light interference effects had to be eliminated to scale IC geometries smaller. To keep scaling going, optical physicists developed very complex optical proximity correction (OPC) technology where geometries drawn on the reticle are distorted to compensate for light interference effects during printing. The patterns on the reticle are quite different from the patterns after they are printed on the wafer. The patterns that get printed on the wafer are the patterns the designers want.

In the early 1990s, it was my "expert" opinion that transistor scaling was going to stop. I did not think gate geometries with a size comparable to or smaller than the wavelength of light could be printed. I thought light interference would distort small geometries and spaces making them unprintable. At the time we were using 248 nm light. I thought we would have to wait until the 193 nm lithography system was ready before we could print 250 nm gates. Boy was I proven wrong! The lithography engineers invented new procedures and processes and blew right past that roadblock (see Fig. 9.16). By the mid-1990s, we were manufacturing transistor gates that were smaller than the wavelength of the light we were using to print them. Optical physicists and photo engineers developed methods to compensate for light interference. Sophisticated optical proximity correction (OPC) programs calculate how light interference distorts the patterns that the circuit designers draw. The program then recalculates and redraws the patterns on the reticle to compensate for light interference. The patterns on the reticle look very different than what the circuit designer drew. The patterns when printed on the wafer look very close to what the designer drew. The optical physicists use light interference to change the distorted patterns on the photomask to patterns on the wafer that the designers want!

Today billions of transistors with gate geometries less than 20 nm long are being manufactured using 193 nm immersion lithography! The scanner, photoresist, and photomask technologies that were invented and developed to make this possible are truly amazing.

(To observe light interference, put the pinky sides of your hands together. From a distance of about one foot, look with one eye through the narrow vertical slit between your pinky fingers. As you slowly close the slit, you will see vertical black lines form. These black lines are due to light interference.)

After the lithographers figured out how to compensate for light interference effects and were able to print transistor gates on the wafer with lengths the circuit designers intended, they discovered that plasma etching also changed the gate lengths (etch loading effects). Resist erodes during gate etch releasing carbon into the etching plasma. Where gate patterns are dense, more carbon is released. Differences in carbon in the gate etching plasma result in differences in polymer formation on the sidewalls of the gate being etched and differences in polysilicon etching rate. This causes differences in transistor gate lengths post gate etch.

In addition to figuring out how to compensate for light interference, transistor gate engineers had to figure out ways compensate for etch loading effects.

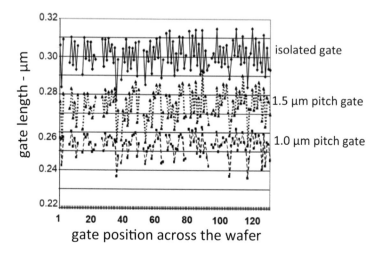

Fig. 9.22 Transistor gate length vs. transistor gate pitch without OPC

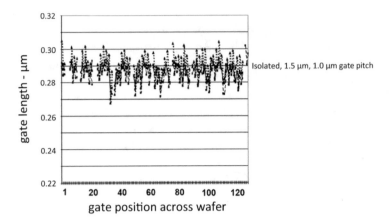

Fig. 9.23 Transistor gate length vs. transistor gate pitch with OPC

9.4.23   Photomasks: Optical Proximity Correction (OPC)

In the early 1990s, while starting development of the 320 nm technology node, we encountered a huge obstacle. The gate geometries were not printing properly because the gate geometries had scaled to a length comparable to the wavelength of the light that we were using to print them (see Fig. 9.22). Previously the solution was to start using a new light source with a shorter wavelength. At the time we were using 248 nm lithography. Our problem was that 193 nm lithography was not yet ready. We discovered that transistors that were drawn with identical gate lengths on the reticle ended up with different gate lengths when the gate pattern was printed on the wafer. The printed gate length depended upon the transistor gate pitch (pitch = transistor gate length plus space between transistor gates) (Fig. 9.22).

Our solution to this problem was to draw these transistors with different gate lengths (optical proximity correction or OPC) on the reticle so that their gate lengths would end up being the same when printed on the wafer (Fig. 9.23).

We soon discovered an additional complication. Gates that were printed with identical gate lengths on the wafer ended up with different gate lengths post etch. This difference depended upon differences in densities of the gate pattern (different gate pitches) across the IC chip (etch loading effect).

Origin of OPC at Texas Instruments

When we discovered optical proximity correction (OPC) was required on our gate pattern, we went to talk with the lithography engineers. They already used OPC on SRAM and DRAM memory cells. We asked them to use their OPC computer programs to apply OPC corrections to our gate reticle. After a week of computing, they managed to correct less than 10% of our gate pattern. Clearly the computer power was insufficient to calculate the OPC corrections needed for our gate pattern. Two engineers from our design group that performed design verification on our circuit patterns proffered they could adjust the length of transistor gates over the active regions as a function of gate pitch by modifying their design verification software. We designed a test chip and built transistors with different gate pitches and with different gate lengths. Using this test chip, we determined what OPC adjustment had to be made to the various pitch transistor gates, so the transistors would end up with identical drive currents after gate pattern and etch. The design verification engineers used this data with their modified design verification software to develop rule-based OPC. It worked. These results were published at the 1996 VLSI Symposium [22]. Their modified design verification software was used to generate the gate patterns for all of Texas Instrument's products for a number of years.

A method had to be developed to distort the pattern on the reticle to compensate for light interference (OPC) and also to compensate for etch loading effects. Very complicated computer programs now calculate optical proximity corrections and etch loading corrections and apply these corrections to the gate geometries that are drawn on the reticle. The gate geometries drawn on the reticle are distorted (Fig. 9.25a) so that after printing and etching, the desired geometries end up on the wafer (Fig. 9.25b). This method is now applied to virtually all the patterned layers in integrated circuits.

For several technology nodes, computers were not yet powerful enough to calculate and apply the needed OPC plus etch loading corrections. As a stop gap measure, TI's OPC team expanded their IC pattern verification software programs to apply OPC plus etch loading corrections to active, contact, via, all interconnect patterns, and some of the implant patterns. Their rule-based OPC software programs became extremely complicated. A new OPC Group with photo engineers, optical physicists, computer simulation experts, and computer programmers was established specifically for this work. In addition to size adjusts, serifs (added patterns) and antiserifs (removed patterns) were added to aid in the printing of the more highly scaled geometries at each new technology node (see Figs. 9.24 and 9.25).

When printed, the serifs reduce end-of-line pull back and corner rounding on outside corners. Antiserfs reduce corner rounding on inside corners (Fig. 9.25).

TI did not apply for a patent on the SSA-OPC methodology. A member of the patent committee said that prior OPC art would render the SSA-OPC methodology unpatentable. Unfortunately, that was bad advice. In 2005, a TI patent attorney contacted me and asked what I knew about the history of SSA-OPC at TI. Another company had patented SSA-OPC and was suing TI for infringement. Since TI was using SSA-OPC on every integrated circuit chip TI was manufacturing, it was going to cost TI tens to hundreds of millions of dollars in royalties if the competitors patent was valid. Conversely, if TI's patent attorney could show TI was using SSA-OPC before the competitors patent was filed, TI would not have to pay royalties. Fortunately, I had more than sufficient documentation to prove TI was using SSA-OPC prior to the filing date on the competitor's patent. The TI patent attorney castigated me severely for not patenting SSA-OPC. He said it would have cost TI a few thousand dollars to secure the SSA-OPC patent, but since we had not, it had cost TI several hundred thousand dollars to fight the SSA-OPC patent infringement lawsuit. In addition, he said TI probably lost 10s or 100s of millions of dollars in patent royalties TI could have collected had TI owned the SSA-OPC patent.

Texas Instruments produced high-performance processor chips for SUN Microsystems servers. At each new node, Texas Instruments developed a custom, high-performance manufacturing process for SUN. Our goal was to have the custom SUN manufacturing process ready by the time SUNs designers sent the patterns for their newly designed server. As soon as we got the pattern designs from SUN, we would rush to make the reticles and to process the SUN lot with top priority. We needed to get completed chips back to SUN as soon as possible, so the SUN engineers could test and tweak their design if needed. SUN would then send us a revised set of patterns and we would repeat the process with SUN until their server chip met all specifications. To process the SUN lot as fast as possible, we would hand carry the lot through the manufacturing line 24 h a day, 7 days a week. It could take from 3 to 6 weeks to process a SUN lot from start to finish. Technicians would pick up the lot from one machine as soon as it completed processing and hand carry it to the next machine. Each machine would be held idle waiting for the SUN lot to arrive. The equipment and process engineers responsible for each process would stand by her or his machine any time of the day or night to immediately resolve any issue should one arise.

One time, SUN sent a metal1 pattern to us. We had debugged our new optical proximity correction OPC on TI designs, but the SUN design was bigger and more complex. We built the first metal1 reticle and checked it out by running a focus/exposure matrix. (Low to high focus in the x-direction and low to high exposure in the y-direction across a wafer.) All die on the focus/exposure matrix had thousands of metal shorts. We trashed that reticle, made adjustments to the OPC, and generated a new reticle. The new reticle still had hundreds of shorts—not nearly as many as the first reticle, but still too many for SUN's server chip to function. While the SUN lot sat idle, we repeated this process three times before we were able to build a metal1 reticle with no shorts. The SUN lot sat idle for almost 1 month while we updated our OPC and generated new reticles. Top-level managers at SUN were upset. Top-level managers at Texas Instruments were livid. Neither could understand why we kept screwing up so badly. They could not comprehend why we could not do it right the first time. My group trashed over a quarter million dollars in reticles before we finally got it right. When we finally did get it right, we restarted the SUN lot with high priority and soon delivered completed IC dies to SUN. SUN's new server design worked. SUN was happy and TI managers were mollified.

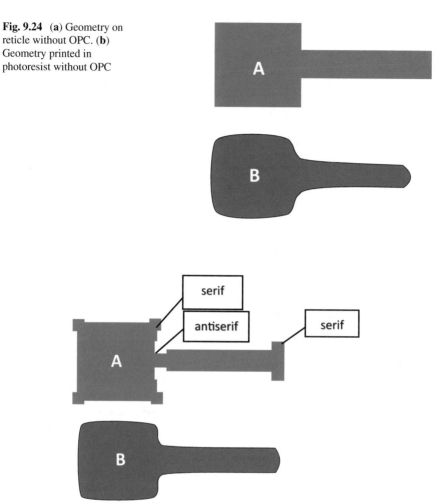

Fig. 9.24 (**a**) Geometry on reticle without OPC. (**b**) Geometry printed in photoresist without OPC

Fig. 9.25 (**a**) Geometry on reticle with OPC, serifs, and antiserifs; (**b**) Geometry printed in photoresist with OPC, serifs, and antiserifs

9.4.24   Dummy Fill Geometries

Dummy fill geometries are added to the patterns that the designers layout to improve manufacturability. Dummy geometries are not functional in the integrated circuit. They are not electrically active. The dummy geometries improve planarization, reduce etch loading variability, improve patterning, and improve matching between transistors.

Dummy geometries were initially added to the active patterning layer to help manufacturing with shallow trench isolation chemical mechanical polish (STI-CMP). Pattern generation engineers wrote a dummy fill program to automatically

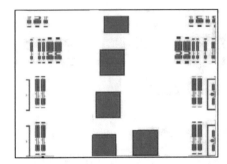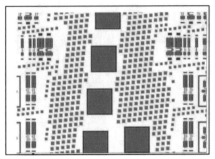

Fig. 9.26 Metal pattern without and with dummy fill geometries. https://semiengineering.com/knowledge_centers/materials/fill/advanced-smart-fill/ [24]

add dummy fill geometries to the computer file (gds file) containing the circuit layout pattern from the design engineers. This file with the circuit layout plus the dummy geometries is sent to the mask shop to make the reticles).

After OPC was introduced, the OPC software program would first process the file with the circuit layout from the design engineers (gds file) before the dummy fill software program added the dummy fill geometries.

At the 180 nm node, dummy assist features were added close to active geometries in the circuit layout such as transistor gates, active areas, and metal lines. These dummy assist features improve the ability of manufacturing to etch geometries such as gates and metal lines uniformly across dies and wafers (reduce etch loading differences) and to print geometries uniformly across dies and wafers. Pattern generation engineers wrote dummy assist computer programs to automatically add these dummy assist geometries to the circuit layout file provided by the circuit design engineers. After these dummy assist features are added, the computer program that automatically applies the optical proximity corrections to the circuit layout also adds optical proximity corrections to the dummy assist features. The dummy fill computer program then fills the large open spaces between the outer edges of the circuit and the scribe streets with additional large dummy fill geometries before the file is sent to the reticle shop.

Texas Instruments first introduced dummy active geometries at the 320 nm node to help manufacturing with chemical mechanical polish (CMP) of shallow trench isolation (STI). STI trenches are etched into the silicon and overfilled with a dielectric such as silicon dioxide. The overfill is removed from the active surface using CMP prior to building transistors (see Sect. 10.2.3 in Chap. 10). Before dummy active geometries were added, the isolation oxide over large active areas underpolished leaving residual oxide and isolation oxide over small active geometries over-polished damaging the single crystal silicon substrate (Fig. 10.3). The addition of dummy active structures between widely spaced geometries made the active area much more uniform (Fig. 10.4).

At the 250 nm technology node, metal dummy structures were added for metal CMP (Fig. 9.26).

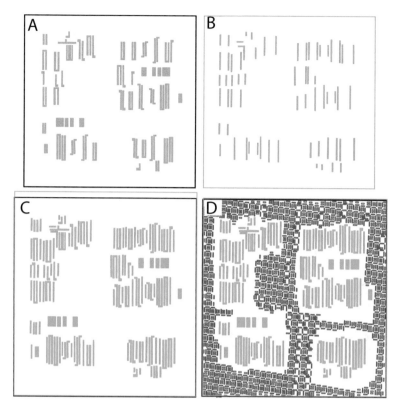

Fig. 9.27 (**a**) Gate pattern from design engineers, (**b**) dummy assist geometries generated by the dummy assist software program, (**c**) gate pattern plus dummy gate pattern before optical proximity corrections (OPC) are applied, (**d**) gate pattern after OPC and after dummy fill. This pattern is sent to the reticle shop

At the 180 nm technology node, polysilicon gate dummy geometries were added to assist gate photolithography and to provide a more planar surface post premetal dielectric (PMD) CMP for improved contact pattern depth of focus.

At the 130 nm node, dummy transistor gates were added next to isolated core transistor gates to improve transistor-to-transistor matching by assisting with printing the gates and minimizing etch loading effects.

Figure 9.27a–d illustrate the placement of dummy poly geometries in the layout of a transistor gate pattern.

Figure 9.27a is the core transistor gate pattern as laid out by the design engineers. In Fig. 9.27b, dummy gate assist features are placed next to the core transistor gates to facilitate gate photolithography and to reduce etch loading effects. Figure 9.27c is the gate pattern with the core transistor gates plus the dummy gate assist features.

OPC corrections are applied to these core transistor gate and dummy gate assist geometries. Figure 9.27d shows the gate layout with large lastly placed dummy fill geometries filling the open spaces between the borders of the circuit and the edge of the dies to facilitate CMP and to reduce etch loading effects by reducing pattern non uniformity across the die.

Polysilicon Gate Dummies

Process integration engineers, process engineers, and manufacturing engineers wanted to add dummy polysilicon structures to improve CMP of the premetal dielectric (PMD) layer covering the transistor polysilicon gates and also to improve transistor matching.

The post etch gate sidewall profile on isolated polysilicon gates is different that the post gate etch profile on closely spaced polysilicon gates. By putting dummy gates close to isolated polysilicon gate, the post gate etch profile could be made the same.

As the holes in the contact pattern get smaller and smaller, the depth of focus range gets smaller and smaller. Any non-planarity of the surface on which the contact pattern is being imaged reduces the depth of focus. The addition of dummy poly structures makes polysilicon geometries more uniform across the wafer and makes the premetal dielectric layer (PMD) that is deposited over the polysilicon geometries more uniform and more flat after CMP.

Designers objected strongly when we first proposed the addition of dummy polysilicon structures. They claimed it would invalidate their SPICE models (circuit simulations) because the parasitic capacitances introduced by the dummy poly structures were not comprehended in their models. They said the cost was too much and time too long to rewrite their SPICE modeling programs and recalibrate them to include dummy poly. I contacted the engineers that laid out our test chips and asked them to place two circuits side-by-side on the next test chip: one without the dummy poly structures and one with them. After processing the test chip, we gave both integrated circuits to the designers. We asked them to characterize both and to let us know which worked best. We told them we were comparing two different OPCs and needed to decide which one to go with in production. The designers said both circuits worked equally well and to go with whichever OPC we preferred. This is how assist dummy poly was first introduced at TI.

9.5 Planarization: Chemical Mechanical Polish (CMP)

9.5.1 Shallow Trench Isolation

Local oxidation of silicon (LOCOS), which was used to grow the silicon dioxide dielectric (SiO_2) isolation between the CMOS transistors, reached its scaling limit by the 320 nm technology node. Figure 10.1 To keep scaling going, shallow trench isolation (STI) in which trenches are first etched into the silicon substrate between the CMOS transistors and then refilled with deposited silicon dioxide dielectric was invented and developed. Figure 10.2a difficulty with STI is that the deposited oxide overfills the trenches (Fig. 9.28). The overfill must be removed from where transistors are to be built. Etch back techniques were not manufacturable.

Chemical Mechanical polish (CMP) of STI was a critical enabler to keeping IC scaling going (see also Sect. 10.2.3 in Chap. 10).

9.5.2   Pre-metal Dielectric Planarization

As ICs scaled smaller, contact holes scaled smaller. As the holes scaled smaller, they became increasingly more difficult to print on a non-planar surface because of the limited depth of focus. If the small contact holes were in focus in the valleys of a non-planar surface, they were not in focus on the hills and would not print.

At the 320 nm technology node, PMD CMP solved this issue enabling scaling to continue. Had the PMD CMP process not been developed, complicated optical proximity corrections plus multiple patterning steps with multiple focus settings would have been required.

9.5.3   Multi-level Metal

As IC area scaled, more and more transistors were crammed onto the IC chip and the number of wires needed to connect them together exploded. This created the demand for additional layers of wiring stacked one on top of another. Planarization became a huge problem. Complicated and expensive manufacturing flows were developed to provide surfaces sufficiently planar to image the wiring layers and sufficiently planar to etch the wiring layers without leaving metal residues that caused metal shorts in valleys. At the 320 nm technology node, Intermetal dielectric (IMD) CMP solved these planarization issues. ICs with a dozen or more wiring layers are now manufactured routinely using CMP.

Fig. 9.28 STI trenches filled with deposited silicon dioxide dielectric. https://pdfs.semantic-scholar.org/ed33/a6084fd959bcd1b4a82ffaffa6beb1563e3b.pdf [25]

Fig. 9.29 Intel 45 nm
PMOS transistor with a
replacement metal gate.
Pre-metal dielectric over a
polysilicon gate is CMP
planarized before
replacement with a metal
gate. https://en.wikichip.
org/wiki/File:intel_45nm_
gate.png [26]

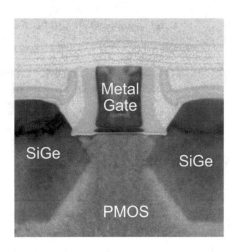

9.5.4 Copper Metallization

As ICs are scaled smaller, the smaller aluminum wires became too resistive. The sputter deposition and plasma etching processes used for aluminum cannot be used for copper. Copper can be sputter deposited but cannot be plasma etched because etching gases such as fluorine and chlorine do not form volatile etch products with copper that can be pumped away.

Copper wiring was introduced at the 180 nm technology node. Copper wiring is manufactured by etching trenches into a silicon dioxide (SiO_2) dielectric layer, electroplating copper to fill and over fill the trenches, and then using copper CMP to remove the copper overfill. Copper CMP makes copper wiring possible.

9.5.5 Metal Replacement Gate Transistors

On some of the most advanced ICs, nitrided silicon dioxide gate dielectric is replaced with a high dielectric constant (High-k) gate dielectric such as hafnium dioxide (HfO_2), and the highly doped polysilicon gates are replaced with metal gates such as titanium nitride (NMOS) and aluminum titanium nitride (PMOS). In some of these manufacturing flows, the MOS transistors polysilicon gates are first manufactured as usual and then are replaced with metal gates using CMP. After the premetal dielectric layer (PMD) is deposited over the polysilicon gates, the PMD surface is CMP'd planarizing it and exposing the tops of the polysilicon transistor gates (Fig. 9.29). The polysilicon gates are then removed and replaced with metal gates using a single damascene gate replacement process.

9.6 Rapid Thermal Processing

9.6.1 Introduction

When IC areas were large and transistor source/drain junctions were large, dopants were thermally diffused into the single crystal silicon to form the source/drain diodes in furnace tubes using gaseous dopant sources at high temperatures. As ICs scaled smaller, diffusion doping was replaced with ion implant doping plus activation anneal. First dopant atoms are implanted into the single crystal silicon substrate to precise depths and then activated (replace silicon atoms in single crystal silicon lattice with dopant atoms) by heating them to a temperature of about 900 °C or more. Prior to the 130 nm node, a batch of dopant implanted wafers were loaded into a furnace tube and annealed for a half hour or longer.

When transistors scaled to the point that these long anneal times were causing dopants to diffuse too much so that the source/drain diodes were larger and deeper than allowed by design rules, a new anneal technology was needed that could heat the wafers high enough to activate the dopants and for a short enough time to prevent the source/drain diodes from diffusing to large and too deep.

9.6.2 Rapid Thermal Anneal

Rapid thermal anneal (RTA) equipment was invented and developed to heat wafers from room temperature to 900+ °C and return them to room temperature again in a manner of a few minutes or less.

As ICs continued to scale activation, anneal times had to keep getting shorter to accommodate the requirement for shallower and smaller source/drain diodes. Rapid thermal anneal times dropped from minutes to seconds and then to milliseconds.

9.6.3 Spike Anneal

Spike anneal equipment and processes were invented and developed to reduce anneal times to hundreds and then tens of milliseconds.

9.6.4 Flash Anneal

Flash anneal equipment and processes additionally reduced anneal times to less than a millisecond. Flash anneal is usually combined with a refined spike anneal. The refined spike anneal temperature is just high enough to regrow the surface amorphous layer into single crystal doped silicon with minimal dopant diffusion. The Flash anneal then raises the temperature to 1000 °C or more for less than a millisecond to activate more dopant atoms without giving them time to diffuse.

9.6.5 Laser Anneal

With continued scaling to 20 nm technology and below, source/drain junctions are even shallower and smaller. To achieve higher dopant activation to additionally reduce series resistance, laser annealing was invented and developed. Laser annealing heats the surface layer containing the dopant atoms from a pre-anneal chuck temperature of about 400 °C up to an annealing temperature as high as 1300 °C and returns the surface layer temperature back to chuck temperature within a few microseconds (millionths of a second) or even tens of nanoseconds (billionths of a second).

The performance of NMOS transistors is most improved if the laser anneal is performed after the refined spice anneal. This provides maximum dopant activation and lowest NMOS source/drain series resistance. The performance of PMOS transistors is most improved if the laser anneal is performed before the refined spice anneal. It is believed that the laser anneal dissolves boron clusters making more boron atoms available for activation when the amorphous doped layer is epitaxially regrown into doped single crystal silicon during the refined spike anneal. Some highly scaled transistor manufacturing flows incorporate a laser anneal before and after the refined spike anneal to boost performance of both NMOS and PMOS transistors.

The challenges in developing rapid thermal processing are daunting. Thermal stresses shattered wafers during initial development. The entire wafer must remain at nearly the same temperature while the temperature is changing by more than 1000°/s to prevent shattering. The surface of the wafer has some patterns that absorb light strongly and other patterns that absorb light less strongly. Keeping the temperate the same so all transistors turn out the same is difficult even on our best days. In some processes, a layer of absorbent material such as amorphous carbon is deposited prior to flash or laser anneal, so the heating is less pattern-dependent.

Had rapid thermal anneal technology not kept pace with IC scaling, IC scaling would have stopped.

9.7 Defect Detection and Defect Analysis

9.7.1 Introduction

Defects cause integrated circuit dies to fail. There are two ways IC dies are usually defective: (1) they are functional but fail to meet electrical specifications such as speed or standby current; or (2) they fail to function due to defects caused by particles, marginal depth of focus, and narrow process windows. Manufacturing processes are tightly controlled so that most functional die meet electrical specifications. Most failures in manufacturing are caused by defects in the wiring levels. Just a few defects at each metal level accumulate to significant yield loss for multiple levels of metal.

The number of particles increases exponentially as particle size gets smaller. (When particle size goes down by 1, the number of particles goes up by 10.) As geometries scaled 1000 times smaller, the number of particles that could kill the IC circuit increased dramatically. If breakthrough defect location and identification technologies had not kept pace with scaling, IC scaling would have stopped.

Analysis of the composition of the particles and of the composition of contaminants is critical to determining the source of the particles and contamination. Once the source is pinpointed, it can be eliminated. Without the detection and identification technologies that kept pace with the IC technology as it scaled, profitable yields could not have been maintained. The sensitivity of some of these techniques is incredible. The composition of contaminants in one monolayer (one layer of atoms or molecules) can be determined. Individual atoms in single crystal semiconductors can be imaged (see Fig. 9.32).

9.7.2   Killer Particles

A particle under a photoresist pattern causes a bump in the photoresist that defocuses the pattern image causing distortion or failure of the geometry to print. A particle on the top of photoresist blocks the photoresist from being exposed by light. This can cause a metal, active, or polysilicon short. This can also cause a missing contact or missing via.

A particle that falls on the pattern after exposure and develop can block etching resulting in active, poly, or metal shorts. A particle can block contact or via etch causing a missing contact or via.

A particle that falls on a pattern after etching can block thin film deposition resulting in voids in shallow trench isolation or voids in metal wires. A particle that falls on a contact hole can block CVD-W fill causing an open contact failure. A particle that falls on a via hole can block copper electroplating causing an open via failure. A particle can fall into an etched metal trench and block copper electroplating causing a metal open failure.

By the mid-1990s, I was convinced we were nearing the end of scaling. Killer defects were becoming so small I could no long find them using our most powerful inspection microscope. The number of killer defects was rising exponentially as critical geometries scaled smaller and got closer together. One small particle could bridge the space between two geometries (Figs. 9.30 and 9.31). I thought this was a showstopper. But then…

I was working in TI's IC prototyping fab trying to find at least one of the defects that was causing a 10+ % yield loss. I had spent a half hour or so scanning the surface of the wafer using a high-power, optical microscope without success. My tech-

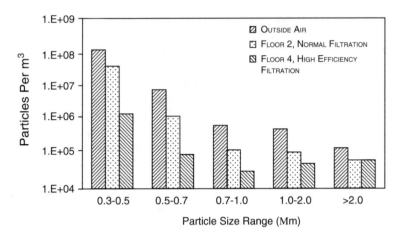

Fig. 9.30 As particle size goes down, the number of small particles rapidly rises. https://www.tandfonline.com/doi/pdf/10.1080/027868200303452 [27]

DRAM Density	4M	16M	64M	256M	1G
Resolution (Mm)	0.65	0.50	0.35	0.25	0.15
Wavelength (nm)	436	365	365/248	248/193	193/157
Criticle Particle Diameter (Mm)	0.13	0.10	0.07	0.05	0.03

1992 1993 1994 1995 1996 1997 1998 1999 2000 2001 ⟩

19042

Fig. 9.31 Critical (killer defect) size versus technology. http://smithsonianchips.si.edu/ice/cd/CEICM/SECTION3.pdf [28]

nician, who was much better than I at finding defects, could not find them either. She heard about a new KLA 21XX automated wafer inspection tool in TI's R&D fab and brought one of our wafers over to them to see if they could locate our killer defects. She returned with a wafer map showing thousands of the killer defects, the exact location (coordinates) of each killer defect on the wafer, and several high-resolution scanning electron micrographs (SEMS) images of the killer defects. I was blown away. We were saved.

I took the wafer back to my high-resolution optical microscope and, knowing exactly what I was looking for, and still was unable to find even one after searching for 10+ min. I then took the wafer to a scope with a computer-driven stage. I entered the coordinates of one of the defects into the computer. The computer drove the stage placing the defect directly under the highest-power lens. Lo and behold, there it was! I could never have found the defect without its coordinates and without the computer-driven microscope stage.

9.7.3 Automated Killer Particle Detection

This new defect inspection tool was a game changer. Since we could now find the killer defects, we could fix them. Without this capability scaling would have stopped. As transistor geometries scaled smaller, the number of particles that became killer defects skyrocketed. Automated wafer inspection tools with higher and higher resolution were developed in pace with scaling. Microscopes with increasing magnification were integrated with computer-driven stages and provided with computer-driven cameras to detect and provide images of smaller and smaller defects. When optical magnification could be pushed no farther, defect inspection tools with scanning electron microscopes (SEMS) were developed and introduced into failure analysis labs for routine automated defect detection.

SEMs with higher and higher magnification were invented and developed to keep pace with the scaling. Scientists and engineers kept pushing electron microscope technology to image smaller and smaller particles. Tunneling electron microscopes (TEMs) were developed when SEMs no longer provided enough magnification. Scanning tunneling electron microscopes (STEMs) provided increased magnification over the TEMS. (We can now see the individual atoms in a single crystal of silicon! (Figure 9.29). I never dreamed this would be possible) (Fig. 9.32).

9.7.4 Killer Particle Analysis

Once a particle is found, the next challenge is to determine where it came from. This requires a detailed analysis of the shape, size, morphology, and composition of the particle.

If the particle is a flake, it most likely is from a thin film that deposited on the inside walls of a plasma deposition or plasma etching tool and then peeled off. If it is round like a sphere, then it most likely is a droplet formed in the gas phase either by gas phase nucleation or condensation (When moist air cools fast due to a pressure drop droplets of fog form). If it is irregularly shaped, depending upon its composition it may be a particle that cracked off a highly stressed film or may be a particle produced by friction between moving parts in the machine or from deterioration of a material such as an O-ring or a plastic wafer carrier.

Amazing material analysis tools were developed to identify and characterize the composition of smaller and smaller particles, thinner and thinner films and finally even determine the composition and bonding of monolayers of contaminate molecules. Most of these techniques were first invented and demonstrated at university labs and research institutes and then developed into the computerized tools purchased by failure analysis labs at semiconductor manufacturing companies and used every day to troubleshoot IC yield problems.

Ion milling tools with integrated optical and scanning electron microscopes dig trenches through dielectric and interconnect layers and take high-resolution pictures

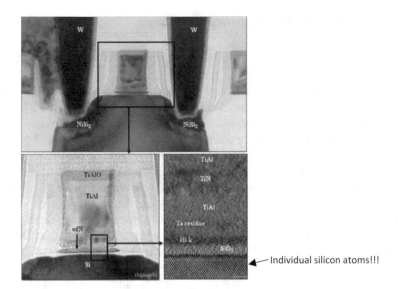

Fig. 9.32 TEM of Intel's 45 nm NMOS transistor. Individual silicon atoms can be seen below the amorphous silicon dioxide (SiO_2) gate dielectric in the bottom right hand TEM. https://en.wikichip.org/wiki/intel/process [26]

of regions of the IC the failure analysis engineers have determined to be defective. Scanning atomic force microscopes map dopants and defects in the silicon crystal lattice. High-resolution thermal microscopes detect circuit hot spots. Acoustic microscopy detects fractures and delamination. X-ray diffraction determines if the particle is made of crystalline, polycrystalline, or amorphous material. X-ray energy spectroscopy (EDX) identifies the chemical composition of particles. Auger electron spectroscopy (AES) determines the composition of surface contaminations. Microprobing equipment assesses functionality of subcircuits in ICs. Sophisticated computer programs with pattern recognition software on the automated inspection tools help to classify the various defects that are detected.

Had we not been able to continue detecting the particles and defects as geometries scaled smaller, scaling would have stopped. Had we not been able analyze the particles and defects so we could determine where they came from and get rid of them, yields would have plummeted so low IC factories would have had to close their doors.

9.8　Functional and Reliability Testing

9.8.1　Introduction

Ensuring integrated circuits are reliable is huge field of discipline by itself. Rule of thumb is all packaged ICs are required to function perfectly for at least 10 years. Tens of thousands of new IC's are designed and manufactured every year - ICs that must endure the harsh environment, temperature, and pressure in oil wells that are miles deep; ICs that will spend their working lives immersed in gasoline tanks; ICs that must survive the heat and vibration of truck and automobile engines; ICs in satellites that must survive the constant harsh bombardment of radiation in space; ICs that run our computers 24 h a day, 7 days a year, for years and years.

Many ICs are so complex and must flawlessly perform so many different tasks (your cell phone IC and computer CPU) that it is impossible to test and verify every function works. IC chips cannot be tested for 10 years to ensure 10-year reliability before they are released to the market. So many chips are manufactured each day that it is impossible to test the reliability of every one before it goes out the door.

Millions of ICs are manufactured each day. Mathematicians and reliability physicists have developed sophisticated statistical sampling programs. With these programs, they accurately predict how many ICs from each production lot must be sampled and tested to ensure that less than one IC out of a billion ICs sold to customers will later fail.

9.8.2 Functional Testing

After a wafer completes manufacturing, it is sent to multiprobe for functionality testing. Test engineers generate complex sets of testing signals (test vectors) that are fed into the inputs of the IC, and they then check the signals on the output pins to verify they are correct. Even with ICs performing billions of calculations a second, it would still take far too long to test all functions. Design engineers build self-testing circuits (BIST) into the ICs whose sole purpose is to test and verify subcircuits function properly. These BIST circuits operate only during multiprobe. While the multiprobe tester is testing the overall functionality of the IC, these on-chip BIST circuits are in parallel testing functionality of multiple subcircuits.

At multiprobe, every IC die on the wafer is tested. Passing die are marked green on the multiprobe wafer map, failing dies are given different colors depending upon what caused them to fail (see Fig. 9.16).

9.8.3 Reliability Testing: Introduction

An IC that functions perfectly during multiprobe testing may fail in a customer's electronic equipment the field months or years later. Months or years after you bought your car a pressure sensor in your car's tire may start giving a false signal and have to be replaced. Since ICs cannot be reliability tested for 10 years to ensure 10-year error-free service, scientists and engineers had to come up with ways to perform accelerated testing to ensure 10-year reliability.

9.8.4   Accelerated Reliability Testing

In accelerated testing, the IC is put under a stress that accelerates the failure mechanism. For example, if the power supply voltage is 1.5 V, the IC may be tested with 5 V stress to accelerate the failure mechanisms. If the IC is expected to operate at 50 °C, the IC may be operated at over 100 °C during reliability testing. If the wiring in an IC is expected survive electromigration with 10 milliamps of current, the wiring may be stressed with 50 milliamps during reliability testing. Typical reliability stress tests include high and low temperatures, high voltage stress, high current stress, drop testing, vibration testing, temperature cycling, harsh environment testing, high humidity testing, plus others as needed.

9.8.5   Accelerated Reliability Modeling

For accelerated testing to be valid, the reliability engineers must understand the failure mechanisms in great detail to ensure the stress they apply accelerates but does not change the physics of how the failure occurs. For example, if a spring is to be stress-tested to ensure it will retain its springiness, the applied stress must not deform the spring. Likewise when a current stress is applied to test the electromigration reliability of an IC wire, the current stress must not heat the wire to the point that the metal atoms move due to thermal motion in addition to electromigration.

Chemists, material scientists, and physicists study the mechanisms and kinetics of each failure. They plus mathematicians and computer scientists generate amazingly complex mathematical models for the failure mechanisms. These models are used to determine what stress to apply during reliability testing and are used to predict reliability lifetimes based upon the reliability stress testing data.

A few examples of mathematical reliability stress models include the following: models for electromigration reliability of metal leads; models for gate oxide dielectric wear out under current stress; models for gate oxide, PMD dielectric, IMD dielectric wear out under voltage stress, models for gate oxide, PMD dielectric, IMD dielectric breakdown under voltage stress; channel hot carrier reliability models; and electrostatic discharge models.

A different model must be developed for each different material. Copper and aluminum wires require different models. Small wires require a different model than large wires because they have different grain structures (see Figs. 11.16 and 11.17). The gate dielectric at each new technology node requires a new model. The channel hot carrier model is different for each transistor in a given technology and must be updated for each new technology node.

The electromigration model for IC wires includes equations that predict how metal atoms move through the interior of a wire through the metal grains and along metal grain boundaries as well as how metal atoms move along the outer surface (skin) of a wire as they are pushed by the electrons in the stressing current. The electromigration model also includes equations that predict changes in electromigration due to current heating the wire and differences in wire heating due to differences in the thermal conductivity of the different dielectric layers encapsulating the wire.

A different dielectric model is generated for each different gate dielectric—core transistors, I/O transistors, SRAM transistors, FLASH transistors, analog transistors, etc. Equations in the dielectric models predict the rate at which chemical bonds in the dielectric are broken depending upon the magnitude of the electric current stress and also predict the rate that electrons are trapped and build up charge (voltage) in the dielectric. The equations predict how long it takes until the dielectric breaks down depending upon the voltage and current stressing level.

The modeling programs are run many thousand times in Monte Carlo simulations to mimic manufacturing variation. For example, an electromigration modeling program is run thousands of times to take into account the variation in the size and composition of the wire die-to-die across wafers and the variation wafer-to-wafer across production lots over months and years of manufacturing. A gate dielectric reliability program is run thousands of times to take into account the variation in the nitrogen profile and thickness of the gate dielectric die-to-die across wafers and the variation wafer-to-wafer across production lots over months and years of manufacturing.

9.8.6   Qualification Reliability Testing

After the manufacturing flow for a new technology has been developed, it is transferred to manufacturing. Qualification testing is performed to determine: (1) Can the new technology be manufactured with sufficient yield to be profitable? And (2) is the new technology reliable?

Once it is determined that the new technology meets yield targets, several (usually 3–6) lots are manufactured and sent for reliability testing. Reliability testing typically takes a minimum of 3 months. If the new technology does not pass reliability testing, changes are made to the manufacturing flow, more reliability qualification lots are run, and the reliability testing is repeated. The first die is sold to a customer only after the technology passes qualification testing.

Reliability testing and prediction is a key technology in part responsible for the incredible success of the IC industry.

Reliability Testing Failure

Our first manufactured IC on a new technology with aluminum wiring and CVD-W via plugs passed reliability qualification with flying colors.

Later, a new IC manufactured using this same new technology failed reliability testing.

It turned out the first IC had a few million vias, whereas the new IC had hundreds of millions of vias. The first part met the part per billion failure qualification specification, but the new IC with many more vias did not.

After tweaking the metal and via process recipes, we were able to pass reliability testing with the new IC.

9.9   Computer Technology; Computer Simulations; Data Analysis

9.9.1   Introduction

The incredible IC scaling that reduced transistor gate lengths by 1,000 and reduced IC area by 1,000,000 would not have happened without the incredible increase in computer power enabled by the scaling. Each new technology node pushed the fastest computers then available to the limit. The faster and more powerful computers made possible by each new technology node were critical to developing the next more advanced technology node which in turn made possible even faster and more powerful computers.

It takes over a month (24 h per day, 7 days per week) to process an integrated circuit from wafer start to metal1 where the transistors can first be tested. It takes 2–3 months to complete manufacture of an IC chip with multiple levels of wiring and test to see if it works. If every process change had to be tested by actually building and testing wafers, the development time for a new technology node would be prohibitive.

Computerized process simulation programs run experiments in a few minutes. Only after a set of process conditions are selected based upon simulation experiments, are wafers actually run in the waferfab to verify or disprove the simulated results.

The incredibly complex integrated circuits with billions of transistors could not be designed if every change to the layout has to be proven out by actually building and testing the integrated circuit. Computerized integrated circuit simulation programs predict in minutes how a design change impacts circuit functionality and performance. Wafers are run to verify or disprove the simulated circuit results.

When a wafer completes manufacturing, it is sent to probe for electrical testing. At parametric probe, electrical data from transistors, diodes, capacitors, and other IC components are acquired by probing specially designed test structures (parametric test structures) in the scribe lanes between the IC dies. At multiprobe, the functionality and performance of the IC dies themselves are tested. Dies that pass all specifications are inked green. Dies that fail are inked with a color code that corresponds to the reason they failed.

Sophisticated data analysis programs pull the electrical data from parametric probe and multiprobe as well as in-line manufacturing data such as when the wafers were processed, in which tools the wafers were processed, in-line electrical data, in-line metrology data such as across wafer gate length and across wafer gate dielectric thickness. These sophisticated data analysis programs look for correlations between failure patterns and the process and metrology data. The fab yield and process integration teams then work together to eliminate the correlation.

These simulation and data analysis programs are upgraded, enhanced, and retuned at each technology node to accommodate higher packing densities, the higher current densities, new IC structures, and the properties of the new materials.

The higher packing density requires a smaller mesh size and increased computing power. High yields in semiconductor fabs would not be possible without these incredibly complex and sophisticated computer programs.

One manufacturing machine in today's IC manufacturing fab can have hundreds of sensors that many times a second measure parameters such as temperature, pressure, and flow rates to monitor if the tool is operating properly. The sensors from one manufacturing machine can generate over a terabyte (one million million bytes or 8,000,000,000,000 bits) of data every day. Powerful computers and sophisticated computer programs are key to turning this enormous flow of data into information, so engineers can evaluate and make adjustments to the machine or process if needed.

Simulated experiments are used in almost all aspects of integrated circuit technology development and are critical to integrated circuit scaling and to designing integrated circuits with increasing complexity.

The integrated circuit fab is computerized. Computers plan what IC wafers are to be started and when they are started. Computers plan when wafers are processed through each manufacturing machine to avoid bottlenecks, to reduce cycle time, and to maximize fab output. Computers gather data from in-line metrology equipment that measure such parameters as transistor gate lengths and metal resistance, plot these data on control charts, and warn operators if a trend is developing and if something goes out of specification. Computers track, store, and analyzes the voluminous data collected on each wafer processed through the fab.

9.9.2   Computer Simulation Programs: Transistor Processing

Extremely sophisticated Technology Computer-Aided Design (TCAD) computer programs such as Suprem II, III, and IV developed through collaboration between scientists from industry and professors at Stanford University provided transistor development engineers the ability to run simulated experiments in a manner of minutes or hours. This eliminated the need to process silicon wafers through the fab taking a month or more. TCAD programs calculate MOS and bipolar transistor parameters based upon the transistor structure and processing conditions. Inputs to these TCAD programs are details of the transistor structure (gate dielectric thickness, gate length, etc.) and processing conditions (implant dose, implant energy implant angle, thermal drive temperatures and times, sidewall dielectric type and thickness, etc.). The TCAD simulation program predicts transistor electrical characteristics such as transistor drive current, transistor OFF current, and transistor turn ON voltages. At the beginning of new technology development, transistor engineers run a design of experiment (DOE) test lot with splits around their initial best guess transistor structure and process conditions. The TCAD engineers tune their process simulation program to match results of this test lot. The TCAD engineers then run simulated experiments by varying transistor structure parameters such as gate length, transistor width, and gate dielectric thickness and by varying transistor processes parameters such as sidewall thickness, source/drain extension implant doping and energy, source/drain implant doping and energy, and anneal time and temperature. Based upon these TCAD simulation results, the technology development engineering teams select target process conditions and run design of experiment (DOE) confirmation lots in silicon. DOE software constructs contour maps of the experimental space by extrapolating between the actual process conditions. A location on the contour maps with transistor parameters closest to the desired specifications is selected as the center point for the next DOE transistor lot to be built in silicon. When the DOE transistor lot comes out of the fab, the TCAD simulation engineers retune parameters in their TCAD simulation program to reproduce the measured results. They then run new TCAD simulation experiments with their retuned TCAD program to determine a new center point for the next DOE transistor silicon lot.

These cycles of transistor simulation followed by DOE verification lots in silicon and retuning the simulator are repeated until the transistors finally meet specifications.

Without TCAD, the rapid advancement in integrated circuits would not have happened. For each new technology, TCAD simulation programs pushed the fastest available computer chips to their limit. Each new technology built faster, more powerful computer chips. As the technology scaled, the distance between the points in the computation grid got closer together and the number of grid points increased. For each new technology node, new materials (new dielectrics, silicide, implant dopants, etc.), new processes (spike anneals, gate dielectric nitridation, new silicide, etc.), and new structures (sidewalls, extensions, gate stack, extended drain transistors, etc.) needed to be characterized and simulated. As technology scaled, physical phenomena that previously could be ignored became important. New physical simulation models had to be developed and included in the TCAD program (carrier scattering off dopant atoms, random dopant fluctuations, dopant supersaturation, dopant deactivation, etc.)

Process Simulation Versus Transistor Guru

Texas Instruments provided custom process flows for SUN Microsystems Server chips. The SUN process flows produced the highest performance transistors at each new technology node. For each technology node, an initial SUN process flow was developed and then a year or so later followed up with a higher performance process flow.

We had just delivered server chips built with the initial SUN process flow to SUN and were ready to start developing the higher performance transistor process flow. We went to the TCAD engineers and asked them to run simulations to tell us what changes we should make to the transistor structure and process to meet the higher performance transistor specifications. They told us how much to reduce the gate length and how much to increase the doping in the channel to keep the transistor OFF current in spec., etc. We ran a DOE transistor lot around their recommended conditions and found the transistor performance went the wrong way! Instead of higher performance, the transistors had poorer performance.

I took these puzzling results to TI's transistor guru. After about 15 min of looking at our process conditions and electrical results, he announced that electrons flowing from source to drain (transistor current) were colliding with dopant atoms in the channel and slowing down (reduced carrier mobility). This was why the transistor performance was going the wrong way. He told me to go back to the TCAD engineers, tell them to add a dopant scattering model to their TCAD deck, and try again. Since I knew at the minimum this would take weeks, I asked the guru if he could give me a set of conditions that I could use to run a lot, so we could learn something while we were waiting. A couple of hours later, based upon back of the envelope calculations, he gave me a starting point for our next DOE transistor lot. We ran the lot and Mr. Guru was spot on. The transistor performance improved as he predicted. TCAD engineers used data from this lot to tune their new simulation deck (that now included a dopant scattering model) and we were off and running again.

At each new technology node, enhanced TCAD programs pushed the new more powerful computer chips to their limits. The most powerful computer chips from the previous technology generation were required to build the next technology node, so even faster and more powerful computer chips could be built to keep scaling going. We were laying track just ahead of the speeding scaling locomotive.

Simulation Helps Solve CHC Problem
During development of one of our next-generation technologies, all transistor specifications were on target except for channel hot carrier reliability (CHC). Instead of a CHC reliability lifetime of years, the CHC lifetime of our transistor was just a few seconds. The hot channel electrons (CHC) were being diverted away from the drain and toward the gate by high electric fields near the transistor drain. These CHC electrons were getting trapped in the gate dielectric causing the transistor to fail. We processed a design of experiments (DOE) lot. The TCAD (technology computer-aided design) engineers tuned their simulation deck to reproduce our DOE results. The TCAD engineers then ran a series of simulated experiments to engineer the flow of the channel current with respect to the position of the peak electric field near the drain. They generated a recommended set of dopant dose and anneal conditions for the drain extension and the halo implants that their simulations said would reduce CHC. We ran a design of experiments (DOE) around the TCAD engineers simulated conditions and were able to find a set of manufacturing conditions that resolved the CHC problem.

9.9.3   Computer Simulation Programs: IC Processing

When designing the manufacturing flow for the next-generation IC, test chips with thousands of test structures duplicating small subcircuits and electrical devices on the IC are built and electrically tested. Computer simulation programs are tuned to replicate these electrical results. These simulation programs are then used by designers to predict if their layout will work. These simulation programs are also used by process engineers to determine if they need to change their materials and/or processes. For example, metal wire test structures with single and multiple wires with different lengths and various spacing are manufactured, tested, and used to tune the metal performance and reliability models. Designers use simulations to determine if the metal wire they are laying out has sufficiently low resistance for signal propagation and sufficiently low current density for reliability. Process engineers use the simulations to determine if reliability and resistance specifications are met across all design rule possibilities. At the 180 nm technology node, these test structures and simulations determined that the previously used Ti/TiN/AlCu/TiN metal stack no longer could pass electromigration reliability testing across the entirety of the design space, so they changed to an $Al_3Ti/AlCu/Al_3Ti/TiN$ metal stack. In a similar manner, these test structures and simulation programs determined the need to replace titanium silicide with cobalt silicide to meet resistance specifications on narrow leads, and the need to replace cobalt silicide with nickel silicide. The test chips and simulations help guide the selection of the various intermetal dielectrics and selection of the thicknesses of the barrier layers and copper seed layers as the copper wires scale.

9.9.4   Computer Simulation Programs: Integrated Circuit Design

Early on, integrated circuit design engineers used bread boards to wire transistors, resistors, and capacitors together to test out new circuit designs. They would then adjust the resistances and capacitances to optimize performance. Today an integrated circuit chip can have billions of transistors switching at over a billion times per second. Today circuit designers test out new designs using integrated circuit simulation computer programs such as SPICE (Simulated Program with Integrated Circuit Emphasis). SPICE was developed by scientists at the University of California Berkley in collaboration with scientists from industry. SPICE modeling programs have evolved and have been continuously updated to keep up with the rapid pace of integrated circuit scaling. HSPICE is a product from Synopsis. PSPICE is a product from Cadence Design Systems.

The SPICE programs are enhanced, upgraded, and retuned at each new node to accommodate the increase in the number of transistors and to accommodate changes in material and device properties. The increasing packing densities require smaller mesh sizes and require increasing computer power. Integrated circuit designers could not design complex integrated circuits and get them to work without these simulation programs.

A detailed model of transistor electrical characteristics such as ON current, OFF current, diode leakage, and gate leakage is constructed to match measured results on wafers processed through the fab. Models for other electrical components such as poly-to-substrate capacitors, poly-to-poly capacitors, metal-to-metal capacitors, implanted resistors, polysilicon resistors, metal resistors, and diodes are also constructed to match wafers processed through the fab. These models also predict variations that occur in the performance of these electrical components due to process variation from manufacturing. Models are also constructed for parasitics that degrade and delay signals in the circuit such as capacitive coupling (metal-to-metal same layer, metal-to-metal different layers, metal-to-gate, metal-to-substrate, etc.). The models also include leakage currents such as gate leakage, transistor off current leakage, and reverse biased diode leakage. Different resistances delay signals by different amounts. (Source/drain, contact plug, via plug, metal wires of differing sizes, etc.) All these various models are used by SPICE to predict what signals come out of an IC based upon what signals are input into the IC. SPICE accurately predicts how integrated circuits in your cell phone (some with over a billion transistors switching over a billion times a second) convert the Wi-Fi signal from a broadcasting station into the voltage and current levels that illuminate the over two million pixels (1920 horizontal by 1200 vertical pixels) on your cell phone screen at 120 Hz (120 times/s) as you stream a movie or facetime with your friend in real time.

9.9.5 Computer Simulation Programs: Thermal Heating in IC Dies and IC Packages

Heat that is generated during the operation of integrated circuits is an important reliability consideration. Designers use thermal simulation programs to model heat generation and to design means of heat removal when laying out their circuits. Heat-generating subcircuits are spaced apart. Metal interconnect and dummy metal geometries are positioned to conduct heat away. Heat spreaders are attached to the IC dies to conduct heat away if needed.

Thermal simulation experiments are run extensively when designing packages for integrated circuits. Circuit layout engineers and package design engineers iterate simulation experiments back and forth to provide IC layout plus package solutions for circuits that generate a lot of heat.

9.9.6   Computer Simulation Programs: Stress in IC Dies and IC Packages

Different materials expand more than others when heated and shrink more than others when cooled (differences in coefficient of thermal expansion—CTE). One common source of killer particles in IC manufacture is when stress from a difference in CTEs causes one thin film to delaminate from an underlying thin film.

One common cause of failure of packaged ICs is when stress causes molding compound to delaminate from the packaged IC die.

Transistor and IC engineers use stress simulation programs to identify where on the IC die stress may be a problem and also to quickly evaluate solutions such as the addition of an adhesion promoting process such as sputter etch, the addition of a stress buffer layer, re-layout of geometries to break up large stress areas, and the addition of dummy geometries.

Transistor engineers use simulation programs to predict the performance of stress-sensitive transistors and of other stress-sensitive electronic components. The performance of analog transistors and devices such as bulk acoustic wave resonators is especially affected by stress.

Packaging engineers use stress simulation programs to design packages for IC dies whose performance is sensitive to stress and to reduce packaged die failures caused by molding compound delamination.

9.9.7   Computer Simulation Programs: Reliability Simulation Programs

Metal electromigration, a primary source of IC die failure due to broken metal wires, is very sensitive to temperature. Design engineers use simulation programs tuned to fab data to evaluate current density and thermal rise in metal wires when designing and laying out metal patterns to avoid reliability problems. These programs are upgraded, enhanced, and retuned at each technology node to accommodate the smaller more closely spaced wires and to accommodate the different dielectrics around them.

Channel hot carrier simulation experiments are run by designers to determine if their transistors will remain in spec. throughout the lifetime of their integrated circuit.

Gate dielectric breakdown simulation experiments are run by design engineers to determine if they can boost the gate voltage on a transistor to improve performance without causing the dielectric to wear out. If the transistor is usually in the OFF state (no voltage on the gate) and the gate dielectric is not stressed very often, this approach can work.

9.9.8   Computer Simulation Programs: Photo Simulation Programs

Optical physicists have written sophisticated modeling programs to predict how light will expose the resist on a wafer after passing through the transparent regions of a reticle. When geometries on the reticle approach the wavelength of light being used to print the geometry in the photoresist, light interference causes severe distortions. Using optical modeling programs, optical physicists calculate how to draw geometries on the reticle so that with light interference, the geometries print correctly on the wafer. These programs are rewritten at each new technology node and retuned using fab generated data to accommodate the smaller and more closely spaced geometries, to accommodate the increasing number of transistors on a reticle, and to accomodate changing etch loading effects.

9.9.9   Computer Simulation Programs: Other Simulation Programs

The number of computer data analysis and simulation programs used in a wafer fab are too numerous to mention. Computer programs are used to simulate virtually every process that is performed in a fab. Equipment simulation programs and process simulation programs help engineers tune processes and improve across wafer uniformity. Fab throughput programs plan which wafers from which lot go through which manufacturing tool at which time to maximize the number of wafers that can be processed through the fab. Control chart programs gather data from processing equipment, metrology equipment, and manufacturing processes and plot them on control charts to make sure everything is in spec. These programs send warning to fab personnel if something goes out of spec. or if a statistically significant trend is developing that will soon cause something to go out of spec.

9.9.10   Data Analysis Programs: Probe Data

After manufacturing is complete, the drive currents on NMOS and PMOS transistors are measured and plotted (see Fig. 9.33). This data is first plotted for all transistors on all wafers in a lot (usually 24 wafers) and then is plotted for each wafer individually. A huge amount of effort is expended to make the distribution of drive currents as narrow as possible. When the distribution is narrow, designers can tighten their design specifications and design higher performance ICs. Data analysis engineers have written sophisticated data analysis programs to pull data from tools

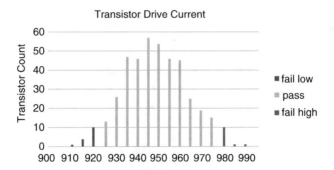

Fig. 9.33 Transistor drive current distribution plot

in the manufacturing fab, so they can run correlations between machine data and transistor data such as (Fig. 9.33) looking for statistically significant correlations. When a statistically significant correlation is found, the process engineers and process integration engineers go to work to try to get rid of the correlation and narrow the drive current distribution. One common correlation is lot-to-lot correlation. This occurs when the drive current on one group of lots is statistically different than the drive current on a different group of lots. These types of correlations usually can be traced to differences between manufacturing tools that the two groups of lots were processed through or to differences in a material such as photoresist used on the two groups of lots or to differences in times the two groups of lots were processed. (One correlation was traced to which way the wind was blowing! When the wind came from the South ammonia fumes from an exhaust on the roof of the fab were sucked into the air intake for the lithography area. Ammonia fumes messed up the chemically amplified resist).

Another common correlation is between wafer order in the lot. Sometimes the drive current on the first few wafers is different because they were the first wafers to run through a tool after the tool was down for an extended time. The fix for this correlation is to run a couple of dummy wafers to season the tool before starting to run wafers from the production lot.

A common correlation between transistor drive current on individual wafers is center-to-edge correlation. The drive current on transistors in the center of the wafer are either hotter or colder than the drive current on transistors on the edge of the wafer. This type of correlation is usually caused by a center-to-edge non-uniformity in one of the single-wafer processes such as gate etch, sidewall deposition, sidewall etch, rapid thermal anneal, contact etch, metal trench pattern, and metal trench etch. In single-wafer processes, the wafer sits in the middle of a process chamber on a wafer chuck. Gases are usually introduced through a shower head above the wafer and exit the chamber through vacuum ports around the edges of the wafer. This configuration frequently results in a difference in deposition rate or etch rate between the center and edge of the wafer.

9.9.11 Data Analysis Programs: Sensor Data and Fault Detection and Classification (FDC) Data

One piece of manufacturing equipment in today's IC manufacturing fab can have hundreds of sensors. These sensors monitor tool parameters such as voltages, currents, and position of valves to ensure the machine is operating properly and monitor process parameters such as temperature, pressure, plasma density, deposition rate to ensure the process is operating properly. The sensors from one manufacturing machine can generate over a terabyte (one million million bytes or 8,000,000,000,000 bits) of data every day. A discipline called Fault Detection and Classification (FDC) has been developed to deal with this enormous flow of data. FDC transforms sensor data into summary statistics and generates models that can be compared to user-defined limits to detect equipment and process excursions.

Computers collect the machine sensor data and constantly monitor the data for indications that something may be drifting and require maintenance. The computers make phone calls to the appropriate fab personnel at any time day or night anything is going amiss.

Computers collect process data and plot the data on graphs for the equipment and process engineers to view and determine if the process is operating in spec. For example, the equipment and process engineers can monitor pressure or temperature of a plasma deposition or etching process. They can view and set limits on rate that temperature and pressure rise in the processing chamber after the wafer is loaded. They can view how long it takes the temperature or pressure to stabilize. They can view and set limits on temperature and pressure throughout processing. They can view and set limits on when the process terminates and the rate at which the temperature and pressure are allowed to ramp down.

Sophisticated statistical models compare FDC data to yield data and to in-line metrology data to identify statistically significant correlations that might be eliminated to increase yield.

References

1. Process Technology History–Intel, WikiChip, Chips and Semi, https://en.wikichip.org/wiki/intel/process. Accessed 4 Mar 2020
2. Intel Microprocessor Quick Reference Guide, WikiChip, Chips and Semi, https://www.intel.com/pressroom/kits/quickrefyr.htm. Accessed 4 Mar 2020
3. Semiconductor Manufacturing Clean Room, Wikipedia, the free encyclopedia, https://en.wikipedia.org/wiki/Cleanroom. Accessed 4 Mar 2020
4. R. Johnson, GlobalFoundries Gears Up for the Next Generations of Chip Manufacturing, Tech Report, 6 Mar 2018, https://techreport.com/review/33337/globalfoundries-gears-up-for-the-next-generations-of-chip-manufacturing/
5. Clean Room Garmet, Wikipedia, the free encyclopedia, https://en.wikipedia.org/wiki/Cleanroom#/media/File:Cleanroom_Garment2.JPG. Accessed 4 Mar 2020

6. A. Holbrook, What Is a Cleanroom?, Critical Environment Solutions LTD., Swindon 2015., https://www.criticalenvironmentsolutions.co.uk/what-is-a-cleanroom/. Accessed 4 Mar 2020

7. P. Moorehead, The U.S. Already Has Bleeding Edge Technology Manufacturing With GlobalFoundries Fab 8 In Malta, NY, (Forbes, 2018), https://www.forbes.com/sites/patrick-moorhead/2018/02/04/the-u-s-already-has-bleeding-edge-technology-manufacturing-with-globalfoundries-fab-8-in-malta-ny/#1ecc2fff23af

8. History of Silicon Wafers used in the semiconductor industry, Chips, Etc., Vintage Computer Chip Collectibles, Memorabilia and Jewelry, Website, https://www.chipsetc.com/silicon-wafers.html. Accessed 4 Mar 2020

9. One Set of 3″ Diameter 25 Group Wafers Carrier Box—SP5-3-25, MTI Corporation, Website, https://www.mtixtl.com/onesetof3diameter25groupwaferscarrierboxsp5-3-25.aspx. Accessed 4 Mar 2020

10. N-Type 5 Inch Polished Monocrystalline Silicon Wafer for Semiconductor, Made-in-China, Connecting Buyers with Chinese Suppliers, Website, https://hncrystal.en.made-in-china.com/product/pXBxfRNGsuYh/China-N-Type-5-Inch-Polished-Monocrystalline-Silicon-Wafer-for-Semiconductor.html. Accessed 4 Mar 2020

11. 150 mm VWS Wafer Shipper, SPI/Semicon, Website, http://www.spisemicon.com/category/products/wafer-containers/. Accessed 4 Mar 2020

12. Wafer Transport Boxes for Semiconductor Contamination Control, Brooks Website (2020), https://www.brooks.com/products/semiconductor-automation/semiconductor-contamination-control/wafer-carriers-boxes/wafer-transport-boxes

13. EM-Tec precision wafer handling tweezers, Micro to Nano Innovative Microscopy Supplies, Website, https://www.microtonano.com/EM-Tec-precision-wafer-handling-tweezers.php. Accessed 4 Mar 2020

14. Wandshop.com: specialist product handling solutions for a range of industries, Wand Shop, Website, https://www.wandshop.com/. Accessed 4 Mar 2020

15. M. Moslehi, Y. Lee, C. Shaper, T. Omstead, L. Velo, A. Kermana, C. Davis, Chapter 6 Single Wafer Process Integration and Process Control Techniques, in *Advances in Rapid Thermal and Integrated Processing*, (Springer-Science+Business, B.V, Berlin, 1996), pp. 166–191, https://books.google.com/books?id=Y37sCAAAQBAJ&pg=PA166&lpg=PA166&dq=MMST+single+wafer+processing&source=bl&ots=fx85m5hPAj&sig=ACfU3U3DRYRDpigpBxF_LdlBSmkLHn-4tQ&hl=en&sa=X&ved=2ahUKEwj_gtuLhaDjAhXFLs0KHS5FBJ0Q6AEwB3oECAkQAQ#v=onepage&q=MMST%20single%20wafer%20processing&f=false

16. TI achieves fastest cycle time for chip making, UPI Archives (1993), https://www.upi.com/Archives/1993/06/30/TI-achieves-fastest-cycle-time-for-chip-making/8926741412800/

17. P. Silverman, The Intel lithography roadmap. Intel Technol. J. **06**(2), 55–64 (2002)., https://www.intel.com/content/dam/www/public/us/en/documents/research/2002-vol06-iss-2-intel-technology-journal.pdf

18. ASML's EUV systems pattern new Samsung and TSMC chip, Optics.org The Business of Photonics, https://optics.org/news/10/4/31

19. J. Hruska, Intel Reportedly Won't Deploy EUV Lithography Until 2021, ExtremeTech (2018), https://www.extremetech.com/computing/276376-intel-reportedly-wont-deploy-euv-lithography-until-2021

20. Nikon ArF Immersion Scanner NSR-S635E, Nikon Corporation, Website, https://www.nikon.com/products/semi/lineup/pdf/NSR-S635E_e.pdf. Accessed 4 Mar 2020

21. S. Dana, P. Ware, A. Tanimoto, Progress Report: 157-nm Lithography Prepares to Graduate, Oemagazine (2003), http://spie.org/news/progress-report-157-nm-lithography-prepares-to-graduate?SSO=1

22. A. Appel, S. Crank, Y. Kim, C. Scharrer, D. Spratt, B. Strong, M. Yao, H. Tigelaar, R. Melanson, A manufacturable 0.30 μm gate CMOS technology for high speed microprocessors. Symp. VLSI Technol. **22**(3), 220–221 (1996)., https://ieeexplore.ieee.org/stamp/stamp.jsp?arnumber=507694

23. A. Thampi, Immersion Lithography, 3rd Seminar M. Sc. Physics, Cochin Univ. (2006), https://www.slideshare.net/anandhus/immersion-lithography
24. Advanced (Smart) Fill, Semiconductor Engineering knowledge center, 13 Dec 2012, SemiEngineering.com, Website, https://semiengineering.com/knowledge_centers/materials/fill/advanced-smart-fill/. Accessed 5 Mar 2020
25. J. Pan, D. Ouma, P. Li, D. Boning, F. Redeker, J. Chung, J. Whitby, Planarization and Integration of Shallow Trench Isolation, 1998 VMIC (1998), pp. 1–6, https://pdfs.semantic-scholar.org/ed33/a6084fd959bcd1b4a82ffaffa6beb1563e3b.pdf
26. Intel 45 nm gate, Wikichip Chips and Wiki, https://en.wikichip.org/wiki/File:intel_45nm_gate.png. Accessed 4 Mar 2020
27. W. Fisk, D. Faulkner, D. Sullivan, M. Mendell, Particle concentrations and sizes with normal and high efficiency air filtration in a sealed air-conditioned office building. Aerosol Sci. Tech. **32**, 527–544 (2000)., https://www.tandfonline.com/doi/pdf/10.1080/027868200303452
28. Chapter 3 Yield and Yield Management, Integrated Circuit Engineering Corporation, Smithsonian chips, pp. 1–20, http://smithsonianchips.si.edu/ice/cd/CEICM/SECTION3.pdf

Chapter 10
The Incredible Shrinking IC: Part 2 FEOL Isolation Scaling and Transistor Scaling

10.1 Introduction

Every 2–3 years, IC manufacturers came out with IC chips that were half as large and more than twice as powerful. As the IC geometries got smaller, the demands on dimension control increased. Equipment was upgraded when possible to meet the more stringent demands. New equipment was required at almost every new technology node. When the current manufacturing methods ran out of gas, completely new manufacturing processes and methods were developed. Completely new manufacturing equipment was invented to implement the new processes and methods. At every new node existing fabs were retooled with tens of millions of dollars of new equipment. Every few nodes completely new IC factories were built costing hundreds of millions of dollars. Today a new IC factory can cost upwards of ten billion dollars.

10.2 Shrinking Transistor Isolation

10.2.1 Overview

As integrated circuit area shrinks, the width of the dielectric isolation (distance between the transistors) shrinks. Several times the isolation technology was pushed to its limit and a new isolation technology had to be invented to keep Moore's law on track.

© Springer Nature Switzerland AG 2020
H. Tigelaar, *How Transistor Area Shrank by 1 Million Fold*,
https://doi.org/10.1007/978-3-030-40021-7_10

10.2.2   LOCOS Isolation

Local Oxidation of Silicon (LOCOS) was used to isolate transistors until the 320 nm technology node when LOCOS could not be scaled any further. In LOCOS isolation, a silicon nitride layer on the surface of the silicon wafer is patterned and etched to form openings in the silicon nitride layer. The silicon nitride layer blocks underlying silicon from oxidizing. Silicon dioxide (SiO_2) isolation dielectric is thermally grown in the openings to electrically isolate the active transistor regions. LOCOS isolation ran out of gas at about the 0.5 µm node in the mid-1990s. At this node, the LOCOS bird's beak (see Fig. 10.1) was causing the gate oxide to be thicker over almost half the transistor channel. This significantly degraded transistor performance. Various process enhancements such as recessed LOCOS, poly buffered LOCOS, and side-wall-sealed LOCOS [3] extended LOCOS isolation to the 320 nm technology node.

10.2.3   Shallow Trench Isolation (STI)

Shallow trench isolation (STI) was invented and developed to enable integrated circuit area to keep scaling. In the STI process, shallow trenches are etched into the silicon substrate between the transistor active areas and filled with silicon dioxide (SiO_2) dielectric (Fig. 10.2). The major challenges in STI isolation are to fill the trenches with dielectric without forming voids as the trenches get narrower with scaling and to planarize the wafer surface by removing dielectric trench overfill.

A chemical vapor deposition (CVD) process is used to fill and overfill the STI trenches with SiO_2 dielectric. This dielectric is then planarized using chemical mechanical polish (CMP) to produce a flat, planar wafer surface. Initially the new chemical mechanical process (CMP) was not manufacturing friendly. Several additional inventions were required to render the CMP process truly manufacturable.

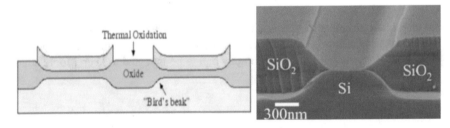

Fig. 10.1 Cartoon and SEM of LOCOS isolation, https://web.stanford.edu/class/ee311/NOTES/ Isolation.pdf   [1];   https://www.researchgate.net/ figure/a-Simulated-profile-of-round-like-LOCOS-waveguide-b-SEM-micrograph-of-round-like_ fig3_47404033 [2]

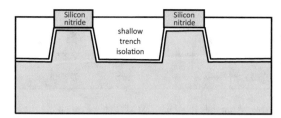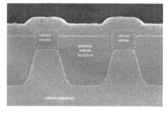

Fig. 10.2 Cartoon and SEM of STI isolation, Journal of Applied Physics 94, 5574 (2003); doi: 10.1063/1.1611287. https://www.google.com/search?q=shallow+trench+isolation+SEM+pictures &tbm=isch&source=iu&ictx=1&fir=fBshWIV23C0FGM%253A%252CnjGggP_6I97jlM%252C _&vet=1&usg=AI4_-kRndP8GQHlnr7RcYuEDriVgEGLuGQ&sa=X&ved=2ahUKEwifyfaEoPP jAhVHA6wKHW2nCnkQ9QEwAHoECAkQBg#imgrc=fBshWIV23C0FGM:&vet=1 [4]

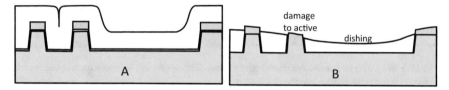

Fig. 10.3 STI trenches without active dummies filled with SiO$_2$. (a) Cartoon on the left is before CMP. (b) Cartoon on the right is after CMP

A first innovation that improved CMP manufacturability was the addition of dummy active geometries to the active pattern. Dummy active geometries are not part of the integrated circuit. They are added to the circuit pattern to facilitate CMP by making the pattern density more uniform. For example, a desired active pattern is shown in Fig. 10.3 (left), with a conformal layer of SiO$_2$ dielectric filling the trenches. CMP planarization of this structure results in dishing of the isolation oxide over wide isolation areas and damage to small active areas adjacent to the widely spaced active areas (Fig. 10.3, right).

Dummy active structures can be added to improve the manufacturability of CMP planarization (the darker active structures in Fig. 10.4a, b). A set of design rules specifically for drawing and placing dummy active geometries were developed and imposed upon the designers by the processing engineers to improve yield. Additional design rules limiting the size of large active geometries such as capacitors (designers had to construct large capacitors by wiring together several smaller capacitors). Designers were required to distribute groups of multiple capacitors to make STI CMP more manufacturing friendly. Despite these additional layout restrictions, new integrated circuit designs satisfying all the design rules (design rule clean) were still difficult to CMP. A customized CMP recipe was required for almost every new product.

To enable multiple new devices to be manufactured using the same CMP recipe, reverse active patterning and oxide etch back processing steps (RPOE) that added

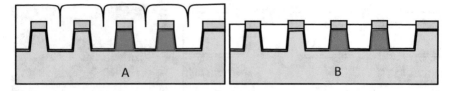

Fig. 10.4 STI trenches with dummy active geometries filled with SiO₂. (**a**) Cartoon on the left is before CMP. (**b**) Cartoon on the right is after CMP

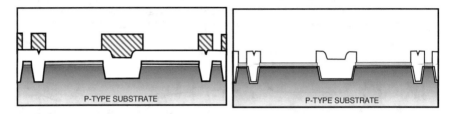

Fig. 10.5 STI with reverse active photoresist pattern. Cartoon on left is before reverse oxide etch. Cartoon on the right is after reverse oxide etch

cost and cycle time were added. The RPOE process removes most of the STI oxide overfill prior to CMP. A reverse active photoresist pattern is formed on the STI oxide with openings over the STI oxide overfill covering the active areas (Fig. 10.5, left). Prior to CMP, this STI overfill oxide is etched away (Fig. 10.5, right). This RPOE process enabled the manufacturing of different devices using the same CMP manufacturing recipe. Even with RPOE, a few new designs still required custom CMP process recipes.

Reverse active patterning and the STI overfill oxide removal etch (RPOE) was used for the 320 and 250 nm technology nodes. By the 180 nm technology node, the CMP development team (CMP equipment engineers, CMP process engineers, CMP slurry chemists) invented and made manufacturable a high selectivity slurry (HSS) process. The CMP development team developed a new CMP machine, new CMP polishing pads, and a new CMP high selectivity slurry that greatly improved CMP manufacturability. Chemicals that etched the surface of the SiO₂ on the hills and valleys were added to the slurry. The CMP polishing pad immediately removed the chemically etched products from the hills but not the valleys. This greatly acceler- ated the removal of the SiO₂ overfill and greatly improved CMP planarization. By combining chemical etching of the STI oxide with mechanical removal of etching products, this team of scientists and engineers was able to deliver a CMP STI oxide planarization process that did not require RPOE patterning and etching. This new HSS-CMP process combined with some additional HSS-CMP specific design rules was manufacturable. It enabled the same CMP recipe to be used for all new devices introduced into manufacturing.

10.2.4   *Void-Free Gap Fill*

As integrated circuits scaled smaller dimensions, the width of the STI trenches got narrower and narrower. This made it increasingly difficult to fill the narrow trench gaps without forming voids. Each time one process could no longer fill these narrow trenches void-free; a new process was invented to keep the scaling going (Table 10.1). Shallow trench isolation was introduced at the 320 nm technology node. The first dielectric used to fill STI trenches was atmospheric CVD (APCVD). For APCVD, the deposition gases are TEOS (tetraethyl orthosilicate) plus ozone (O_3). APCVD SiO_2 was able to fill the narrowest STI trenches at the 320, 250, and 180 nm nodes without voids (Fig. 10.6). Because APCVD oxide could not fill the STI trenches in the 130 nm technology node void-free, it was replaced with high-density plasma (HDP) SiO_2 deposition (Fig. 10.7). HDP was also able to fill STI trenches in the 130 nm and the 90 nm technology nodes. HDP was extended to fill the even smaller STI trenches in the 65 nm technology node by periodically halting the deposition and inserting etch back steps (dep/etch/dep).

HDP Dielectric Deposition
The HDP dielectric deposition process simultaneously deposits and etches the dielectric. Gases used are silane (SiH_4), oxygen (O_2), and argon (Ar). Silane and oxygen are deposition gases. The argon atoms are ionized and accelerated toward the depositing dielectric using bias. The argon ions sputter etches the depositing film. The deposition and sputter etch rates are balanced, so the more film is deposited than is etched. The dielectric film etches faster on the flat surface than down in the narrow trenches. As a result, the dielectric on the flat surface thickens slowly while the narrow trenches fill faster and fill void-free. The HDP deposited dielectric is almost planar, is void-free, and is easily planarized (Fig. 10.7).

A high aspect ratio dielectric deposition process (HARP) was developed to meet gap fill requirements for the 45 nm node. The HARP process is a TEOS plus ozone CVD process that does not use plasma. A high ozone (O_3) flow rate slow deposition process with good gap fill is used to initially fill narrow gaps. Gas ratios are adjusted as the deposition proceeds to speed up the deposition rate to meet both gap fill and deposition rate targets. An enhanced HARP (eHARP) process that adds water vapor (H_2O) during the deposition extended the eHARP process to fill the gaps in the 32 and 28 nm nodes.

Table 10.1 Evolution of isolation dielectrics versus technology node

Node nm	Dielectric	Gases	Pressure Torr	Temp °C	Comments
>320	Thermal SiO$_2$	O$_2$; H$_2$O	760 or >>750	~1000	LOCOS
320 250 180	APCVD	TEOS; O$_3$	760	300– 450 ~380	Shallow trench isolation
130 90	HDP	SiH$_4$; O$_2$; Ar	<10 mT	400– 600	Deposition and sputter etch—plasma
65	HDP	SiH$_4$; O$_2$; Ar	<10 mT	400– 600	Dep./etch/dep.
45	HARP	TEOS; O$_3$	600	~450	No plasma—flow rates adjusted during dep.
32 28	eHARP	TEOS; O$_3$; H$_2$O	600	~450	Flow rates adjusted during dep. plus H$_2$O
22 14	Flowable CVD	Deposit silizane + O$_3$ anneal	?	<650	Fill trenches with flowable silizane-based liquid. An O$_3$ anneal converts Si-H and Si-N bonds to Si-O

Fig. 10.6 STI trenches filled with SACVD-deposited silicon dioxide. https://pdfs.semantic-scholar.org/ed33/a6084fd959bcd1b4a82ffaffa6beb1563e3b.pdf [5]

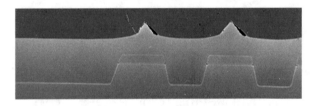

Fig. 10.7 STI trenches filled with HDP-deposited silicon dioxide

A flowable CVD dielectric [6] process is being developed to fill smaller gaps in the 22 and 14 nm nodes. https://www.eetimes.com/applied-flows-into-flowable-cvd/ [7].

The first 180 nm lot we ran with a new product design had zero yield. Failure analysis engineers determined that the root cause was polysilicon filling voids in the STI and causing shorts. The designers followed all the design rules. Their layout was design rule clean. The failure analysis (FA) engineers determined that the STI voids occurred at only one location on the die—a narrow STI trench between two closely spaced wide active regions (Fig. 10.8).

We had four choices: (1) Buy a new multimillion-dollar deposition tool that could fill the narrow trenches without voids. (2) Increase the size of the integrated circuit by re-laying out the pattern with a slightly larger STI trenches that would fill void-free with the current processes. (3) Tweak the current equipment and process to eliminate the void, and (4) Tweak the pattern to widen the STI trench, so it would fill void-free using the current process.

The first option was too expensive. The second was also too expensive and would take far too long. We tried very hard to implement the third option but failed. One of TI's layout gurus wrote a computer program which located each instance of the narrow STI trench between two wide active geometries in the layout and replaced them with a new pattern with a slightly wider trench. This solution was fast, cheap, and it worked! (This solution cheated the active overlap of contact design rule at this location. High-resolution SEM cross-sectional micrographs at this location showed the active overlap of contacts with this fix was actually o.k. in spite of the design rule violation).

There are a number of other isolation challenges that were encountered and conquered such as introducing isolation for high-voltage transistors onto a low-voltage circuits, squeezing the n+ diode to p+ diode space across the well boundary in SRAM cells to reduce the size of the SRAM memory, isolating transients in high-power circuits from disturbing their low-power circuit neighbors, etc. ...

Fig. 10.8 Voids in STI isolation voids SEM courtesy of Texas Instruments

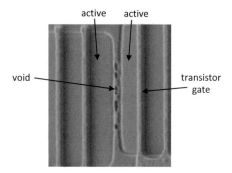

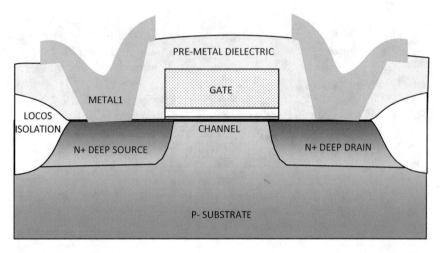

Fig. 10.9 NMOS transistor

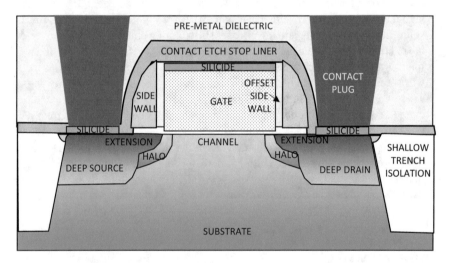

Fig. 10.10 NMOS transistor

10.3 Shrinking MOS Transistor

10.3.1 Overview

The NMOS transistor (Fig. 10.9) is simply an electric switch formed on a capacitor between two diodes—a gate- to- channel capacitor between a source-to-substrate pn-diode and drain-to-substrate pn-diode. When a sufficiently high positive voltage (above the turn on voltage = V_{TN}) is applied to the gate, electrons in the p-type substrate are attracted to the surface of the p-type substrate immediately under the gate (transistor channel). The excess of electrons in the transistor channel turns the p-type substrate in the channel to n-type silicon (inverts the channel doping = inversion). The n-type channel connects the n-type source diode to the n-type drain diode allowing current to flow freely through the n-type channel from the n-type drain to the n-type source.

Down to about the 1.5 µm technology node, the MOS transistor structure was as illustrated in Fig. 10.9. The transistors are isolated from each other with LOCOS isolation (Fig. 10.1). Deep source and drain pn-diodes are implanted self-aligned to the transistor gate. A pre-metal dielectric is deposited over the transistor and is smoothed by thermal reflowing at high temperature. Aluminum alloy metal1 is sputtered into contact openings etched through the pre-metal dielectric.

The changes that were implemented to improve the performance of these transistor switches as they scaled about 1000 times smaller are summarized in Fig. 10.10. Shallow trench isolation (STI) replaced LOCOS isolation. Offset sidewalls are added on the gate, so source and drain extensions can be implanted self-aligned to them. The source and drain extensions are added to connect the deep source and drain pn-diodes to the transistor channel. Halos covering the ends of the source and drain extensions are added to combat short channel effects. Sidewalls are added to space the deep source and drain diodes the required minimum distance apart. The deep source and drain diodes are implanted self-aligned to these sidewalls. Silicide is formed on the gate, source, and drains to lower resistance. A contact etch-stop liner (CESL) is deposited before the pre-metal dielectric (PMD) to facilitate contact etch. The pre-metal dielectric is planarized using chemical mechanical polish. Contact holes are etched through the PMD and CESL layers and filled with chemical vapor deposited tungsten metal (CVD-W). The reason for these enhancements and the technology nodes at which they were introduced are briefly described below.

The objective is to turn the transistor OFF and ON as quickly as possible. The speed at which the transistor can switch is limited by the resistance to the current flow through the transistor and is also limited by the amount of current (electrons) needed to charge capacitors in the transistor. During scaling, the goal is to engineer capacitance and resistance to be low, so the transistors are fast (see Chap. 4). A secondary objective is to engineer the transistor so that when it is OFF, leakage current from source to drain is low, leakage current through the gate dielectric is low, and reverse diode leakage current from the drain to substrate is low. This is desirable for long battery life.

Transistor resistances are illustrated in Fig. 4.4 and listed in Table 4.1.

Transistor capacitances are illustrated in Fig. 4.6 and listed in Tables 4.2 and 4.3.

The substrate surface of the n-type doped source and drains is doped as high as possible for low resistance. To avoid high reverse-biased diode leakage and low diode breakdown voltage, the doping under the surface is graded from high at the surface to low deeper down. The lighter doping at the bottom of the source/drain pn-junction diode plus the lightly doped p-type substrate forms a wide depletion region with low capacitance.

The surfaces of the source and drains are silicided to provide a low-resistance interface between the surface of the source and drains and the contact plug. The surface of the transistor gate is silicided to lower the resistance of polysilicon word

Table 10.2 Summary of the history of transistor scaling at Intel

Node, nm	Year	Power supply	Gate length, nm	Gate diel.	Gate diel. tkns, nm	Microprocessor clock speed, cycles/s
10,000	1972	5.0	10,000	SiO$_2$	100	200,000
6000	1976	5.0	6000	SiO$_2$	70	2,000,000
3000	1979	5.0	3000	SiO$_2$	40	8,000,000
1500	1982	5.0	1500	SiO$_2$	25	12,000,000
1000	1987	5.0	1000	SiO$_2$	--	33,000,000
800	1989	4.0	800	SiO$_2$	15	33,000,000
500	1993	3.3	500	SiO$_2$	8	100,000,000
350	1995	2.5	350	SiO$_2$	6	133,000,000
250	1998	1.8	200	SiO$_2$	4	300,000,000
180	1999	1.6	130	Nitrided SiO$_2$	2	700,000,000
130	2001	1.4	70	Nitrided SiO$_2$	~1.4	1,330,000,000
90	2003	1.2	50	Nitrided SiO$_2$	~1.2	3,060,000,000
65	2005	1.0	35	Nitrided SiO$_2$	~1.2	3,800,000,000
45	2007	1.0	25	SiON and high-K	~1.0	3,000,000,000
32	2009	0.75	30	High-K	~1.0	3,460,000,000
22	2011	0.75	2.6	High-K	~0.9	3,900,000,000
14	2014	0.70	1.8	High-K	–	4,500,000,000

https://en.wikichip.org/wiki/intel/process [8]

lines that span memory arrays and turn the memory transistors ON and OFF. Without silicide the word line resistance would slow the memory requiring the IC to wait for the word from the memory to arrive. The resistance of nickel silicide is about 15 μΩ-cm ($\mu = 10^{-6}$). The resistance of the source/drain without silicide is about 15 Ω-cm (~a million times higher).

Between 1972 and 2014 transistor gate lengths scaled over 5000 times shorter. This increased computer processing unit (CPU) speed by over 20,000 times (Table 10.2).

As the gate dielectric scaled thinner and thinner, the power supply voltage (voltage on the gate) was reduced to prevent excessive gate current from leaking through the thinner gate dielectric and breaking it down. Hundreds of improvements to existing equipment were invented, innovated, and implemented as the transistors scaled. At each node, the transistor structure had to be modified to meet new aggressive transistor specifications required by the designers, so they could meet the new circuit performance targets demanded by their customers.

Originally sources/drains and wells were formed by thermally diffusing dopants into the silicon substrate using gaseous sources. Phosphine (PH_3) was used for n-type phosphorus and diborane (B_2H_6) for p-type boron. Dopant gas from heated liquid sources such as (phosphoryl chloride ($POCl_3$) for n-type phosphorus and boron tribromide (BBr_3) for p-type boron) were also used. The dopant gases thermally decompose releasing dopant atoms. The dopant atoms diffuse into the exposed (not covered with a silicon nitride layer) surface of the wafer doping it heavily n-type or p-type. The wafers are then annealed at high temperatures for extended periods of time to thermally drive the dopants into the wafer to form WELLs, sources, and drains (pn-junction diffusions) with the desired depth.

As the transistor area scaled, WELLs and source/drain diffusions scaled smaller and shallower. To scale WELLs and pn-junctions shallower, diffusion doping was replaced with implant doping and an hour or more long furnace anneal was replaced with a few seconds of rapid thermal anneal.

Ion Implantation
An ion implant machine ionizes dopant atoms (gives them a charge) and then with electric fields accelerates the ionized dopant atoms to high speeds (near the speed of light) and shoots them (implants them) into the single crystal silicon. Higher energies implant the dopant atoms deeper into the silicon wafer. The ionized atoms are neutralized just before they hit the surface of the wafer to prevent the wafer from building up charge. Implantation energies range from a few thousands of electron volts (kilovolts or Kev) for shallow implants such as source and drain extensions to hundreds of millions of electron volts (megavolts or MeV) for deep wells such as high voltage wells, and for deep buried diffusions (DUFs).

10.3.2   Scaling Source/Drain Junction Diodes: Implant Angle

Initially dopants were implanted at 7° (Fig. 10.11, left) to avoid channeling of the dopant atoms through the single crystal silicon lattice. When dopant atoms are implanted at 0° (perpendicular to the surface), most of the dopant atoms hit silicon atoms and stop near the surface. Some dopant atoms, however, travel (channel) through the spaces between the silicon atoms in the crystal and go deep into the wafer. Channeling makes the pn-junction less well defined. After dopant implantation, a thermal drive is used to activate the dopant atoms (replace silicon atoms in the single crystal lattice with dopant atoms) and to diffuse the pn-junctions to the desired depth. When dopant is implanted at 7°, the transistor gate causes a dopant shadow next to the gate (Fig. 10.11, left). Prior to the 600 nm technology node, the thermal drive of the source/drain anneal was sufficient to connect source/drain extensions implanted at 7° with the transistor channel (Fig. 10.11, right).

This worked until with scaling thermal drives got shorter and source/drain extensions became shallower. At the 600 nm node, the thermal drive was no longer sufficient to connect the source/drain extensions implanted at 7° to the transistor channel.

The dopant implant angle was changed from 7° to 0° to eliminate the gap (Fig. 10.11, left). To prevent the dopant atoms from channeling, an amorphous silicon dioxide (SiO_2) layer was deposited before implanting the dopant. Silicon and

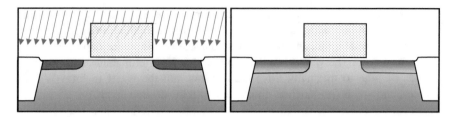

Fig. 10.11 Cartoon on left is after 7-degree source/drain extension is implanted self-aligned to the gate. Note gap between extension and gate on left side after implant. Cartoon on right is of the extension implants after annealing. Note in this instance the thermal drive is sufficient to close the gap

oxygen atoms are randomly distributed in the amorphous SiO_2 layer, so implanted dopant atoms scatter off silicon and oxygen atoms in the amorphous SiO_2 layer and remained close to the surface (no channeling).

10.3.3   Scaling Source/Drain Junction Diodes: Source/ Drain Extensions

Prior to about the 1.5 μm technology node, source and drain extensions were not needed. As transistor gates scaled to smaller and smaller dimensions, the deep source/drain junctions got closer together. At the 1.5 μm technology node, leakage current under the transistor channel from source to drain exceeded acceptable limits. A parasitic NPN bipolar transistor is formed under each NMOS transistor. (emitter = n-type source; base = p-type substrate, collector = n-type drain) (Fig. 10.12, right, and Chap. 5). To prevent this parasitic NPN bipolar from turning on and destroying the MOS transistor, engineers added a punch through implant that increased the doping between the source and drain under the channel (Fig. 10.12, left). This increases the doping in the base of the parasitic NPN to keep it from turning on. This worked for a few scaling nodes before it too ran out of gas.

In 1982, Bill Hunter, a colleague at Texas Instruments, invented a transistor structure that enabled engineers to keep scaling. His invention is still used on virtually every transistor manufactured today (US Patent 4,356,623) [9].

Bill's idea was to implant shallow, source/drain extensions self-aligned to the transistor gate as shown in Fig. 10.13. Next, he formed dielectric sidewalls on the vertical sides of the transistor gate and implanted the deep source and drains self-aligned to the dielectric sidewalls (Fig. 10.14). This process turned out to be inexpensive and manufacturable. It kept the deep source and drain diffusions sufficiently far apart to reduce leakage current under the channel to within the specification limits. It also reduced the gain of the underlying parasitic NPN bipolar transistor by increasing the length of the base keeping it from turning ON.

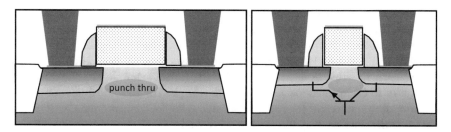

Fig. 10.12 NMOS transistor with punch through implant doping under the channel, between the deep source and drain junctions. The punch through implant increases the base doping of the parasitic NPN bipolar transistor under the channel keeping it from turning on (right cartoon)

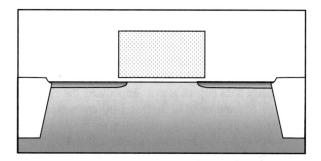

Fig. 10.13 NMOS transistor with source and drain extensions

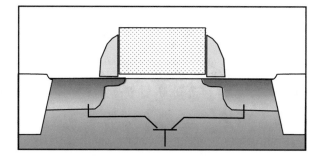

Fig. 10.14 NMOS transistor with source and drain extensions and with deep source and drains

10.3.4   Engineering Source/Drain Extension Resistance and Capacitance

Next to gate etch, source/drain extension resistance and capacitance are the most critical factors for transistor performance. As the end of the extension is moved closer to the channel, the contact resistance between the end of the extension and the transistor channel is reduced. This improves transistor speed. However, as the end of the extension moves under the edge of the gate, capacitive coupling between the transistor gate and the extension increases slowing the transistor (capacitance b in Fig. 4.6). Maximum transistor speed is a delicate trade-off between channel/extension resistance and gate/extension capacitance. To position the end of the extension optimally, in some transistor manufacturing flows process integration engineers either thermally grow or deposit thin sidewalls (offset sidewalls) on the transistor gates prior to extension dopant implantation (Fig. 10.10). In some transistor manufacturing flows, different thickness offset sidewalls are formed on the NMOS and PMOS transistors so that the dopant activation anneal simultaneously diffuses NMOS and PMOS transistor extensions to their optimal location.

As transistors scaled, extensions had to get shallower and shallower to reduce leakage from the drain to source under the channel. Transistor engineers had to keep coming up with ways to reduce extension resistance as they scaled shallower. They also had to keep coming up with ways of making the extension pn-junctions shallower and more sharply defined. The transistor engineers annealed the dopants at temperatures high enough to activate the dopants but for shorter and shorter times as scaling continued (Sect. 9.5). To additionally lower the resistance of the extensions as the extensions got shallower, engineers discovered ways to supersaturate the single crystal silicon with dopant atoms. Implanting the surface of the NMOS extension with a high concentration of arsenic (As) and annealing it with a laser anneal (900+ °C for a few millionths of a second) gives the As atoms sufficient time to replace silicon atoms in the single crystal lattice (activate) but virtually no time to diffuse and increase the depth of the extension junction. Supersaturation produces a much higher concentration of activated As (much lower resistance) than is possible under equilibrium conditions.

Supersaturation

At room temperature only a precise amount of salt can dissolve. If a lot of salt is poured into water at room temperature, only a fixed amount of salt will dissolve. The rest of the salt will remain as solid salt crystals in the bottom of the container. The salt solution is in equilibrium with the salt crystals at room temperature.

If the salt solution is heated, more salt dissolves and a new equilibrium condition is reached between the salt solution and the salt crystals at the higher temperature. If this solution is very carefully cooled to room temperature, it forms a super saturated salt solution. There is more salt in solution than the room temperature equilibrium condition. If a small crystal of salt is added to the supersaturated salt solution, the excess salt quickly precipitates out and falls to the bottom of the container. The salt continues to precipitate out until the super saturated solution again becomes an equilibrium saturated salt solution.

Under equilibrium conditions, only a given number of arsenic or phosphorus atoms can replace silicon atoms in the silicon single crystal. Single crystal silicon super saturated with arsenic atoms is formed by implanting more arsenic atoms than the equilibrium condition and then thermally activating them for such a short time (microseconds); they do not have time to diffuse away from the region of high concentration. This produces supersaturated silicon with low resistance. When silicon supersaturated with arsenic is heated, arsenic atoms deactivate and diffuse away from the supersaturated region until the equilibrium condition is reached (single crystal silicon saturated with arsenic).

Eventually transistors scaled to the point where the dopant implant energy was so low that a large fraction of the implanted dopant remained in the layer of amorphous silicon dioxide (SiO_2) being used to prevent channeling. Process engineers removed the amorphous SiO_2 layer and implanted source and drain extension dopants directly into the surface of the wafer at zero degrees. To prevent channeling, the single crystal silicon surface was first amorphized by implanting large atoms such as silicon and germanium to break up the crystal lattice. After the dopants are implanted, a thermal process was engineered to simultaneously convert the doped amorphous silicon layer back into single crystal silicon (epitaxial regrowth) and to activate the dopant atoms.

As transistors continued to scale, to better define the pn-junction transistor engineers also implanted carbon (not a dopant). Carbon atoms occupy interstitial spaces between the silicon atoms in the single crystal silicon. The interstitial carbon retards dopant atom diffusion resulting in shallower and more sharply defined pn-junctions.

In highly scaled PMOS source and drain extensions, implanted indium dopant atoms were found to create a PMOS flat band shift allowing reduced channel doping and improved transistor performance. Nitrogen implanted into highly scaled PMOS transistors at the same time the PMOS extension dopant atoms are implanted was found to reduce boron penetration through the thin gate dielectric. Fluorine implanted along with PMOS source/drain extension dopants was found to improve negative bias temperature instability (NBTI).

10.3.5   Gate Etch

The gate critical dimension (CD) (gate length) is the parameter that most impacts transistor performance. The need to print gate geometries with shorter gate lengths drove lithography scientists and engineers to develop photolithographic systems with shorter and shorter wave lengths. As transistor gate length scaled, the challenge of etching the transistor gates with vertical sidewalls and with exactly the same gate length (CD) across a die and across the wafer drove constant improvement to the gate etching tools and to gate etching chemistries. When transistor gate length is 1000 nm, a ±5 nm variation per side is no big deal. When transistor gate length scaled to 30 nm, a ±5 nm variation per side is 30% of the gate CD. When gate dielectric thickness was 20 nm, the selectivity of the polysilicon gate etch to the underlying silicon dioxide gate dielectric was not critical. When the gate dielectric scaled to less than 2 nm, the gate etch team had to develop an etch with sufficiently high selectivity to prevent the gate etch from penetrating the thin gate dielectric and etching trenches into the silicon substrate. The gate etch is not 100% uniform across a wafer. The gates can be completely etched in the center of the wafer but not yet completely etched at the edge of the wafer. Gate etching (gate over etch) must continue until all the gates across the entire wafer are completely etched. Tens of seconds of over etch may be required. The gate etch-selectivity to the underlying dielectric must be high enough to withstand the gate over etch with a comfortable margin. The transistor gate team had to constantly refine the gate etch equipment and the gate etching chemistry as transistors scaled and tolerances got tighter.

At one point in the transistor scaling process, lithographic limitations put a stop to the transistor engineer's ability to print transistor smaller gate lengths. This was a showstopper. The transistor gate team came up with incredibly inventive solutions. In one solution, the gate lithography engineers pattern the transistor gates with a thicker resist. The gate etch engineers developed a resist ashing step that very uniformly etches photoresist across the wafer. (Ashing is an oxygen plasma process used to strip photoresist patterns from the wafer.) Their resist etching process etched photoresist isotopically (both vertically and horizontally) very uniformly across the entire wafer, independent of pattern density. Using their resist etch, they reduced the gate length of the photoresist to whatever gate length the transistor engineers wanted. Gate patterns could be printed at 70 nm and reduced to 40 nm prior to the polysilicon gate etch. After the resist trimming step, the gates are etched anisotropically (vertical but not horizontally) to form narrow gates with straight sidewalls.

The gate etch is a series of plasma etching steps. The gate team worked continuously for decades constantly improving the gate etch. The top surface of polysilicon is rough (grain boundaries between the grains of polysilicon). Anything (residual photoresist, oxidized polysilicon, or …) in these grain boundaries blocks the polysilicon gate etch. The first step in the gate etch is engineered to be non-selective—it etches polysilicon, photoresist, oxidized polysilicon, etc. at about the same rate. This clears the surface of anything that might cause micro masking which blocks the etch and leaves etch residues. The etch chemistry is then changed to etch the polysilicon anisotropically (vertical etching only, no horizontal etching) to produce vertical transistor gate sidewalls. This is accomplished by adjusting the gas phase etching chemistry to form and deposit a polymer on the sidewall of the gate as the polysilicon is being etched. This polymer protects the polysilicon gate from being etched horizontally while it is being etching vertically. A bias on the wafer accelerates plasma ions to impinge on the horizontal surface and remove polymer before it can build up and block the etch. When the gate etching nears the underlying gate dielectric, the etching chemistry and plasma conditions are again changed to etch polysilicon with extremely high selectivity to the underlying gate dielectric. (Etches polysilicon and does not etch silicon dioxide.) This prevents the gate etch from punching through the underlying thin transistor dielectric and etching trenches into the substrate. Polymer deposition on the sidewalls of the gate in this final gate etching step must be carefully controlled to match the polymer deposition of the previous gate etching steps. Too little polymer will result in insufficient protection of the gate sidewall from horizontal etching resulting in notching at the base of the gate during gate over etch (Fig. 10.15a) (Gate length too short and transistor too leaky when turned OFF). Too much polymer will result footing at the base of the transistor gate. (Gate too long and the transistor too slow when turned ON) (Fig. 10.15c). The polysilicon etching rates are never completely the same across the wafer. The polysilicon etch may clear the surface of the transistor dielectric in the center of the wafer before it clears at the edge of the wafer. Transistor gates completely etched at the center of the wafer must not change dimension or sidewall profile while gate etching continues until all gates across the wafer are completely etched.

The gate etching chemistry had to be carefully retuned and tweaked for every new technology node.

10.3.6   Gate Dielectric Scaling

The gate dielectric of choice is thermally grown silicon dioxide for all technology nodes until 180 nm. At the 180 nm technology, the gate dielectric became so thin that boron from the PMOS transistor gate penetrated through the thin PMOS gate dielectric doping the channel. This added dopant in the channel and changed the PMOS transistor turn on voltage (V_{TP}). In addition, gate leakage current (I_G) through the thin gate dielectric between the grounded gate and the drain at V_{DD} (power supply voltage) caused excessive IC standby current. At the 180 nm node, nitrogen was incorporated into the SiO_2 gate dielectric by growing the gate dielectric at high temperature (~900 °C) in a nitrogen/oxygen (N_2/O_2) mixture or by annealing the SiO_2 after growth in nitrous oxide (N_2O) or nitric oxide (NO). Adding nitrogen to the gate dielectric suppresses boron penetration through the gate dielectric on PMOS transistors. Adding nitrogen also increases the dielectric constant of the gate dielectric. The increased dielectric constant allows a thicker gate dielectric (less leaky) to be used while still maintaining capacitive coupling control of the transistor channel. Annealing in N_2O or NO for gate dielectric nitridation was used temporarily. It put too much nitrogen at the channel/dielectric interface. Nitrogen at the channel/dielectric interface reduces carrier mobility degrading transistor performance. To better control the nitrogen distribution in the thin gate dielectric, a remote plasma nitridation (RPN) process was invented to replace the N_2O and NO anneals. The

A: gate notching   B: desired gate   C: gate footing

Fig. 10.15 Polysilicon gate profiles after gate etch. (**a**) gate notching, (**b**) desired gate profile, (**c**) gate footing

RPN process uses a plasma to generate highly reactive nitrogen atoms and deliver them to the gate oxide without exposing the gate oxide to the plasma. The RPN nitridation process enabled the gate dielectric team to engineer the nitrogen profile in the gate oxynitride dielectric. (High nitrogen concentration at gate/dielectric interface to block boron diffusion. Low nitrogen concentration at gate/substrate interface to improve CHC without degrading carrier mobility.)

To better control nitrogen profile when the gate dielectric got thinner at the 130 nm node decoupled plasma nitridation (DPN) was introduced and followed by a rapid thermal anneal (RTA) in $N_2 + O_2$. These nitridation steps were performed after the SiO_2 gate dielectric was grown and before the gate polysilicon was deposited. The DPN process is more uniform and more scalable than the RPN process [9, 10].

At the 90 nm technology node, AMAT introduced the Centura[R] cluster tool with multiple chambers to grow in situ steam generation (ISSG) silicon dioxide (SiO_2) dielectric, add nitrogen using DPN, perform rapid thermal oxidation (RTO), and deposit gate polysilicon sequentially in separate chambers without breaking vacuum. This integrated process significantly improved the ability to engineer the nitrogen profile in the gate stack, reduced defects, and improved reliability.

At the 28 nm node on high-performance ICs, high-k dielectrics such as hafnium oxide (HfO_2, k ~ 25) are replacing the nitrided gate oxide dielectric and metal gates such as titanium nitride (TiN on PMOS) and (TiAlN on NMOS) are replacing the heavily doped polysilicon gates.

Silicon Dioxide Gate Dielectric Equipment

When I started working at Texas Instruments, SiO_2 gate dielectric was thermally grown at high temperatures in horizontal quartz furnace tubes. Wafers stood vertically next to each other in slots in a quartz wafer boat. As wafers increased in size, a statistically significant trend in transistor drive current was detected from wafer top to wafer bottom. This was traced to a small but consistent difference in SiO_2 thickness from wafer bottom to wafer top. This was the result of a small difference in temperature between the bottom of the furnace tube and the top. Horizontal furnaces were replaced with vertical furnaces to eliminate this correlation. In a vertical furnace, wafers are stacked horizontally one above the other. The temperature is the same across each wafer. Soon a statistically significant trend was discovered between the top and bottom wafers in the vertical furnace. This trend was traced to reactant gases being depleted by the growing SiO_2 on wafers as the reactant gases rose in the vertical furnace. Thermal profiling in which the top of the vertical furnace was made slightly hotter than the bottom of the vertical furnace was introduced to compensate for reactant gas depletion in some vertical furnaces. Injection tubes that added back some of the reactant gas that was being depleted at various locations up the vertical furnace tube was added in other vertical furnaces. As scaling continued, batch vertical furnaces were replaced with single wafer gate dielectric growth tools. Single wafer cluster tools with multiple chambers allow the transistor engineers to grow the gate dielectric, nitride the gate dielectric, anneal the nitrided gate dielectric, and deposit the gate polysilicon sequentially without breaking vacuum.

10.3.7 Silicide

Nitrogen at Channel/Gate Dielectric Interface

A small amount of nitrogen is desirable at the channel/gate dielectric interface to reduce channel hot carriers. The problem is that as nitrogen concentration at this interface increases, carrier mobility decreases. High nitrogen concentration at the dielectric/polysilicon interface is desirable to block boron penetration on PMOS transistors.

Silicon-hydrogen bonds at the channel-gate dielectric interface are easily broken by channel hot carriers (CHC). This results in trapped charges in the gate dielectric which changes the turn ON characteristics of the NMOS and PMOS transistors over time. Replacing the silicon-hydrogen bonds with silicon-nitrogen bonds or with silicon-fluorine bonds improves CHC.

Incorporation of nitrogen at the gate/channel interface exacerbates negative bias temperature instability (NBTI) in PMOS transistors. The negative bias (gate at ground and NWELL at positive voltage = V_{DD}) on PMOS transistors attracts positively charged holes to the channel/gate dielectric interface. Some of these positively charged holes get trapped at the interface forming a permanent positive charge that increases over time. This raises the turn ON voltage of the PMOS transistor over time reducing the PMOS transistor drive current. Reducing the number of silicon-nitrogen bonds at the PMOS channel/gate dielectric and replacing silicon-nitrogen bonds with silicon-fluorine bonds improves NBTI. Implanting PMOS transistors with fluorine improves NBTI.

Around the mid-1980s simulation programs such as SPICE were telling designers that their memory speed (read and write access times) was being limited by the time it was taking for an electrical signal to propagate across the memory. The integrated circuits that the designers were laying out with the next-generation technology were having to wait for the electrical pulse to travel across the memory array to a memory cell and turn it ON. The pulse travels near the speed of light along a polysilicon word line less than a millimeter long! Designers needed word lines with lower resistance. Equipment and process engineers first came up with a process for depositing molybdenum silicide ($MoSi_2$) or tungsten silicide (WSi_2) on top of polysilicon and then etching the silicide/polysilicon gate stack. The resistivities of $MoSi_2$ (10–50 $\mu\Omega*cm$) and WSi_2 (60–80 $\mu\Omega*cm$) are about 10^6 (one million) times lower than the resistivity of heavily doped polysilicon (~15 $\Omega*cm$). Problem solved.

Deposited silicides such as $MoSi_2$ and WSi_2 reduce the resistance of the polysilicon word lines in memory arrays, but do not reduce the resistance of bit lines (long source and drain junctions in the wafer surface). Scientists and engineers invented and developed self-aligned silicide technologies that simultaneously reduced the resistance of the polysilicon gate word lines and the source/drain diffusion bit lines. In the self-aligned silicide formation process, a refractory metal such as titanium, cobalt, or nickel is deposited on the surface of the wafer and then heated. Where silicon is exposed (top of gates and source/drain diffusions), the metal reacts with the silicon to form a self-aligned silicide. After silicide formation, the unreacted metal (i.e., covering dielectrics such as STI and transistor sidewalls) is dissolved away in acid. (Refractory metals are soluble in acid but their silicides are not.) The silicide is then annealed at a higher temperature to convert it to a low resistance crystalline phase. Titanium silicide was introduced at the 1000 nm node and was used until the 65 nm node. After formation at a lower temperature, the $TiSi_2$ must be annealed at a higher temperature to convert it from the C49 crystalline phase to the lower resistance C54 crystalline phase. This conversion became increasingly difficult as silicide gate lengths decreased. On 65 nm wide word lines not all the C49 crystals converted to C54 crystals. The mixture of C49 and C54 crystals resulted in too much variation in resistance from one-word line to the next. A new silicide process had to be developed.

Cobalt silicide ($CoSi_2$) replaced $TiSi_2$ for the 65 nm node. As transistors scaled from 65 to 45 nm, new problems were discovered with $CoSi_2$ requiring the transistor team to develop yet another new silicide to replace $CoSi_2$ at the 45 nm technology node. Cobalt silicide formation is more difficult on the narrower word lines and perimeter diode leakage (drain/STI interface and especially drain/transistor channel interface) is too high with $CoSi_2$.

The scientists and engineers developed a nickel silicide ($NiSi_2$) process to replace cobalt silicide at the 45 nm node. $NiSi_2$ resolved $CoSi_2$ problems but introduced processing issues of its own. Microscopic residues on the surface of the source/drains generate crystal imperfections (dislocations or mismatched crystal planes) as nickel silicide first forms. Nickel metal diffuses along these crystal imperfections siliciding the dislocations and causing them to propagate. Silicided dislocations (called silicide pipes) were shorting transistor sources to transistor drains through the substrate (see Fig. 10.16) and shorting source and drain diodes to the substrate depressing yield. Prior to nickel metal deposition, a wet etch containing hydrofluoric acid was employed to etch away the dielectric residues that caused nickel pipes to form. Using the best possible HF preclean, nickel silicide pipes still formed and killed ICs. The nickel silicide pipe problem was finally fixed when a new in situ dry plasma preclean process called Siconi© etch was invented. The Siconi© preclean process simultaneously exposes the silicon surface to plasma decomposition products of NH_3 and NF_3 (Siconi patent US8501629) [11].

Fig. 10.16 Nickel silicide pipe shorting source to drain. SEM is courtesy of Texas Instruments

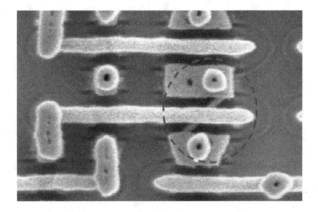

10.4 Transistor Performance Enhancement Using Stress

10.4.1 High-Performance PMOS Transistors with Compressive Channel Stress

The mobility of holes in PMOS transistors is about three times slower than the mobility of electrons in NMOS transistors. To compensate, the width of PMOS transistors in an CMOS inverter is approximately three times wider than NMOS transistors. For high-performance CMOS inverters, compressive stress can be applied to the channel of the PMOS transistors to improve hole mobility. Improved hole mobility improves PMOS transistor performance.

Figure 10.17 shows the CMOS inverter after the source/drain extensions are formed. (Chap. 6, Fig. 6.32) In this process flow a first silicon nitride epi block layer is deposited on top of the gates before the gates are etched. A second silicon nitride epi block layer is deposited, patterned, and etched blocking NMOS transistor areas (Fig. 10.18). Trenches are etched into the exposed silicon source/drain areas of PMOS transistors (Fig. 10.18). Single crystal silicon germanium (SiGe) is epitaxially grown refilling the PMOS source/drain trenches (Fig. 10.19). The Ge atom is larger than the Si atom, so single crystal SiGe is larger than single crystal Si. The larger SiGe crystalline material in the PMOS source/drains pushes against the PMOS channel applying compressive stress which enables holes to move faster. After the SiGe epi is grown, the silicon nitride epi block layer is removed (Fig. 10.20).

Figure 10.21 are a high-resolution cross-sectional TEMs of Intel PMOS transistors at two different technology nodes with SiGe performance enhancement.

10.4.2 High-Performance NMOS Transistors Using Tensile Stress Memorization

The performance of NMOS transistors can be improved by applying tensile stress to the channel of the NMOS transistor. Tensile stress improves the mobility of electrons in the channel, so they travel from source to drain faster. After source/drain

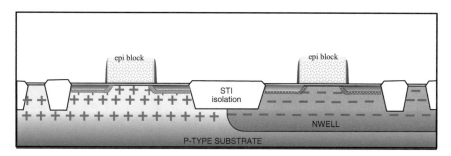

Fig. 10.17 CMOS inverter with source/drain extensions and halos

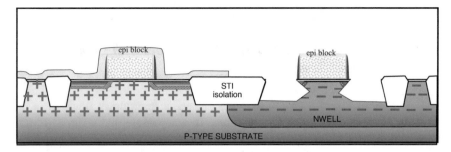

Fig. 10.18 CMOS inverter with Epi Blocking Nitride over NMOS transistors and with trenches etched into PMOS source and drains

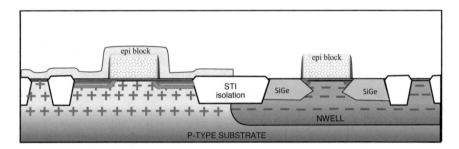

Fig. 10.19 CMOS inverter with Epi Blocking Nitride over NMOS transistors and with epitaxial silicon germanium (SiGe) filling PMOS source and drain trenches. SiGe applies compressive stress to PMOS channel enhancing mobility of hole carriers

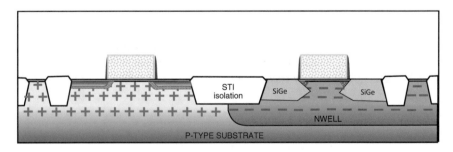

Fig. 10.20 CMOS inverter with Epi Blocking Nitride over NMOS transistors and with epitaxial silicon germanium (SiGe) filling PMOS source and drain trenches. SiGe applies compressive stress to PMOS channel enhancing mobility of hole carriers

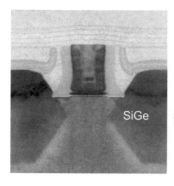 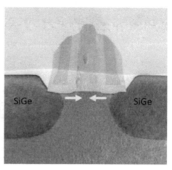

Fig. 10.21 Intel 45 nm PMOS transistor with silicon germanium source and drain. https://en.
wikichip.org/wiki/File:intel_45nm_gate.png [12]

dopant implant and before source/drain anneal (Fig. 10.22), a silicon nitride film
with high tensile stress is deposited (Fig. 10.23). In some process flows, the tensile
silicon nitride film is patterned and etched to remove it from PMOS transistor areas
to avoid degrading the PMOS transistor performance (Fig. 10.23).

During the source/drain anneal silicon grains in the polysilicon gate grow in size
and rearrange in response to the high tensile stress. The polysilicon in the NMOS
polysilicon gate "memorizes" the tensile stress. After the silicon nitride film is
removed, the NMOS polysilicon gate continues to apply tensile stress to the NMOS
transistor channel enhancing the NMOS transistor performance.

After source/drain anneal, the tensile stress memorization layer is removed prior
to silicidation.

10.4.3   High-Performance NMOS Transistors Using Contact Etch-Stop Applied Stress

When contact etch-stop liner (CESL) technology was first implemented (see Sect.
7.2.1 and Fig. 10.24), the transistor device engineers saw a difference in transistor
performance between CESL deposited in one manufacturing tool versus CESL deposited in another manufacturing tool. The NMOS transistors with CESL from the first
tool performed better than NMOS transistors with CESL from the second tool. PMOS
transistor performance was just the opposite. It was discovered that the cause was stress
in the two CESL films was different and that the CESL deposition process needed better tool-to-tool CESL composition uniformity to avoid these differences. The transistor device engineers also realized they had another knob they could use to enhance
transistor performance. The transistor device engineers collaborated with the equipment and process engineers and developed equipment and processes to deposit silicon
nitride (SiN) contact etch-stop liner with high tensile stress to enhance NMOS transistor performance and with high compressive stress to enhance PMOS transistor
performance.

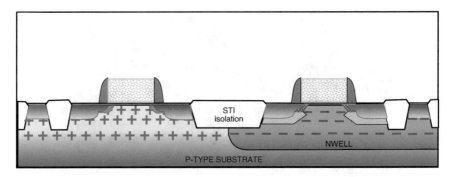

Fig. 10.22 CMOS inverter after deep source/drain dopant implant, but before deep source/drain anneal

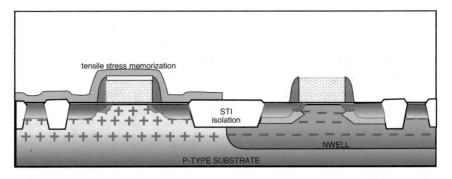

Fig. 10.23 CMOS inverter with a tensile memorization layer covering NMOS transistor areas. Tensile memorization layer is removed from PMOS transistor areas to prevent degrading the PMOS transistor

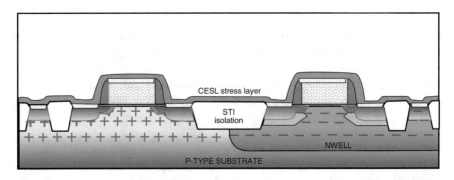

Fig. 10.24 CMOS inverter with a contact etch-stop stress layer covering the CMOS inverter transistor areas. If stress is tensile, NMOS transistor performance is improved, and PMOS transistor performance is degraded. If stress is compressive, PMOS transistor performance is improved and NMOS transistor performance is degraded

10.4.4   Single Contact Etch-Stop Liner Stress Technology

Equipment and process engineers modified the silicon nitride (SiN) contact etch-stop liner (CESL) deposition equipment to deposit SiN films with stress ranging from high tensile stress to high compressive stress. This gave the transistor engineers flexibility to enhance NMOS transistors using high tensile CESL and enhance PMOS transistors using high compressive CESL. The problem is that stress that enhances the performance of one transistor type, degrades performance of the other. To save cost, a CESL film with stress that provides best performance for the IC being manufactured is usually selected.

10.4.5   Dual Contact Etch-Stop Liner Stress Technology: DSL Technology

Dual stress liner (DSL) technology was developed to enhance the performance of both the NMOS transistors and the PMOS transistors using CESL stress (Fig. 10.25) in high performance CMOS ICs dispite the higher cost. A first CESL layer with high tensile stress is deposited over the NMOS transistors and then patterned and etched away from the PMOS transistors. A second CESL layer with high compressive stress is deposited over the PMOS transistors, patterned and etched away from the NMOS transistors. This process adds significant cost to the manufacturing flow but is mandatory to meet specifications for some high-performance IC circuits. The (DSL) process enhances NMOS transistor performance by applying tensile stress only in the NMOS transistor areas and enhances PMOS transistor performance by applying compressive stress only in the PMOS transistor areas (Fig. 10.25).

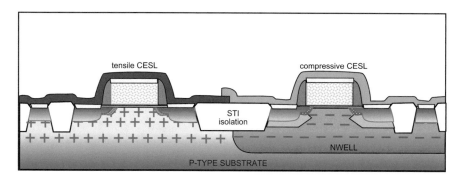

Fig. 10.25 CMOS inverter with a tensile contact etch-stop stress layer covering the NMOS transistor and a compressive contact etch-stop stress layer covering the PMOS transistor. The DSL enhances performance of both NMOS and PMOS transistors

References

1. K. Saraswat, Integrated circuit isolation technologies, EE311 Lecture Notes, Stanford University, https://web.stanford.edu/class/ee311/NOTES/Isolation.pdf. Accessed 4 Mar 2020
2. B. Desiatov, I. Goykhman, U. Levy, Demonstration of submicron square-like silicon waveguide using optimized LOCOS process. Opt. Express **18**(18), 18592–18597 (2010)., https://www.researchgate.net/figure/a-Simulated-profile-of-round-like-LOCOS-waveguide-b-SEM-micrograph-of-round-like_fig3_47404033
3. P. Smyes, Chapter 2 Local Oxidation of Silicon for Isolation, Ph.D. Thesis, Stanford University (1996), https://web.stanford.edu/class/ee311/NOTES/isolationSmeys.pdf
4. V. Senez, A. Armiglaito, I. De Wolf, G. Carnevale, R. Balboni, S. Frabboni, A. Benedetti, Strain determination in silicon microstructures by combined convergent beam electron diffraction, process simulation, and micro-Raman spectroscopy. J. Appl. Phys. **94**(9), 5574–5583 (2003)., https://www.researchgate.net/publication/234893353_Strain_determination_in_silicon_microstructures_by_combined_convergent_beam_electron_diffraction_process_simulation_and_micro-Raman_spectroscopy/figures?lo=1
5. J. Pan, D. Ouma, P. Li, D. Boning, F. Redeker, J. Chung, J. Whitby, Planarization and integration of shallow trench isolation, VMIC (1998), pp. 1–6, https://pdfs.semanticscholar.org/ed33/a6084fd959bcd1b4a82ffaffa6beb1563e3b.pdf
6. Y. Yan, B. Zhang, H. Deng, L. Chen, L. Xiao, B. Zhang, Y. Chen, Flowable CVD process application for gap fill at advanced technology. ECS Trans. **60**(1), 503–506 (2014)., http://ecst.ecsdl.org/content/60/1/503.abstract
7. M. LaPadus, Applied flows into flowable CVD, EE Times, 20, https://www.eetimes.com/applied-flows-into-flowable-cvd/
8. Process History—Intel, WikiChip Chips and Semi, https://en.wikichip.org/wiki/intel/process. Accessed 4 Mar 2020
9. W. Hunter, Fabrication of semiconductor devices, US Patent 4356623 (1982), http://www.pat2pdf.org/patents/pat4356623.pdf
10. H. Tseung, Y. Jeon, P. Abramowitz, T. Luo, L. Hebert, J. Lee, J. Jiang, T. Pl, G. Yeap, M. Moosa, J. Alvis, S. Anderson, N. Cave, T. Chua, A. Hegedus, G. Miner, J. Jeon, A. Sultan, Ultra-thin decoupled plasma Nitridation (DPN) Oxynitride gate dielectric for 80-nm advanced technology. IEEE EDL **23**(12), 704–706 (2002)., https://ieeexplore.ieee.org/abstract/document/1177959?section=abstract
11. J.Tang, N. Ingle, D. Yang, Smooth SiCoNi Etch for Silicon Containing Films, US Patent 8501629 (2013), https://patentimages.storage.googleapis.com/02/f9/32/ad817ef855bdb4/US8501629.pdf
12. Intel 45 nm Gate, WikiChip Chips and Semi, https://en.wikichip.org/wiki/File:intel_45nm_gate.png. Accessed 4 Mar 2020

Chapter 11
The Incredible Shrinking IC: Part 3 BEOL Aluminum Alloy Single and Multilevel Metal

11.1 Introduction

As the number of transistors on chips exploded (from about 2000 on Intel's first 4004 CPU in 1971 to almost 2 billion on Intel's Core i7 CPU in 2014—see Table 11.1), the demands upon the wiring resources required to route signals between the transistors exploded. Chip area stayed about the same as the number of transistors on the chip skyrocketed. The wires connecting the transistors got smaller (higher resistance) and closer together (higher capacitance). In 1987, with over a quarter million transistors on a chip, the number of wires demanded no longer could fit on one wiring layer. In 1987 IC designers added a second level of wiring. By 1998 the wires had scaled to such a small diameter that the resistance of the aluminum wires was limiting the speed of the IC circuits. Scientists and engineers developed equipment and processes to replace the aluminum wires with copper wires on high-performance ICs.

The metallization team (scientists, metal and dielectric equipment engineers, metal and dielectric deposition engineers, metal and dielectric etching engineers, metal and dielectric CMP engineers, chemists, material scientists, lithography engineers, etc.) worked together to invent and reduce to practice methods to wire the transistors together using multiple levels of metal stacked on top of one another (Table 11.1 and Figs. 11.1 and 11.2). In 1987 at the 1000 nm technology node a second wiring level was added. As the integrated circuits (ICs) continued to scale, more and more levels of wiring were needed. Today ICs with a dozen or more wiring levels are routinely manufactured.

© Springer Nature Switzerland AG 2020
H. Tigelaar, *How Transistor Area Shrank by 1 Million Fold*,
https://doi.org/10.1007/978-3-030-40021-7_11

Table 11.1 Summary of CPU technology development at Intel

Node	Approx. year	Power supply	Levels of metal	Metal	Intel CPU	Approx. # transistors in CPU
10,000	1971	5.0	1	AlSi	4004	2,300
6000	1976	5.0	1	AlSiCu	8080	6000
3000	1979	5.0	1	AlSiCu	8088	29,000
1500	1982	5.0	1	AlSiCu	80,286	134,000
1000	1987	5.0	2	AlSiCu	386SX	275,000
800	1989	4.0	3	AlSiCu	486DX	1.2 million
500	1993	3.3	4	AlSiCu	IntelDX4	1.6 million
320	1995	2.5	4	AlSiCu	Pentium	3.3 million
250	1998	1.8	5	Cu	Pentium2	7.5 million
180	1999	1.6	6	Cu	Pentium3	28 million
130	2001	1.4	6	Cu	Celeron	44 million
90	2003	1.2	7	Cu	Pentium4	125 million
65	2005	1.0	8	Cu	Pentium4	184 million
45	2007	1.0	9	Cu	Xenon	410 million
32	2009	0.75	10	Cu	Core i7 G	1.17 billion
22	2011	0.75	11	Cu	Core i7 IB	1.4 billion
14	2014	0.70	12	Cu	Core i7 BU	1.9 billion

https://en.wikichip.org/wiki/intel/process [2]
https://www.intel.com/pressroom/kits/quickrefyr.htm [3]

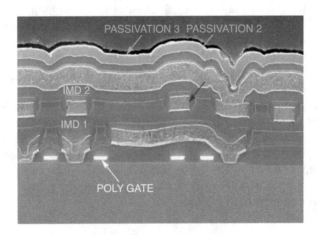

Fig. 11.1 Multiple levels of aluminum alloy wiring in an integrated circuit. https://docplayer.net/54982636-Programmable-logic-devices.html [1]

Applications: Interconnection

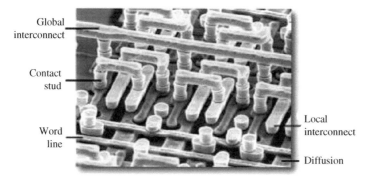

Global interconnect

Contact stud

Word line

Local interconnect

Diffusion

Fig. 11.2 Integrated circuit with multiple levels of metal. Dielectric layers are etched away [22]

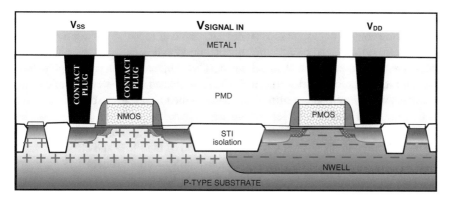

Fig. 11.3 CMOS Inverter with first level of metal wiring

11.2   Pre-Metal Dielectric (PMD)

11.2.1   Overview

After the transistors gates, sources, and drains have been silicided, a pre-metal dielectric (PMD) stack is deposited over the transistor gate topography to electrically isolate the transistors from the IC wires above (Fig. 11.3). The PMD is planarized to facilitate patterning and etching of contact holes and patterning and etching of metal1 wires. Contact holes are etched through the PMD stack to establish electrical contact between the tops of the transistor gates and metal1 wires and between transistor source/drain diffusions and metal1 wires. Metal fills the contact holes and electrically connects the transistor gates and source/drain diffusions to the metal1 wires. The metal1 wires provide a signal path from the inverter to other inverters and transistors and also provides electrical connection to the power supply voltage and ground.

11.2.2   Pre-metal Dielectrics and PMD Stack

The pre-metal dielectric stack, usually comprised of several layers of dielectric, prevents the transistors from shorting to the first metal1 wiring layer (Fig. 11.3).

PMD engineers faced two big challenges. The first big challenge was to fill the ever-shrinking spaces (gaps) between the transistor gates without forming voids. When a void forms in the PMD between closely spaced transistor gates (Fig. 11.4, left), source/drain (S/D) contacts next to the gate are shorted together causing the IC to fail (Fig. 11.4 right). As CVD-W is filling S/D contact holes, the CVD-W metal is also filling the void.

The second big challenge for the PMD engineers was to provide a sufficiently flat surface, so lithography engineers could simultaneously keep all the contact holes on the contact pattern in focus (adequate depth of focus). This was a challenge when the first integrated circuit was manufactured and became an ever-increasing challenge until chemical mechanical polish (CMP) planarization was introduced at the 320 nm technology node. Prior to CMP, complicated, multi-step manufacturing processes were developed to sufficiently smooth the top surface to be able to manufacture contacts and metal1 leads with acceptible yield. Figure 11.5 shows the metal1 wires in an IC before CMP planarization was introduced. Figure 11.6 shows the metal1 wires in an IC after CMP planarization of the PMD and chemical vapor deposition of tungsten (CVD_W) to fill contact holes were introduced.

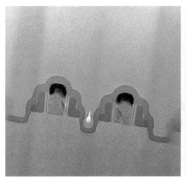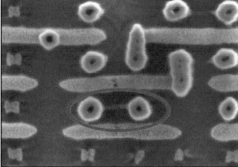

Fig. 11.4 Left: Void in PMD between closely spaced transistor gates. Right: Tungsten filament formed in the void is shorting two adjacent contact plugs together. SEMs are courtesy of Texas Instruments

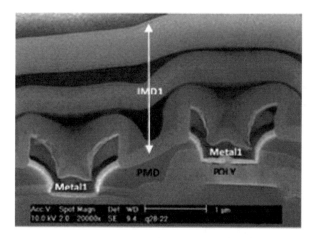

Fig. 11.5 Metal1 lead on left contacts S/D through hole in smoothed PMD. Lead on right contacts transistor gate. PMD smoothed using reflow. http://portal.unimap.edu.my/portal/page/portal30/ Lecture%20Notes/KEJURUTERAAN_MIKROELEKTRONIK/Semester%201%20Sidang%20 Akademik%2020132014/EMT%20357%20-%20Fundamental%20of%20Microelectronic%20 Fabrication/Lecture%20Notes/Deposition%20Process%20Technology.pdf [4]

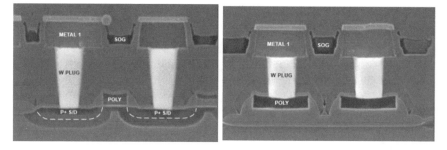

Fig. 11.6 Metal1 leads on BPSG PMD planarized using chemical mechanical polish (CMP). Note the difference in height between source/drain plugs and gate plugs. http://docplayer.net/40707559-Amd-vantis-machl-v-eepla.html [5]

At almost every new technology node, pre-metal dielectric (PMD) deposition equipment and processes were upgraded or replaced to fill the smaller and smaller space (gap) between transistors without a PMD void (Table 11.2).

One of the dielectric layers in the PMD stack is usually phosphorus silicate glass (PSG, phosphorus doped SiO_2). Phosphorus in the silicate glass performs two functions. First it lowers the glass transition temperature enabling the PSG to reflow at high temperatures smoothing the surface. Second, phosphorus getters (renders immobile) mobile ions such as sodium and potassium preventing them from diffusing to the transistor channel and changing the transistor turn on voltage (V_T).

Mobile Ions

Mobile ions such as sodium (Na+) and potassium (K+) readily diffuse through the silicon substrate and through silicon dioxide. Mobile ions accumulate at the gate dielectric/substrate interface giving it a positive charge. For example, when a PMOS transistor is turned on, the negative voltage on the gate attracts the positively charged mobile ions to the channel. The positive charge from the mobile ions can accumulate under the PMOS gate at the dielectric/substrate interface raising the PMOS turn ON voltage causing the integrated circuit to fail.

Mobile ions continued to be a concern for many years. Phosphorus doped glass was used and is still often used to getter mobile ions in PMD. Makers of materials used in semiconductor manufacture now supply materials that are free of mobile ions.

Cleanup engineers developed wet cleans that can remove 100% of the mobile ions from the surface of wafers. As long as all mobile ions are removed from the surface of the wafers before the temperature of the wafer is raised (annealed), mobile ions are not a problem. High selectivity CMP slurries bathe wafers in potassium mobile ions (K+) but the cleanup engineers remove each and every one of them before the wafers are annealed.

Filling the ever-narrowing gap or space between adjacent transistors as gates kept getting closer together required improvements in PMD gap fill process at almost every new technology node (Table 11.2).

As transistors scaled, PMD deposition processes with lower deposition temperatures were required and had to be developed to prevent excessive diffusion of dopants in the transistor source/drain junctions and to prevent silicide sheet resistance from degrading.

11.2.3   Pre-metal Dielectric Planarization: Reflow

Prior to the 500 nm technology node PMD consisted of a thick layer of phosphorus doped silicate glass (PSG) on a thin layer of undoped low pressure chemical vapor deposited (LPCVD) TEOS. The phosphorus doping in the silicate glass performs two functions. First the PSG reflows at about 1000 °C smoothing the topography (Fig. 11.7). Secondly the P atoms trap (getter) mobile ions such as sodium. Boron (B) doping was added at the 320 nm technology node (BPSG) to additionally lower the reflow temperature to about 900 °C. LPCVD gases used to deposit PMD are silane (SiH_4) and oxygen (O_2). Phosphine (PH_3) is used for phosphorus (P) doping and diborane (B_2H_6) is used for boron doping. LPCVD gases used to deposit PMD

Table 11.2 Pre-metal dielectric vs. technology node

Node nm	Pre-metal dielectric	Deposition gases	Pressure, Torr	Temp, °C	Comments regarding PMD planarization
500	PECVD APCVD + dopant	TEOS; O_2 SiH_4; O_2	<10 760	700 300–450 ~380	Resist etch back planarized thick doped APCVD on undoped LPCVD oxide
320	PECVD APCVD + dopant	TEOS; O_3 SiH_4, O_2	<10 760	700 300–450 ~380	CMP planarized BPSG. APCVD oxide on thin undoped PECVD oxide
250	PECVD APCVD + dopant	TEOS; O_3 SiH_4, O_3	<10 760	700 300–450 ~380	CMP planarized BPSG. APCVD with ozone provides better gap fill.
180	PECVD SACVD + dopant PECVD	TEOS; O_3 TEOS; O_2	<10 ~600	300–450 300–450	CMP planarized BPSG capped with PECVD-TEOS SACVD improves gap fill. TiN adheres better to undoped PECVD-TEOS cap than to SACVD BPSG
130 90	HDP-PSG	SiH_4; O_2; Ar + dopant	<10 mT	400–600	SiN contact etch stop liner (CESL) plus HDP-PSG. Simultaneous deposition + sputter etch followed by CMP
65	HDP-PSG	SiH_4; O_2; Ar + dopant	<10 mT	400–600	Dep + etch/etch-only / dep + etch HDP-PSG dielectric dep. followed by CMP
45	HARP™ USG	TEOS; O_3	~600	~450	No plasma. dep. on SiN contact etch stop liner. Flow rates adjusted during dep. to first fill gaps then increase dep. rate, CMP
32 28	eHARP™ USG	TEOS; O_3; H_2O	~600	~450	No plasma. dep. on SiN contact etch stop liner. Flow rates adjusted during deposition to first fill gaps then increase dep. rate, CMP

at a lower temperature are tetraethysiloxane (TEOS) and ozone (O_3). Trimethylphosphate (TMPO) and trimethylphosphite (TMPI) are used for phosphorus doping, and triethylborate (TEB) and trimethylborate (TMB) are used for boron doping at lower temperatures.

As transistors scaled smaller, the high temperatures needed to reflow the PMD could no longer be tolerated. Transistors had scaled to the point where the temperature required to reflow the BPSG diffused the junctions too far laterally and too deeply to meet the new die area target. Designers would have had to space the transistors farther apart to accommodate the temperature required to reflow BPSG. This would have made IC chips larger and more expensive.

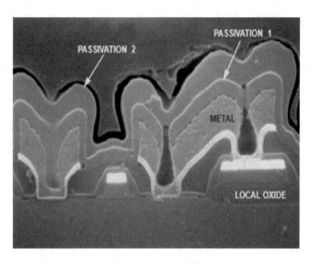

Fig. 11.7 TiW/AlCu single-level metal on reflowed BPSG PMD. http://smithsonianchips.si.edu/ice/cd/9504_408.pdf [6]

11.2.4   Pre-metal Dielectric Planarization: Resist Etch back

Resist etch back (REB) planarization was introduced at the *500 nm technology node* to smooth the PMD at a lower temperature. In REB, a thick layer of PE-TEOS (1 μm or more) is deposited. A thick reflowable resist is coated over the PE-TEOS topography. The resist is heated to a temperature where it becomes liquid and flows producing a more planar surface. Resist reflows at a much lower temperature (100–130 °C) than PSG (950+ °C) or BPSG (850+ °C). A plasma etch that etches the resist and the TEOS with equal rates is then employed to etch away the resist and at the same time etch away the hill tops on the TEOS layer. A capping layer of PSG or BPSG is usually then deposited using APCVD $SiH_4 + O_2$ + dopant gases and densified (not reflowed) prior to contact patterning.

11.2.5   Pre-metal Dielectric Planarization: Chemical Mechanical Polish

Chemical Mechanical Polishing (CMP) planarization of PMD was introduced at the *320 nm technology node*. CMP solved the planarization problem but did not solve the gap fill problem. At each new node, as transistors got closer together, a PMD dielectric with improved gap fill was needed. CMP planarization of PMD introduced a new problem. By removing the topography over the transistor gates, the PMD thickness over the PMD gates was now much less than the PMD thickness over the source/drain diffusions. The contact etch would open the contact holes to the gates long before opening the contact openings to the source/drains. Overetching in the contact holes on the gates damaged the gates and the gate dielectric. A contact etch stop layer (CESL) was added and new contact etching processes and equipment were invented and developed to resolve this issue.

11.2.6   Pre-metal Dielectric Gap Fill

At the *500 nm technology node*, a thick first pre-metal TEOS dielectric layer (PMD1) is deposited using PECVD TEOS + O_2. After smoothing the topography using resist etch back (REB), a second pre-metal dielectric BPSG layer (PMD2) is deposited using atmospheric pressure CVD SiH_4 + O_2 plus boron and phosphorus dopant gases.

A PECVD TEOS process using TEOS plus ozone (O_3) was developed at the *320 nm technology node* to meet the more stringent gap fill requirements. First a thin layer (~100–200 nm) of undoped PECVD TEOS + O_3 is deposited followed by a thick (>1 μm) layer of APCVD BPSG (SiH_4 + O_2 + dopant gases). The BPSG is CMP'ed. to about half the thickness and another BPSG layer is deposited prior to contact pattern.

As scaling continued, at the *250 nm technology node*, APCVD TEOS + O_2 was replaced with APCVD TEOS + ozone (O_3) to meet gap fill requirements. In addition, at the 250 nm node, a thin (20–40 nm) silicon nitride (SiN) contact etch stop liner (CESL) is deposited first on the source, drains, and gates to enable simultaneous etching of shallow contact holes through the thin PMD over the transistor gates and deep contact holes through the thick PMD over the transistor sources/drains (Fig. 11.6).

At the *180 nm technology node*, APCVD TEOS + O_3 is replaced with SACVD-TEOS + O_3 to meet the more stringent gap fill requirements. Sub atmospheric pressure reduces the collisions of the reactant molecules in the gas phase so that they pretty much only react when they hit the surface of the wafer (sidewalls and bottoms of trenches). This results in improved step coverage and gap fill. The penalty paid for improved gap fill is lower deposition rate.

At the *130 nm technology node*, to satisfy still more severe gap fill demands, SACVD-TEOS + O_3 was replaced with high-density plasma (HDP) dielectric deposition using silane (SiH_4) and oxygen (O_2) plus argon (Ar) with biasing. Argon ions with bias sputter etch the surface of the silicon dioxide while it is depositing. By adjusting the deposition rate versus the sputter etch rate, narrow/deep gaps can be filled free of voids. The depositing dielectric layer is sputter etched much more efficiently from an exposed flat surface than from the inside a gap or trench. This produces an almost planar surface that is easily planarized with CMP (Fig. 11.8). The HDP process was extended to the 65 nm technology node by interrupting sputter + etch deposition with sputter-only steps. This dep + etch/etch-only/dep + etch/etch-only/dep + etch process is known as dep/etch/dep.

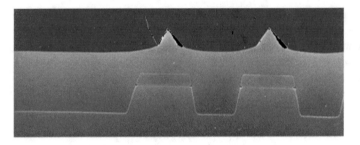

Fig. 11.8 HDP BPSG deposited over 1-μm-wide active geometries. https://pdfs.semanticscholar. org/ed33/a6084fd959bcd1b4a82ffaffa6beb1563e3b.pdf [7]

At the *45 nm technology node*, a high aspect ratio oxide deposition process (HARP™) was introduced to replace the HDP process whose gap fill capability had run out of steam. HARP™ employs thermal decomposition of TEOS + O_3 (no plasma) and manipulates the O_3/TEOS ratio throughout the deposition to first fill small trenches using a low deposition rate and then to fill larger trenches with a higher deposition rate.

Addition of water vapor (H_2O) to the HARP™ deposition process (eHARP™) extended the 45 nm process to fill smaller gaps at the *32 nm technology node*.

At the *22 nm technology node* and beyond a flowable CVD [8] process is being developed to replace eHARP™. The reflowable siloxane-based material $(SiH_2NH)_n$ is deposited into the narrow trenches using a CVD technique. This reflowable siloxane flows and completely fills very narrow trenches. An ozone (O_3) anneal then converts the Si-H and Si-N bonds to Si-O bonds.

11.3   Contact Process

11.3.1   Overview

Contact plugs electrically connect the transistor source/drains and gates to the first level of metal wiring (metal1) (Fig. 11.3). A contact hole pattern with contact hole openings is formed on the surface of the PMD layer. Contact holes are etched anisotropically (vertical etch only) through the PMD layer exposing the surfaces of the transistor source/drains and transistor gates. A barrier layer is deposited on bottom and the sidewalls of the contact holes prior to filling them with the metal contact plugs.

11.3.2   Contact Pattern

Contact holes are small and difficult to pattern. Keeping the millions of holes on a contact pattern in focus is a challenge. Before CMP planarization, bilayer resists were used in some processes. A first layer of planarizing resist was deposited and reflowed to provide a more planar surface before the second layer of imaging resist was deposited. A common IC failure mechanism is missing contacts. Some contacts have less depth of focus than others. The image of these contacts can be blurry. If an insufficient number of photons exposes a contact hole, developer cannot dissolve the resist out of the contact hole and the plasma etch is blocked. When one particular contact or set of contacts is found to always be missing, the lithography engineers and optical physicists tweak their optical proximity software to increase the depth of focus for these contacts, so they always print.

11.3.3   Contact Etch

The contact etch equipment engineers and contact etch process engineers had to constantly make improvements to meet demands imposed by scaling. The dielectrics in the PMD stack changes at almost every node requiring new etching processes to be developed. Until the 250 nm technology node, the PMD stack was composed of layers of undoped and doped deposited silicon dioxide (SiO_2) plus layers of spin-on-glass (SOG). At the 250 nm node, when chemical mechanical planarization (CMP) was introduced, a silicon nitride (SiN) contact etch stop layer (CESL) was added to the PMD stack. The CESL layer required significant changes to the contact etch equipment and the contact etching processes.

With CMP planarized PMD, the depth of the contact holes to the top of the transistor gates is much less than the depth of the contact holes to the transistor source/drains (Figs. 11.3 and 11.6). The contact hole etch opens the gate contact holes first. Without the CESL layer, the contact etch hammers on the top of the transistor gates damaging them while the contact etch continues until the source/drain contacts are also open. The plasma etch engineers developed a plasma etch that etches silicon dioxide (SiO_2) but does not etch the silicon nitride (SiN) CESL layer. Using the new contact etch, the gate contact first opens and stops on the SiN. The source/drain contacts continue etching until they too stop on the SiN CESL layer. The plasm etch engineers also developed a plasma etch that etches SiN but does not etch silicide. When both the gate contact and S/D contact holes are opened down to the CESL layer, the plasma chemistry is changed to etch through the silicon nitride CESL layer and stop on the silicide. Gate contact holes and S/D contacts holes etch through the CESL layer simultaneously.

Resist erodes during the contact plasma etch releasing carbon into the etching plasma. Carbon alters the contact etching chemistry. Less carbon is released into the plasma in regions of high contact density. To prevent carbon from affecting contact etch, some manufacturers deposit a hard mask such as titanium nitride (TiN) before they pattern the contacts. They form the contact pattern on the TiN layer and etch the contact pattern into the TiN. They then strip the resist and etch the contacts using the TiN hard mask. With the resist gone, there is no resist erosion, no carbon introduced into the contact plasma etch, and dense and isolated contact holes etch pretty much the same.

11.3.4   Contact Barrier: Overview

Contact openings must be lined with a barrier material to prevent the aluminum metal from reacting with the silicon on the surface of S/D diodes and shorting the S/D diodes to the substrate. Titanium/tungsten (TiW) or titanium nitride (TiN) barrier layers are used to prevent aluminum from spiking the S/D diodes. A TiN barrier layer is used to prevent fluroine gas from chemical vapor deposited tungsten (CVD-W) from spiking the S/D diodes and from etching the silicon dioxide (SiO_2) PMD sidewalls of the contact holes. As the contact holes scaled smaller and the aspect ratio (depth to width ratio) became greater, new technologies had to be invented and developed to cover the bottom and sidewalls of the contact holes with uniform barrier layers with no pinholes.

At almost every new technology node, metal thin film deposition processes and deposition equipment were upgraded or replaced to deposit thinner and more uniform adhesion/barrier layers on the bottom and sidewalls of the smaller contact holes.

Thin film metal deposition processes that were invented, developed, and used for various technology nodes included DC sputtering, RF sputtering, DC magnetron sputtering, reactive sputtering, columnated sputtering, ion metal plasma (IMP) sputtering, and ion beam sputtering (IBS). See Chap. 2, Section 2.1.5 for brief descriptions of these sputter deposition processes.

11.3.5   Contact Barrier Material and Deposition

A continuous layer (no holes!) of barrier material must be deposited on the bottom and sidewalls of the contact opening to block reaction of metal from the aluminum wires with silicon from the source/drain junctions. Aluminum reacts with silicon from the source/drain diodes causing the diodes to short to the substrate. Prior to the 800 nm technology node, a thin layer of titanium + titanium/tungsten alloy (Ti/TiW) was sputtered into the contact holes to line the contact holes with a barrier that prevented contact between the aluminum alloy metal1 and the source/drain diodes.

At the *800 nm technology node*, columnated sputtering was implemented to provide better step coverage in the smaller contact holes.

For the *500 and 320 nm technology nodes*, the Ti/TiW barrier layer was replaced with a titanium/titanium nitride barrier layer. This barrier layer combination provided a thinner barrier layer with reduced resistance. Columnated sputtered titanium and columnated reactive sputtered titanium nitride were used to deposit the Ti/TiN barrier layers.

> **Columnated Sputtering**
> In the early 1980s while still a sputter deposition engineer, I submitted a patent suggestion for what I called shadow mask sputtering (another word for columnated sputtering). The patent committee decided not to patent this idea believing it was not manufacturable. They reasoned the shadow mask would accumulate sputtered metal and require frequent opening of the sputter chamber and cleaning of the shadow mask. In 1998 Applied Materials applied for and got the patent for columnated sputtering. US6362097 [9] Columnated sputtering was used in the manufacture of integrated circuits for a couple of technology nodes. Columnated sputtering was finally replaced when IMP sputter deposition was invented and developed. It turns out the patent committee was correct. Columnated sputtering was not a very manufacturable process, but until manufacturable IMP titanium and titanium nitride processes became available it was the only game in town.

At the *250 nm technology node*, CVD-TiN replaced columnated TiN to provide a more uniform barrier coating on the contact hole sidewalls.

Columnated titanium (Ti) sputtering was replaced by ion metal plasma (IMP) Ti deposition at the *180 nm technology node*.

At the 180 and 130 nm technology nodes, the IMP-Ti and the CVD-TiN were deposited in different tools. The Ti adhesion layer was annealed in a rapid thermal anneal (RTA) tool after the IMP-Ti deposition and before the CVD-TiN deposition. (The RTA forms titanium silicide in the bottom of the contact hole lowering contact resistance.) At the *90 nm technology nodes*, the AMAT 5500 Endura[R] multi-chamber deposition tool provided an IMP-Ti deposition chamber, an RTA chamber, and a CVD-TiN deposition chamber on the same mainframe. This improved the contact adhesion/barrier process by sequentially depositing the IMP-Ti adhesion layer, annealing it, and depositing the CVD-TiN barrier layer without breaking vacuum.

At the *45 nm technology node*, the Endura[R] multichambered contact barrier tool was also used to deposit the Ti/TiN barrier layer.

11.3.6 Contact Fill and Metal1

At the *1000 and 800 nm technology nodes*, contact openings were large enough for a continuous layer of aluminum alloy (AlCu) to be sputtered into them. As shown in Fig. 11.9, at the 800 nm technology node, marginal step coverage caused the sputtered metal1 to be barely continuous as it enters the contact hole. This thinned AlCu is susceptible to electromigration failure in customers equipment (see Sect. 11.4.3).

A thin layer (~50 nm) barrier layer of titanium/tungsten ($Ti_{0.3}W_{0.7}$) is first sputtered into a contact hole before aluminum/silicon/copper alloy is sputter deposited.

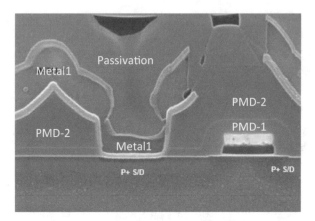

Fig. 11.9 Smoothed BPSG PMD using high-temperature reflow. http://smithsonianchips.si.edu/ice/cd/9712_573.pdf [10]

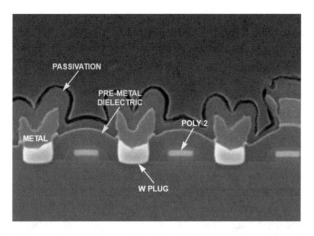

Fig. 11.10 Aluminum-copper leads on etched back CVD-W contact plugs. http://smithsonianchips.si.edu/ice/cd/9612_517.pdf [11]

The titanium layer lowers the contact resistance. The TiW barrier layer prevents the aluminum alloy from contacting the silicon and spiking (shorting) the source/drain diodes. In some manufacturing flows, beginning at about the 500 nm technology node, a layer of CVD-tungsten metal (chemical vapor deposition) was deposited onto the TiW in the contact holes and etched back to provide improved contact filling and a more planar surface for the AlCu metal1 wires (Fig. 11.10).

At the *800 nm technology node*, some manufacturers kept the CVD-W layer, then patterned and etched it. They used it for local interconnect (metal0) or for the first wiring layer (metal1). Because it is difficult to print a pattern on the rough grainy surface of as deposited CVD-W, prior to patterning surface roughness is removed using resist etch back (A coat of resist is spin-coated on the rough surface and reflowed to have a smooth surface. An etch that etches the resist and tungsten metal with the same etch rate was developed and used to etch the resist and the tungsten forming a smooth tungsten surface) (Fig. 11.22).

11.3.7   Contact Plugs: Tungsten Chemical Vapor Deposition

At the *half micron (500 nm) technology node*, sputtered aluminum no longer formed a continuous layer when sputtered into the contact hole (Fig. 11.11).

To enable scaling to continue, a tungsten chemical vapor deposition process (CVD-W) was developed to fill small aspect ratio (deep and narrow) contacts with a plug of tungsten metal. The CVD-W process completely fills contact holes with tungsten metal without forming voids. Tungsten hexafluoride (WF_6) gas molecules contact the hot pre-metal dielectric (PMD) surface and decompose into tungsten metal and fluorine (F_2) gas. The tungsten metal fills the contact holes void free. The fluorine gas is pumped away. Fluorine gas aggressively etches silicon and silicon dioxide. The F_2 gas must be blocked from attacking the source/drains in the bottom of the contact holes and blocked from attacking the silicon dioxide pre-metal dielectric (PMD) on the sides of the contact holes. A thin barrier layer of titanium plus titanium nitride (TiN) is deposited on the bottom and sidewalls of the contact holes and on the surface of the PMD prior to CVD-W. The titanium reacts with residues on the surface of the source/drains lowering contact resistance. Titanium forms an adhesion layer on the sidewalls of the contact holes for the TiN. TiN is a conductive barrier layer that F_2 gas cannot penetrate.

After filling the contact holes, the CVD-W overfill plus the TiN/Ti barrier/adhesion layers are removed from the surface of the PMD using either etch back or

Fig. 11.11 Void formed when aluminum is sputtered into a 500-nm contact hole. http://acceleratedanalysis.com/?page_id=140 [12]

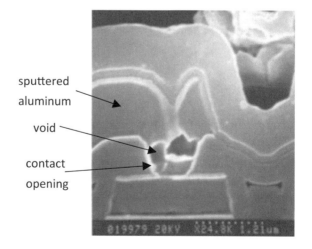

sputtered
aluminum

void

contact
opening

Fig. 11.12 Contact filled with CVD-W as deposited. https://core.ac.uk/download/pdf/11465298.pdf [13]

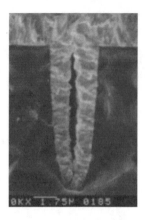

Fig. 11.13 Contact filled with CVD-W after etching back. https://www.sciencedirect.com/science/article/pii/S0026271405001976 [14]

CMP. At the 500, 320, and 250 nm technology nodes, the CVD-W overfill is removed using contact plug plasma etch back. From the 180 nm technology node on, it is removed using chemical mechanical polish (CMP). CMP provides a planar PMD surface with the tops of the CVD-W contact plugs exposed (Figs. 11.12 and 11.13).

11.4   Single-Level Metal

11.4.1   Single-Level CVD-W Metal (SLM)

Some manufacturers did not remove the CVD-W and used it for local interconnect or metal1 (Fig. 11.14). When CVD-W instead of AlCu is used for metal1, the designers must take into account the higher resistance of CVD-W when laying out long metal1 leads.

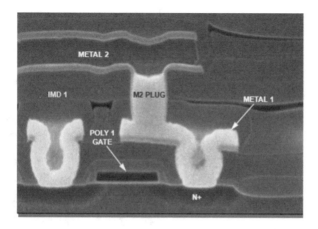

Fig. 11.14 CVD-W metal1. http://smithsonianchips.si.edu/ice/cd/9702_529.pdf [15]

11.4.2   Single-Level Aluminum Alloy Metal

At the 1000 nm technology node, 1–2% (atomic percent) of silicon was added to the aluminum sputtering targets to reduce the solubility of silicon in the aluminum wires. Over time when the IC is in use silicon from the source/drains migrates into the aluminum leads under current stress and forms aluminum spikes which short the source/drain pn-junctions to the substrate. The added silicon significantly reduces junction spiking. At the 500 nm technology node, the Ti/TiW/AlSiCu barrier/metal1 stack (bottom to top) was replaced with Ti/TiN/CVD-W/TiN/AlCu stack.

11.4.3   Aluminum Metal1: Electromigration

As transistors scaled, transistor drive current stayed about the same or even increased a bit. Driving the same current through smaller and smaller wires soon led to reliability problems. As electrons flow through the wires, they bump into metal atoms pushing them out of the way (electromigration). Over time electromigration forms voids in the metal wires that cause the IC to fail (Fig. 11.15). Material scientists repeatedly had to come up with new materials and new processes to combat electromigration as wire sizes scaled smaller.

At the 800 nm technology node ICs with failures caused by electromigration were being returned by customers to the IC manufacturors.

Up to 4% (atomic percent) copper is added to aluminum sputter targets to reduce electromigration failures. Aluminum electromigration occurs primarily along the grain boundaries between aluminum crystals. Voids formed primarily at grain

Fig. 11.15 Electromigration caused hillock and void on aluminum alloy leads. https://www.mwrf.com/ active-components/13-key-considerations-aerospace-rfmicrowave-devices [16]

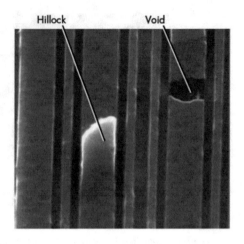

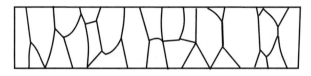

Fig. 11.16 Grain boundaries between multiple aluminum grains in wide leads

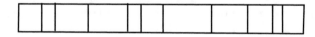

Fig. 11.17 Grain boundaries between two aluminum grains in narrow leads (bamboo grain structures)

boundary triple junctions (between three or more aluminum crystals) (Fig. 11.16). Copper diffuses to and concentrates along aluminum grain boundaries and in grain boundary triple junctions. The copper atoms retard the movement and transport of electro migrating aluminum atoms. Copper doping significantly improves the electromigration reliability of aluminum wires.

When the width of the aluminum wires scaled to the 500 nm technology node, the wire width became smaller than the size of the aluminum grains. Grain boundary triple junctions do not form in the narrow wires. The grain structure in the narrow wires is a bamboo grain structure (Fig. 11.17).

In aluminum leads with bamboo grain structures about 0.5% copper doping is sufficient to suppress electromigration within the aluminum wire. Electromigration along the surface of the aluminum wire became the dominate mode of electromigration failure.

At the 500 nm technology node, to reduce electromigration of aluminum atoms along the surface of the aluminum wires, the Ti/TiW/AlSiCu(2%) metal1 stack (bottom to top) was replaced with a TiN/AlCu(0.5%)/TiN. The bottom TiN layer improves adhesion of the aluminum to the dielectric and also suppresses aluminum atom electromigration along the aluminum wire bottom surface. The top TiN layer performs a dual function. It improves electromigration, by suppressing aluminum electromigration void and hillock formation. Figure 11.15 The top TiN layer also functions as an antireflection coating (ARC) layer for metal1 patterning. The TiN/AlCu/TiN metal stack was used for the 500, 320, and 250 nm technology nodes.

Hillocks
As electrons flow through the aluminum wires, electrons bump into aluminum atoms causing them to move. Voids form in the aluminum wires causing the wires to fail. The aluminum atoms that are moved out of the void area must go somewhere. They accumulate and form bumps (hillocks) on the surface of the aluminum wire (Fig. 11.15). Hillocks sometimes cause shorts between adjacent wires. Hillocks also cause bumps in resist patterns above the hillock that result in defocus and failure of the overlying pattern to print correctly.

At the 180 nm technology node, the higher current densities through the scaled TiN/aluminum/TiN wires failed electromigration reliability testing. A new Ti/aluminum/Ti/TiN stack was introduced. The titanium layers underlying and overlying the aluminum lead react with the aluminum forming titanium aluminide (Al_3Ti). Aluminum alloy leads encased on top and bottom with Al_3Ti pass electromigration reliability testing. The top TiN layer suppresses hillocks and functions as a bottom antireflective coating (ARC) for patterning the aluminum (see Sect. 9.4.14 and Fig. 9.19).

11.5 Multilevel Aluminum Metal

Prior to the 1000 nm technology node, a single level of aluminum wiring (wires about 100 times smaller than a human hair) provided enough wires to route electrical signals between the increasing numbers of transistors. While transistor area shrank by 50% every couple of years, the size of the integrated circuit chip stayed about the same. At each new technology node, circuit engineers designed increasingly powerful computer chips by packing on more and more transistors with faster and faster switching speeds. In 1976, Intel's 8080 computer chip had about 6000 transistors that switched about two million times per second. Just 10 years later, Intel's 386SX computer chip had over 250 million transistors that switched about 30 million times per second! (Table 11.2). By the 1000 nm technology node the transistor count had increased to the point that a second level of wiring was required to wire the transistors together.

Adding another level of wires presented enormous challenges. The topography of intermetal dielectric stack (IMD2) deposited over the metal1 wires is severe (Fig. 11.9). Without planarization, printing via1 and metal2 photoresist patterns is impossible. The uneven topography provides inadequate depth of focus for the photolithography printers. As the spacing between wires scaled smaller, filling the smaller spaces (gaps) with dielectric void-free became increasingly difficult. At almost every technology node, improved dielectric deposition processes were needed at a minimum and usually new deposition processes and equipment were also required. Even though the planarization problem was fixed with the introduction of chemical mechanical polish (CMP), filling the increasingly higher aspect ratio gaps (narrow and deep) between the wires without leaving voids continued to present huge technical challenges.

The primary challenges for multilevel aluminum alloy scaling are as follows: (1) void-free gap fill in IMD layers as wires get closer together and (2) electromigration reliability as current densities increase as the wires get smaller.

11.5.1   Multilevel Aluminum Metal: IMD1 Resist Etch Back Planarization

At the *1000 nm technology node*, resist etch back planarization (REB) of IMD1 was used (Fig. 11.18). In REB a layer of PE-TEOS dielectric is deposited over the metal1 topography. A thick layer of resist is then coated on the PE-TEOS layer and heated to melt the resist so that it reflows filling the valleys and thinning over the hills to form a planarized surface. This reflowed resist is then etched back (REB) using a plasma etch that etches resist and PE-TEOS at about the same rate. Most of the resist and the tops of the PE-TEOS hills are etched away smoothing the IMD1 surface. Resist that remains in the valleys is ashed off. A capping layer of PE-TEOS is deposited prior to via1 patterning. In some manufacturing flows the REB process is repeated multiple times to achieve acceptable planarization.

Via1 openings are patterned on the smoothed IMD1 surface and etched through the IMD1 stack stopping on the underlying metal1. The via1 openings at the 1000 nm node are large enough to accommodate sputtered TiW/AlSi metal2 (Fig. 11.18).

11.5.2   Multilevel Aluminum Metal: Tungsten Filled Via Plugs

At the *800 nm technology node*, a third level of metal (TLM) was needed to accommodate the increased transistor count and IC complexity. By the 800 nm node, the via openings had become so small that sputtered aluminum became almost discontinuous, when filling the via holes (Fig. 11.19). At this node, some manufacturers began to fill the via holes with CVD-tungsten plugs (Fig. 11.20).

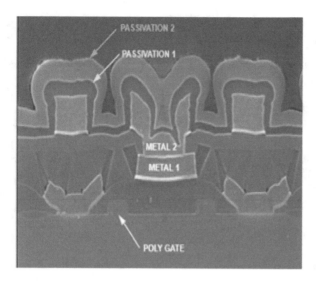

Fig. 11.18 Double-level metal with resist etch back planarization. http://smithsonianchips.si.edu/ice/cd/9512_444.pdf [17]

Fig. 11.19 Double-level metal (DLM) with sputtered TiN/AlSiCu/TiN metal and with SOG etch back planarization of IMD. https://www.entrepix.com/docs/papers-and-presentations/050509_ENTR_NSTI2009.pdf [18]

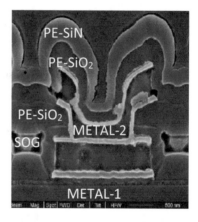

For the CVD-W via plug process, a titanium/tungsten (TiW) barrier layer is deposited to cover the bottom and sides of the via openings. The TiW barrier layer prevents fluorine gas generated by the thermal decomposition of WF_6 during CVD-tungsten (CVD-W) deposition from attacking and etching the aluminum metal in the bottoms of the via holes and from attacking the PMD (silicon dioxide). The CVD-W deposits conformally filling the via holes and covering the surface of the IMD layer. A TiW/AlCu metal2 stack is deposited on the CVD-W. Prior to the metal2 deposition, the CVD-W is etched back to remove it from the surface of the

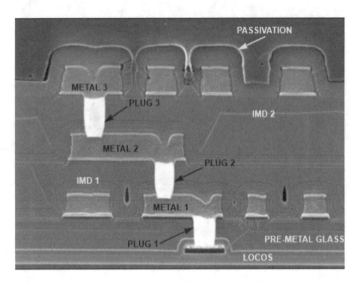

Fig. 11.20 Triple-level metal (TLM) with via1 and via2 filled with etched back CVD-W. Metal layers are TiN/AlCu/TiN. IMD layers are planarized using CMP. Integrated Circuit Engineering (ICE) Report SCA 9711-567 (Figure 9). http://smithsonianchips.si.edu/ice/cd/9711_567.pdf [19]

IMD1 leaving CVD-W plugs filling the via holes (Figs. 11.20 and 11.22). Some IC fabs added a titanium nitride capping layer on the aluminum (Figs. 11.21 and 11.22). The TiN cap provided the dual benefits of providing an antireflection coating (ARC) for aluminum patterning and also suppressing electromigration hillock formation.

11.5.3   Multilevel Aluminum Metal: Spin-on-Glass Etch Back Planarization

At the 500 nm technology node resist etch back (REB) no longer provided a sufficiently smoothed surface. After resist etch back the resist is removed from the valleys in the IMD by ashing leaving gaps too small to fill. Resist etch back was replaced with spin-on-glass (SOG) etch back. After SOG is etched back, SOG remains on the wafer filling the small gaps and providing a smoother surface. The IMD1 stack at the 500 nm technology for several manufacturers is PETEOS/SOG/etch-back/PETEOS (bottom to top) or PETEOS/SACVD/SOG/etch-back/PETEOS (Figs. 11.21 and 11.22).

SOG is coated with a liquid such as hydrogen silsesquioxane (HSQ) suspended in ethyl acetate and isopropyl alcohol solvent. The liquid HSQ flows off the top of hills, filling valleys and forming a somewhat planar surface. After SOG is coated on the wafer, the SOG is annealed to drive off solvents and convert the SOG into silicon dioxide (SiO_2) dielectric. SOG has poor structural stability and needs to be removed from where via holes are to be etched. The SOG is etched off using a

Fig. 11.21 TiN/AlCu/TiN
wiring with SOG etch back
planarization at metal1.
http://smithsonianchips.
si.edu/ice/cd/9504_403.
pdf [20]

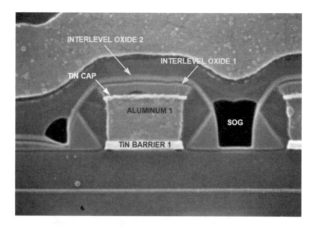

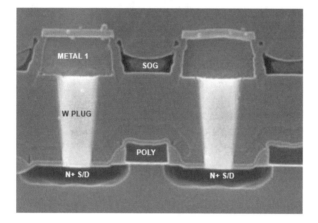

Fig. 11.22 Aluminum alloy metal with SOG etch back planarization at metal2. http://docplayer.
net/40707559-Amd-vantis-machl-v-eepla.html [5]

plasma etch that etches the underlying PE-TEOS or SACVD-TEOS layer at about
the same rate as the SOG. The SOG is completely etched off the hills where vias are
formed but remains in the valleys between the hills (see Figs. 11.21, 11.22, and
11.23). After SOG etch back, a capping layer of PE-TEOS is deposited. In some
manufacturing flows, a layer of resist is coated on the SOG and reflowed prior to
etch back for additional planarization. In these manufacturing flows the plasma etch
back step is tuned to etch resist, SOG, and PE-TEOS and/or SACVD-TEOS at
about the same rate.

This REB planarized IMD and metallization scheme was repeated over and over
to add more levels of aluminum alloy wires (Fig. 11.23).

HSQ (hydrogen silsesquioxane) SOG has a dielectric constant of about 3.0.
Replacing the hydrogen atoms (H-) in hydrogen silsesquioxane with methyl groups
(CH_3-) to form methyl silsesquioxane SOG reduces the dielectric constant to
about 2.7.

Fig. 11.23 Five layers of TiN/AlCu/TiN wiring with tungsten filled contact and via plugs planarized using resist etch back and SACVD-TEOS plus SOG etch back. SEM is courtesy of Texas Instruments

At the 180 nm technology node, the TiN/AlCu/TiN aluminum alloy stack was replaced with an $Al_3TiN/AlCu/Al_3TiN$ aluminum alloy stack to meet electromigration reliability specifications on the smaller metal leads.

11.5.4   Multilevel Aluminum Metal: CMP Planarization

At the *320 nm technology node*, the ILD2 dielectric stack was changed to accommodate the new chemical mechanical polish (CMP) planarization process. A thin layer (~100 nm) of PE-TEOS is deposited over the metal1 topography. A layer of SOG is spun on to fill the small gaps. The SOG and then annealed to drive off solvent and convert it to silicon dioxide and etched back to remove SOG from the tops of metal1 leads where vias are to be formed. A thick layer (>1000 nm) PE-TEOS is then deposited and planarized using CMP. The surface is essentially flat post CMP. This provides excellent depth of focus for via1 and metal2 patterns. The long metal over etch previously required to clear metal filaments from valleys in the IMD also is no longer needed. A capping layer of PE-TEOS is deposited to hit the IMD2 dielectric thickness target.

Via1 openings are then patterned and etched. Via1 plugs are formed in the via1 holes using a columnated sputtered Ti/TiN adhesion/barrier, CVD-W deposition, and tungsten plug etch back. The metal2 stack for the 320 nm node is TiN/AlCu/TiN.

This IMD + TiN/AlCu/TiN metallization manufacturing process was repeated to add more levels of aluminum alloy interconnect for the 250 and 180 nm technology

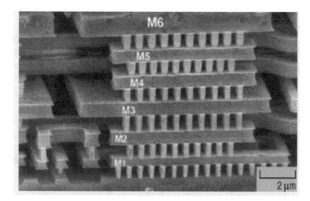

Fig. 11.24 Six layers of TiN/AlCu/TiN wiring with tungsten filled contact and via plugs planarized using chemical mechanical polish (CMP) and with PMD and IMD planarized CMP. https://www.dbstalk.com/community/index.php?threads/digital-swm-theory-and-speculation.209678/page-2 [21]

nodes. Columnated sputtering was first replaced with IMP sputter deposition and then IMP sputter deposition was replaced by chemical vapor deposited TiN to provide barrier layer integrity as the via hole size scaled.

This CMP planarized IMD and TiN/AlCu/TiN metallization scheme was repeated over and over to form additional levels of aluminum wires (Fig. 11.24).

At the 130 nm technology node, the TiN/AlCu/TiN metallization was replaced with $Al_3Ti/AlCu/AL_3Ti$ metallization to meet electromigration reliability specifications on the smaller metal leads.

Because aluminum wiring manufacturing is less costly than copper wiring manufacturing, upper levels of wiring where looser design rules are used can be aluminum alloy. Even on high performance ICs upper levels of aluminum alloy wires are frequently used to distribute power supply voltage and ground across the IC chip. Low resistance is not needed when current flow is low. Large busses that carry high current are also sometimes made in the upper levels of metal using thick and wide aluminum alloy wires.

ICs that are not high performance or are not high density can be manufactured with multilevel aluminum alloy wires to reduce cost.

References

1. Chapter 4 Programmable Logic Devices, in *Integrated Circuit Engineering Corporation*, DocPlayer, pp. 1–18, https://docplayer.net/54982636-Programmable-logic-devices.html
2. Process Technology History—Intel, WikiChip Chips and Semi, https://en.wikichip.org/wiki/intel/process. Accessed 4 Mar 2020
3. Intel Microprocessor Quick Reference Guide, https://www.intel.com/pressroom/kits/quickrefyr.htm. Accessed 4 Mar 2020

4. A. Ayub, *Class Slides—Deposition Processes (Thin Films)—CVD & PVD* (Institute of Nano Electronic Engineering, Institute of Nano Electronic Engineering, Universiti Malaysia Perlis, Arau)., http://portal.unimap.edu.my/portal/page/portal30/Lecture%20Notes/ KEJURUTERAAN_MIKROELEKTRONIK/Semester%201%20Sidang%20Akademik%20 20132014/EMT%20357%20-%20Fundamental%20of%20Microelectronic%20Fabrication/ Lecture%20Notes/Deposition%20Process%20Technology.pdf
5. Construction Analysis—AMD/Vantis MACHL V EEPLA, ICE Report SCA 9709-554, http:// docplayer.net/40707559-Amd-vantis-machl-v-eepla.html
6. Construction Analysis—Atmel AT27C010-45DC 1Mbit EPROM, ICE Report SCA 9709-408, http://smithsonianchips.si.edu/ice/cd/9504_408.pdf
7. J. Pan, D. Ouma, P. Li, D. Boning, F. Redeker, J. Chang, J. Whitby, Planarization and Integration of Shallow Trench Isolation (STI), VMIC (1998), pp. 1–6, https://pdfs.semantic-scholar.org/ed33/a6084fd959bcd1b4a82ffaffa6beb1563e3b.pdf
8. M. LaPadus, Applied flows into flowable CVD, EE Times (2010), https://www.eetimes.com/ applied-flows-into-flowable-cvd/
9. R. Demaray, D. Deshpandey, R. Pethe, Collimated Sputtering of Semiconductor and Other Films, USP 6362097 (2002), https://patents.google.com/patent/US6362097
10. Construction Analysis—Lattice ispLSI2032-180L CPLD, ICE Report SCA 9712-573, http:// smithsonianchips.si.edu/ice/cd/9712_573.pdf
11. Construction Analysis—AMD AM27C010 1M UVEPROM, ICE Report SCA 9612-517, http://smithsonianchips.si.edu/ice/cd/9612_517.pdf
12. Oxide Etch Polymer, Accelerated Analysis, http://acceleratedanalysis.com/?page_id=140
13. J. Schmitz, A. Hasper, On the mechanism of the step coverage of blanket tungsten chemical vapor deposition. J. Electrochem. Soc. **140**(7), 2112–2116 (1993)., https://core.ac.uk/down-load/pdf/11465298.pdf
14. J. Kim, K. Kim, S. Jeon, J. Park, Reliability improvement by the suppression of keyhole generation in W-plug vias. Microelectron. Reliab. **45**(9–11), 1455–1458 (2005)., https://www.sci-encedirect.com/science/article/pii/S0026271405001976
15. Construction Analysis—Hitachi HM5293206FP10 8Mbit SGRAM, ICE Report SCA 9702-529, http://smithsonianchips.si.edu/ice/cd/9702_529.pdf
16. J. Lisle, "13 Key Considerations for Aerospace RF/Microwave Devices, Microwaves & RF (2015), https://www.mwrf.com/active-components/13-key-considerations-aerospace-rfmicro-wave-devices
17. Construction Analysis—QLogic ISP1000 SCSI Processor, ICE Report SCA 9512-444, http:// smithsonianchips.si.edu/ice/cd/9512_444.pdf
18. R. Rhodes, New Applications for CMP: Solving the Technical and Business Challenges, NSTI Conference, May 5, 2009, https://www.entrepix.com/docs/papers-and-presentations/050509_ ENTR_NSTI2009.pdf
19. Construction Analysis—NEC 79VR5000 RISC Microprocessor, ICE Report SCA 9711-567, http://smithsonianchips.si.edu/ice/cd/9711_567.pdf
20. Construction Analysis—Actel A1440 FPGA, ICE Report SCA 9504-403, http://smithson-ianchips.si.edu/ice/cd/9504_403.pdf
21. Digital SWM Theory and Speculation Forum, DBS Talk, p. 2, #25 out of 241, https://www.dbstalk.com/community/index.php?threads/digital-swm-theory-and-speculation.209678/page-2
22. Hong Xiao, U. of Texas, Metallization, Chapter 11; http://apachepersonal.miun.se/~gorthu/ ch11.pdf

Chapter 12
The Incredible Shrinking IC: Part 4 BEOL Low-K Intermetal Dielectrics and Multilevel Copper Metallization

12.1 Introduction

By the 250 nm technology node, the size of aluminum wires had scaled so small that aluminum wire resistance was degrading the speed of high-performance ICs. ICs with large memory arrays were having to wait for data traveling through high resistance aluminum wires. Lower resistance copper wiring was developed to replace aluminum alloy wires. As ICs area scaled, copper wires got closer together increasing capacitive coupling. This slowed the speed of electric signals through the copper wires limiting the speed of the IC. New low dielectric constant (low-k) intermetal dielectrics were developed to reduce capacitive coupling. As the wires scaled closer together, the gap-fill capability of the dielectrics had to be continuously improved.

In 1998, both IBM (https://www-03.ibm.com/press/us/en/pressrelease/2486. wss) [1] and Motorola (https://www.eetimes.com/document.asp?doc_id=1188138) [2], working on a joint development project at the 250 nm technology node introduced versions of the PowerPC™ with copper wiring. In 2001, Intel introduced copper metallization in it's 130 nm technology node computer chip (Table 12.1) (https://www.intel.com/pressroom/archive/releases/2000/cn110700.htm) [3]

A revolutionary technology was developed for copper wiring. Unlike aluminum wires, copper wires cannot be etched using a plasma. Copper does not form gaseous etching products. Development of a chemical mechanical polish (CMP) process to physically remove copper metal was the breakthrough process technology that makes copper wiring possible. Copper metal is electroplated to fill and overfill via holes and trenches etched into the intermetal dielectric (IMD). Copper CMP removes copper overfill leaving copper wires filling the trenches.

A cross section for multilevel copper wiring is shown in Fig. 12.1, left. A projection view of an IBM computer chip with the intermetal level dielectric (IMD) layers removed is shown in Fig. 12.1, right.

Today some ICs have 15 or more wiring levels.

© Springer Nature Switzerland AG 2020
H. Tigelaar, *How Transistor Area Shrank by 1 Million Fold*,
https://doi.org/10.1007/978-3-030-40021-7_12

12.2   Low-K Intermetal Dielectrics (Low-K IMDs)

The time delay (RC delay) of signals propagating along a wire in an IC is proportional to the resistance (R) of the wire multiplied by the capacitance (C) between that wire and surrounding wires. (The unit of resistance times capacitance is frequency = dt = delta time.)

Table 12.1 Summary of CPU technology development at Intel

Node	Approx. year	Power supply	Levels of metal	Metal	Intel CPU	Approx. # transistors in CPU
10,000	1971	5.0	1	AlSi	4004	2300
6000	1976	5.0	1	AlSiCu	8080	6000
3000	1979	5.0	1	AlSiCu	8088	29,000
1500	1982	5.0	1	AlSiCu	80,286	134,000
1000	1987	5.0	2	AlSiCu	386SX	275,000
800	1989	4.0	3	AlSiCu	486DX	1.2 million
500	1993	3.3	4	AlSiCu	IntelDX4	1.6 million
320	1995	2.5	4	AlSiCu	Pentium	3.3 million
250	1998	1.8	5	AlSiCu	Pentium2	7.5 million
180	1999	1.6	6	AlSiCu	Pentium3	28 million
130	2001	1.4	6	Cu	Celeron	44 million
90	2003	1.2	7	Cu	Pentium4	125 million
65	2005	1.0	8	Cu	Pentium4	184 million
45	2007	1.0	9	Cu	Xenon	410 million
32	2009	0.75	10	Cu	Core i7 G	1.17 billion
22	2011	0.75	11	Cu	Core i7 IB	1.4 billion
14	2014	0.70	12	Cu	Core i7 BU	1.9 billion

https://en.wikichip.org/wiki/intel/process;   https://www.intel.com/pressroom/kits/quickrefyr.htm [6, 7]

Fig. 12.1 Multilevel copper on IBM computer chips. Left is cross section. Right is projection view with dielectric etched away. https://www.electrochem.org/dl/interface/spr/spr99/IF3-99-Pages32-37.pdf [4]; https://www.ibm.com/ibm/history/ibm100/us/en/icons/copperchip/ [5]

$$\tau \ \left(\text{signal time delay} \right) = R \ \left(\text{resistance} \right) * C \ \left(\text{capacitance} \right)$$

$$\tau = \frac{V \ \left(\text{voltage} \right)}{I \left(\text{current} \right)} * \frac{Q \ \left(\text{charge} \right)}{V \ \left(\text{voltage} \right)}$$

$$\tau = \frac{Q \ \left(\text{charge} \right)}{I \left(\text{current} \right)} = \frac{Q * dt \ \left(\text{time} \right)}{Q} = dt \ \left(\text{time} \right)$$

With the introduction of lower resistance copper metallization, wire-to-wire capacitance became a dominant factor in RC delay. When wire-to-wire capacitance causes a signal on one wire to induce a signal on an adjacent wire a logic failure can be the result. New equipment and processes were developed to replace dielectrics that had higher dielectric constants (high-k) with new dielectrics having lower dielectric constants (low-k). Some of the dielectrics that were developed and introduced into manufacturing as the IC metallization scaled are listed in Table 12.2.

The next section describes the evolution of copper wiring manufacturing through the various technology nodes. The technology node where each of these low-k dielectric materials is introduced is described.

12.3   Damascene Copper Metalization

The damascene copper process for forming copper wires on integrated circuits gets its name from damascene jewelry. Damascene jewelry is made by etching trenches into a piece of jewelry and then refilling the etched-out trench area with metal of a different color.

Table 12.2 Dielectric constants of dielectrics used to isolate wires in copper metallization

Dielectric	Dielectric constant
Silicon nitride Si_3N_4 (often written in the literature as SiN)	~7.0
Silicon oxynitride SiON	~5.0
Silicon carbon nitride SiCN	~4.8
Silicon oxycarbide SiCO:H	~4.65
BloK™ SiCOH	~4.5
Silicon dioxide SiO_2	~4.0
FSG Fluoro silicate glass F-SiO_2	~3.7
(OSG) Organo silicate glass SiOCH	~2.9
ULK ultra low dielectric constant	~2.5
Aerogels-porous SiO_2	1.1–2.2

12.3.1   Single Damascene Copper Metalization

A single damascene process is used for the first layer of copper wires that are formed on top of the PMD and on top of the tungsten-filled contact plugs. After the tungsten overfill is removed using tungsten CMP, a first intermetal dielectric layer (IMD) is deposited. A metal1 trench pattern is formed on the IMD layer and metal1 trenches are etched through the IMD layer exposing the tops of the tungsten contact plugs. Copper metal is electroplated filling and overfilling the metal1 trenches. Copper CMP removes the copper overfill leaving copper wires in the IMD trenches.

12.3.2   Dual Damascene Copper Metalization

A dual damascene copper process is used to simultaneously fill the metal2 trenches in the IMD1 and to fill the via1 holes that electrically connect the bottoms of the metal2 trenches to the metal1 wires below. In the dual damascene copper process, using a first pattern plus etch, metal2 trenches are etched part way through the intermetal dielectric stack (IMD1). Using a second pattern plus etch process, via1 holes are etched through the IMD1 in the bottom of the metal2 trenches to connect the bottom of the metal2 trenches with the top of an underlying metal1 copper wire. The via1 holes and the metal2 trenches are then simultaneously filled with electroplated copper. Since both metal2 trenches and via1 holes are simultaneously filled, the process is called dual damascene.

12.4   180 nm Copper Multilevel Metal

12.4.1   180 nm Single Damascene Copper Metal1

After the PMD is planarized and the tungsten contact plugs are formed using tungsten CMP, the first intermetal dielectric stack (IMD1) is deposited. IMD1 for the 180 nm technology node is (SiN/FSG1). A thin (~30 nm) silicon nitride (SiN) metal1 trench etch stop layer (M1TESL) is first deposited. A thick layer of PECVD fluorine-doped silicon glass (FSG1) deposited on top (Fig. 12.2a). The thickness of the IMD1 layer determines the thickness of the metal1 copper wires. FSG1 has a lower dielectric constant (3.7) than the PETEOS (4.0) used in previous technology nodes (Table 12.2). Metal1 trenches are patterned (Fig. 12.2b) and etched through the FSG1 stopping on the M1TESL. The etching chemistry is changed from silicon oxide etch to silicon nitride etch to etch the silicon nitride M1TESL (Fig. 12.2c). After the metal1 trenches are etched, a thin (~25 nm) barrier layer of IMP sputtered tantalum nitride (TaN) is deposited to cover the surfaces of the IMD1. Copper readily diffuses through silicon dioxide and silicon and will change the transistors performance if it reaches them. The TaN barrier layer seals the copper lead and

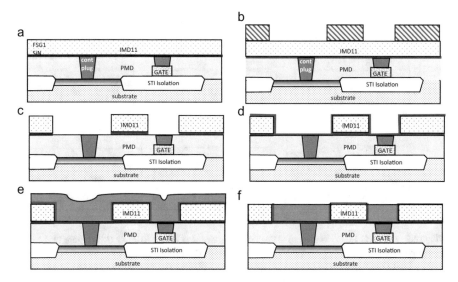

Fig. 12.2 (**a**) M1ESL and IMD1 on PMD and tungsten contact plugs, (**b**) metal1 trench pattern on IMD1, (**c**) metal1 trenches etched through IMD1 layer, (**d**) tantalum nitride liner and copper seed layers covering metal1 trenches, (**e**) electroplated copper overfilling the metal1 trenches, (**f**) single-level damascene copper wires after copper CMP

blocks copper from diffusing into the IMD. A thin layer (~100 nm) of copper seed is IMP (ion metal plasma) sputtered on top of the TaN (Fig. 12.2d). The copper seed layer provides a low resistance path for the copper electroplating current. Copper is then electroplated to fill and over fill the metal1 trenches (Fig. 12.2e). The copper overfill and the TaN barrier layer are removed from the IMD1 surface using CMP (Fig. 12.2e). This leaves copper metal1 wires in the metal1 trenches.

12.4.2   180 nm Dual Damascene Copper Metal2

The IMD2 dielectric stack for high-performance 180 nm copper metallization is SiN/FSG2/SiN/FSG3 (Fig. 12.3a). The bottom layer of silicon nitride (SiN) is a via1 etch stop layer (V1ESL). The first layer of FSG2 electrically isolates metal2 from metal1. The second SiN layer is metal2 trench etch stop layer (M2TESL). The metal2 trenches are formed in the top FSG3 layer. After the ILD2 stack is deposited, a via1 photoresist pattern is formed on the top FSG3 layer and via1 holes are etched through the FSG3 layer and stop on the SiN M2TESL. The via1 etch chemistry is then changed to etch through the SiN M2TESL (Fig. 12.3a). The via1 pattern is stripped and a metal2 trench photoresist pattern, aligned to the via1 holes, is printed on the FSG3 layer. Because the metal2 pattern is aligned to the via1 holes, the metal2 leads must be sufficiently wide to completely cover the via1 holes including the manufacturing misalignment tolerance (Fig. 12.3b). Metal2 trenches are etched through the FSG3 layer stopping on the M2TESL (Fig. 12.3c). The partially etched vias are at the

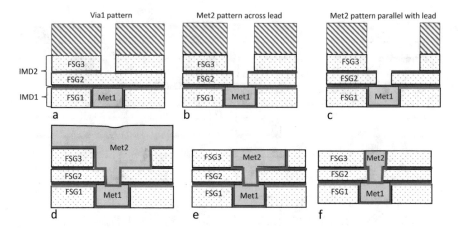

Fig. 12.3 (**a**) FSG IMD2 on metal1. Post via1 pattern and etch, (**b**) FSG IMD2 metal2. Metal2 trench pattern and etch (perpendicular to lead), (**c**) IMD2 with metal2 trench patterned and etched (parallel with lead), (**d**) metal2 trench and via1 filled with electroplated copper (parallel with lead), (**e**) dual damascene metal2 and via1 post copper CMP (parallel with lead), (**f**) dual damascene metal2 and via1 post copper CMP (perpendicular to lead)

same time etched through the FSG2 layer and stop on the underlying V1ESL. The plasma etching chemistry is then changed to etch SiN instead of FSG. The V1ESL in the bottom of the via1 holes is etched through to expose the underlying metal1 lead. At the same time the SiN M2TESL with its high dielectric constant is etched from the bottoms of the metal2 trenches. A thin layer of copper seed is (ion metal plasma) IMP sputtered on top of a thin IMP sputtered TaN barrier layer. Electroplated copper fills the via1 holes and over fills the metal2 trenches (Fig. 12.3d). The copper overfill and the TaN barrier layers are removed from the IMD2 surface using CMP. Copper metal2 wires remain filling the via1 holes and filling the metal2 trenches (dual damascene) (Fig. 12.3e, f). This dual damascene copper wire process is repeated for additional levels of wiring at the 180 nm technology node.

12.5   130 nm Copper Multilevel Metal

The IMD1/metal1 and IMD2/Via1/metal2 processes for 130 nm node are almost the same as for the 180 nm technology node. Since the materials are the same and the structure is the same only scaled smaller, the manufacturing flow is essentially the same. The manufacturing equipment and manufacturing processes are updated versions of 180 nm equipment and processes. They are updated to meet the tighter deposited thin film thickness tolerances and tighter tolerances of metal trench and via hole geometries post etch. An improved copper electroplating solution and improved copper electroplating procedure was introduced to fill the smaller vias and trenches void-free. At the 130 nm technology node, in a number of fabs 300 mm (12 inch) wafers were introduced for the first time. This required an entirely new set of manufacturing equipment and usually the construction of an entirely new wafer fab.

12.6   90 nm Copper Multilevel Metal

12.6.1   90 nm Single Damascene Copper Metal1

By the 90 nm technology node, the copper wires were smaller and higher resistance. In addition, the copper wires were closer together with higher capacitance. Signals were being slowed by the increased resistance∗capacitance (RC delay) and were degrading IC performance. New dielectric materials with lower dielectric constants (low-k) were introduced to reduce the capacitance that was slowing circuits.

The IMD1 dielectric stack for 90 nm copper metallization was changed to SiCN/OSG/SiN to reduce capacitance (Figs. 12.4 and 12.5a). The silicon carbon nitride (SiCN) metal1 trench etch stop layer (M1TESL) has a lower dielectric constant (4.8) than the silicon nitride (SiN) (7.0) previously used. The organosilicate glass (OSG) has better step coverage and has a lower dielectric constant (2.9) than FSG (3.7). The top SiN layer provides improved structural integrity over the OSG, functions as a bottom antireflection coating (BARC) for metal1 trench pattern, and is a copper CMP stopping layer. Much of the SiN dielectric layer with its relatively high dielectric constant (~7.0) is removed during copper CMP.

In the 90 nm technology node, a thin (~10 nm) layer of tantalum (Ta) metal is deposited on the TaN barrier layer. The Ta layer improves adhesion of copper (Cu) to TaN and also blocks copper diffusion. The tantalum nitride barrier layer thickness is reduced to about ~10 nm to lower resistance of the copper lead. The copper seed layer at the 90 nm node is thinned to ~100 nm and is deposited using self-ionized plasma (SIP) sputter deposition to provide more uniform coverage. (The copper seed layer was deposited using IMP sputter deposition at the previous technology node.) An improved copper electroplating solution and improved copper electroplating procedure was introduced to fill the scaled smaller trenches void-free.

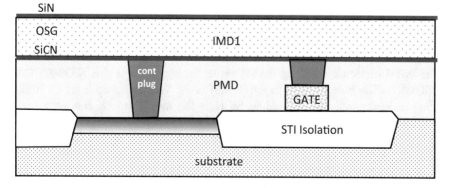

Fig. 12.4   90 nm IMD1 stack (SiCN/OSG/SiN) on PMD and contact plugs

Low-k Organo Silicate Glass (OSG)

https://www.svmi.com/custom-film-coatings/low-k-films/ [8]

Black Diamond II™ (BD) is an Applied Materials low-k dielectric. BD is deposited near room temperature using CVD with an organosilane precursor plus oxygen and nitrous oxide. Various versions of BD have dielectric constants ranging from 2.7 to 3.3.

Coral™ (Lam Research) is deposited with a CVD-like process using tetramethyl silane and tetramethylcyclotetrasiloxane. Coral™ has a dielectric constant of about 2.9.

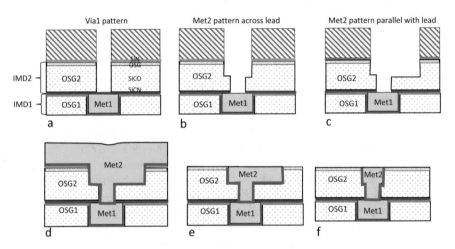

Fig. 12.5 (**a**) OSG IMD2 on metal1. Via1 pattern and etch, (**b**) OSG IMD2 Metal2 trench pattern and etch (perpendicular to lead), (**c**) IMD2 with metal2 trench patterned and etched (parallel with lead), (**d**) metal2 trench and via1 with TaN barrier and electroplated copper (parallel with lead), (**e**) dual damascene metal2 and via1 post copper CMP (parallel with lead), (**f**) dual damascene metal2 and via1 post copper CMP (perpendicular to lead)

12.6.2   90 nm Dual Damascene Copper Metal2

The IMD2 dielectric stack for the 90 nm technology node is (SiCN/OSG/SiCN/OSG/SiN). The bottom SiCN layer ($k{\sim}4.8$) is the via1 etch stop layer (V1ESL) (Fig. 12.5a through f). The middle SiCN is the metal2 trench etch stop layer (M2TESL). The top SiN layer caps the OSG with improved structural intregity, is an antireflection coating (ARC) for via1 and metal2 patterning, and also is a stopping layer for copper chemical mechanical polish. The high dielectric constant SiN ($k = 7.0$) is mostly removed during CMP (Fig. 12.5e, f). To reduce the resistance of the scaled copper leads, the higher resistance TaN/Ta barrier layers are both thinned to about 10 nm. The manufacturing flow follows similar manufacturing steps as the previous nodes. After the via1 etch stops on the SiCN M2TESL, the via1 photoresist pattern is removed and the metal2 trench patten is applied. The first step in the metal2 trench etch is to etch the top SiN layer and to simultaneously etch the M2TESL from the bottom of the via1 holes. The second step in the metal2 trench etch is to simultaneously etch the OSG3 layer in the metal2 trenches stopping on the M2TESL and etch the OSG2 layer from the via1 holes stopping on the V1ESL. The last metal2 trench etch step is to change etch chemistry and simultanously remove the V1ESL from the bottom of the via1 holes and remove the M2TESL from the bottom of the metal2 trenches. Manufacturing equipment and processes are upgraded to process the new materials and to meet the more stringent gap fill, side wall coverage, and uniformity requirements. For example, different manufacturing equipment and processes are required to deposit and etch the SiCN and the OSG. An improved copper plating solution and copper plating manufacturing process is used to fill the scaled via1 holes and metal2 trenches.

12.7   65 nm Copper Multilevel Metal

12.7.1   65 nm Single Damascene Copper Metal1

The IMD1 dielectric stack at the 65 nm technology node is the same dielectrics as for the 90 nm technology node but with different thicknesses. The equipment and processes are upgraded to meet the tighter thin film deposition, etch, and uniformity specifications. In addition, the barrier layer was changed from the bilayer TaN/Ta barrier to a single thin layer of Ta (~20 nm) to reduce resistance of the scaled copper wires (more low resistance copper/less high resistance tantalum).

12.7.2   65 nm Dual Damascene Copper Metal2

A major innovation was introduced into the metal2/IMD2 stack at the 65 nm node. A timed metal2 trench etch replaced the previously used metal2 etch that stopped on the M2TESL. This innovation significantly reduced manufacturing cost by eliminating several manufacturing steps. It also significantly lowered the dielectric constant of the IMD2 dielectric stack by eliminating the SiCN M2TESL ($k = 4.5$).

Various IMD2 dielectric stacks with compromises between structural integrity and low dielectric constant were used at the 65 nm technology node. Dielectric stacks used by various manufacturers include SiCN/OSG/USG/SiN, SiCN/SiCO/ OSG/SiN, SiCN/PETEOS/OSG/PETEOS, SiCN/OSG/FSG/SiN, or…. The bottom SiCN is a barrier ($k = 4.8$) to copper diffusion from the metal1 wires and also is the etch stop layer for the via1 etch (V1ESL). OSG is the bulk of the IMD2 dielectric

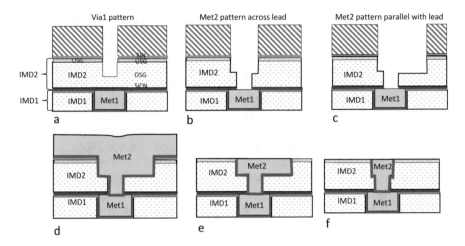

Fig. 12.6 (**a**) OSG IMD2 on metal1. Via1 pattern and etch, (**b**) OSG IMD2 metal2. Metal2 trench pattern and etch (perpendicular to lead), (**c**) IMD2 with metal2 trench patterned and etched (parallel with lead), (**d**) metal2 trench and via1 filled with electroplated copper (parallel with lead), (**e**) dual damascene metal2 and via1 post copper CMP (parallel with lead), (**f**) dual damascene metal2 and via1 post copper CMP (perpendicular to lead)

stack with $k = 2.9$. Undoped silicate glass (USG) is a thin undoped CVD TEOS + O_3 layer ($k = 4.0$) that caps the OSG and provides structural integrity. The top SiN layer ($k \sim 7.0$) provides structural intregity, is an antireflection coating (ARC) for via1 and metal2 patterning, and also is a stopping layer for copper CMP. Much of the high dielectric constant SiN layer ($k = 7.0$) layer is removed during copper CMP planarization (Fig. 12.6e, f).

In Fig. 12.6a the via1 pattern is formed on the SiN antireflective coating (ARC) and etched to the via1 etch stop layer (V1ESL). In Fig. 12.6b, the metal2 pattern is aligned to the via1 holes and etched into the IMD2 layer. The time of the etch determines the depth of the metal2 trenches. A tantalum (Ta) barrier layer (~20 nm) covers the surfaces of the via1 holes and metal2 trenches to prevent copper diffusion into the IMD. A thin copper seed layer is deposited on the barrier layer. The barrier and copper seed layers are deposited using a self-ionized plasma (SIP) process that improves step coverage and uniformity. The TaN barrier layer is omitted to lower the resistance of the copper wires. An improved copper electroplating solution and copper electroplating process are used to fill the smaller via1 holes and metal2 trenches (Fig. 12.6d). Chemical mechanical polish (CMP) is used to remove copper overfill and to remove most of the high dielectric constant (~ 7) SiN ARC layer from the surface of the IMD2 (Fig. 12.6e, f).

Additional levels of copper wires are added by repeating the metal2 dual damascene process.

Timed Metal Etch Eliminates Metal Etch Stop Layer

The metal trench etch team (equipment engineers, etch engineers, thin film deposition, lithography engineers, cleanup engineers, etc.) were able to pull off what I thought was impossible. Their team developed a metal2 trench etch process that etched trenches to the same depth across each die and across the entire wafer without a metal2 trench etch stop layer (M2ESL). Within a chip there are narrow metal2 trenches and wide trenches. Reactive species that etch IMD2 and reaction products generated by the etch have more difficulty diffusing into and diffusing out of narrow trenches than diffusing in and out of wide trenches. In my experience, narrow trenches always etch slower than wide trenches. Without a M2TESL narrow trenches are always shallower than wide trenches. Somehow, they fixed this.

It also was my experience that metal2 trenches the center of the wafer etched a little bit faster (or slower) than metal2 trenches at the edge of the wafer. A M2TESL was needed for the metal2 wires at the edge of the wafer to have the same depth (resistance) as metal2 wires at the center of the wafer. Somehow, they fixed this too! Somehow, they were able to develop a metal trench etch with exactly the same etch rate across the wafer and independent of pattern density.

By eliminating the M2TESL they eliminated the M2TESL deposition and M2TESL etching manufacturing steps. Removing the M2TESL also reduces the dielectric constant of the IMD2. (M2TESL with $k = 4.5$ is replaced with OSG with $k = 2.9$).

12.8 45 nm Copper Multilevel Metal

12.8.1 45 nm Single Damascene Copper Metal1

At the 45 nm technology node, organo silicate glass (OSG, $k \sim 2.9$) is replaced with ultra-low-k dielectric (ULK, $k \sim 2.5$) and the SiCN ($k = 4.8$) via1 etch stop layer (V1ESL) is replaced with SiCO ($k \sim 4.65$). A titanium nitride (TiN) hard mask is also added to reduce etch loading due to photoresist erosion during the metal2 trench etch. The metal2 trench photoresist pattern is etched into the TiN hard mask layer and then the photoresist pattern is removed. The metal2 trench plasma etch erodes the photoresist introducing carbon and hydrogen into the plasma. This locally changes the etch rate and polymer formation resulting in locally non uniform metal trench depths and metal trench profiles. The TiN hard mask eliminates this source of etching nonuniformity. The metal2 trenches are etched using the TiN hard mask pattern.

> **ULK Dielectric**
> Black Diamond™ is a modified organo silicate glass (OSG). The OSG is
> modified with thermally labile organic groups. After the OSG film is depos-
> ited using a variation of PECVD, the OSG film is heated cleaving the ther-
> mally labile organic groups. A UV light cure removes these groups leaving
> pores in the ULK. The pores reduce the dielectric constant to about 2.5 [9].

The IMD1 stack for the 45 nm technology node is SiCN/ULK/SiCN. The bottom
SiCN is the metal1 trench etch stop layer (M1TESL). The top SiCN provides
structural stability for the ULK and also provides a stopping layer for the metal1
CMP. The tantalum (Ta) barrier layer thickness is reduced to about 10 nm to lower
resistance of the scaled copper leads.

12.8.2   45 nm Dual Damascene Copper Metal2

The IMD2 stack for the 45 nm technology node is SiCO/ULK/SiCN/TiN. The SiCO
is the stopping layer for the via1 etch (V1ESL) with a lower dielectric constant (k
~4.65) than the previously used SiCN (k ~5.1). The SiCO is deposited using PECVD
with octamethylcyclotetrasiloxane (OMCTS) and He. The top SiCN layer adds
structural stability for the ULK film and also provides a stopping layer for copper CMP.
   In Fig. 12.7a, the metal2 trench photoresist pattern is formed on the TiN hard
mask layer. The metal2 trench pattern is etched into the TiN hard mask. The metal2

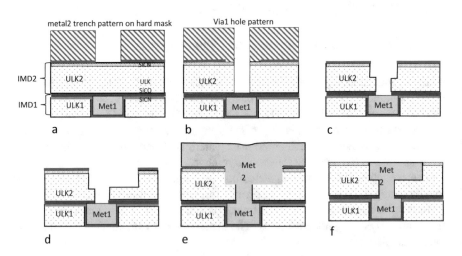

Fig. 12.7 (**a**) ULK IMD2. Metal2 pattern etched into TiN hard mask, (**b**) Via1 patterned on ULK2
and etched, (**c**) Metal2 etch using hard mask (perpendicular to lead), (**d**) Metal2 trench etch using
TiN hard mask (parallel with lead), (**e**) metal2 trench and via1 filled with electroplated copper
(parallel with lead), (**f**) dual damascene metal2 and via1 post copper CMP (parallel with lead)

trench photoresist pattern is removed. (Not all manufacturers use the TiN hard mask.)

In Fig. 12.7b, the via1 photoresist pattern is aligned to the metal1 trench pattern and the via1 hole is etched through the IMD2 layer stopping on the V1ESL.

In Fig. 12.7c (perpendicular to metal2 lead) and 12.7d (parallel through metal2 lead), the metal2 trenches are etched into the IMD2 layer using a timed etch (no etch stop layer!).

In Fig. 12.7d, Ta barrier and Cu seed layers are deposited onto the walls of the via1 holes and the metal2 trenches. An improved copper plating solution and improved copper plating process are used to electroplate copper filling and overfilling the scaled via1 holes and the metal2 trenches.

In Fig. 12.7e, copper chemical mechanical polish (Cu-CMP) removes the copper overfill and the TiN hard mask from the surface of the IMD2 layer. Cu-CMP stops on the SiCN layer.

Additional layers of copper wiring are added by repeating the IMD2/VIA1/METAL2 processing steps.

12.9   32 nm Copper Multilevel Metal

For the 32 nm technology node, the metallization team replaced the SiCN ($k = 4.8$) metal1 trench etch stop layer (M1TESL) with BLOk™ (SiCOH) [10] ($k = 4.5$). (Not at all companies.) BLOk™ is deposited using PECVD with trimethyl silane ((CH_3)$_3$SiH). Two additional new processes were introduced at the 32 nm technology node: first, the metallization team introduced a process that self-aligns the via1 pattern to the metal2 trench hard mask pattern and second, the metallization team introduced a self-formed copper diffusion barrier process to reduce the resistance of the copper leads.

Self-Formed Copper Diffusion Barrier

By the 32 nm technology node, the copper wires had scaled so small that the thickness of the barrier layers was adding significant resistance to the copper wires. By replacing part of the tantalum barrier thickness with copper, the resistance of the copper wires can be reduced. At 32 nm the metallization team introduced a manganese-doped copper self-formed diffusion barrier. First, they deposit the thinnest possible Ta barrier layer—just thick enough to prevent copper from coming into contact with ILD2 dielectric during copper seed deposition. They then deposit a copper seed layer doped with about two atomic percent manganese (Mn) and anneal it. During the anneal, the Mn atoms are rejected from the growing copper crystals. The Mn atoms migrate to the IMD1 interface where they react with SiO_2 forming a $MnSi_xO_y$ barrier layer. This self-formed $MnSi_xO_y$/Ta copper diffusion barrier layer is much thinner than the previously used 10 nm Ta barrier layer and therefore contributes less resistance to the scaled copper wires (Fig. 12.8).

Fig. 12.8 Dual damascene
copper metal with
self-formed $MnSi_xO_Y$
barrier layer [11]

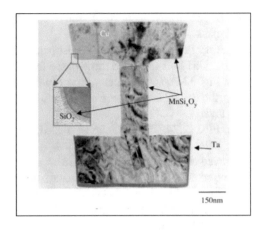

12.9.1   32 nm Single Damascene Copper Metal1

The IMD1 dielectric stack for 32 nm technology node is (BLOk™/ULK/SiCO/
TiN). The bottom BLOk™ layer (SiCOH) is the metal1 trench etch stop layer
(M1TESL). The top SiCO layer (deposited using PECVD of octamethylcyclotetrasiloxane) provides structural stability for the ULK and is a stopping layer for
metal1 CMP. The TiN is a hard mask layer, so metal1 can be etched avoiding the
problems of etching nonuniformity due to resist erosion. The copper diffusion barrier layer is a self-formed $MnSi_xO_Y$/Ta barrier.

12.9.2   32 nm Dual Damascene Copper Metal2

At the 32 nm technology node, the metallization team introduced a via1 self-aligned
to metal2 process. Self-aligning the via1 to metal2 eliminates the metal2 overlap of
via1 design rule. Metal2 leads no longer have to be laid out wider to allow for misalignment of via1 to metal2 during manufacturing. (When the metal2 wire does not
completely cover the via1 plug, current crowding occurs and causes electromigration reliability failures.) Elimination of this design rule enables the width of tens of
thousands of metal leads to be reduced and the integrated circuit chip size to be
reduced. (Manufacturing can now build more chips on each wafer.)

Figure 12.9a through h illustrate the SELF-ALIGNED VIA1 PROCESS. Key to
the via1 self-aligned process is the titanium nitride (TiN) hard mask. In Fig. 12.9a,
b, the metal2 trench photoresist pattern is formed on and etched through the titanium nitride (TiN) hard mask. The metal2 trench photoresist pattern is then stripped.

In Fig. 12.9c and d, the via1 photoresist pattern is formed on the TiN hard mask
and is aligned to the metal2 hard mask pattern. In Fig. 12.9c even with misalignment the metal2 lead will always cover via1. As shown in Fig. 12.9d, the size of the
via1 hole is determined by the width of the metal-2 trench opening in the TiN hard
mask. Even with misalignment, metal2 always completely covers via1.

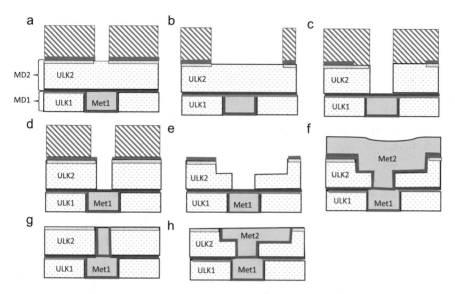

Fig. 12.9 (**a**) ULK IMD2. Metal2 pattern on TiN hard mask. Etched perpendicular to Cu wire, (**b**) ULK IMD2. Metal2 pattern on TiN hard mask. Etched parallel to Cu wire, (**c**) ULK IMD2. Via1 pattern and IMD2 etched parallel to Cu wire, (**d**) via1 pattern on hard mask. Via1 etched using hard mask. Self-aligned perpendicular to Cu wire, (**e**) metal2 timed trench etches to target trench depth using hard mask, (**f**) electroplated copper parallel to metal2 wire, (**g**) copper overfill and hard mask removed using CMP (perpendicular to wire), (**h**) copper overfill and hard mask removed using CMP (parallel to wire)

In Fig. 12.9e, the metal2 trench is etched part way through the ILD2 using a timed etch and the TiN hard mask pattern.

In Fig. 12.9f, the self-formed Ta/CuMn copper diffusion barrier layer is formed and the copper seed layer is deposited. The via1 holes and metal2 trenches are filled with electroplated copper.

In Fig. 12.9g and h, the electroplated copper overfill and the titanium nitride (TiN) hard mask are removed with CMP.

Improved copper plating solution and copper plating procedure is used to fill the scaled via1 holes and trenches.

12.10 Copper Metallization Challenges Going Forward

Resistance and electromigration of the smaller and smaller copper wires is becoming a bigger and bigger problem. One proposed solution is to replace the $MnSi_xO_Y$/Ta barrier with a $MnSi_xO_Y$/Co barrier. A thin layer of cobalt (Co) is deposited on the surfaces of the via holes and sides of the metal trenches before a Mn-doped copper seed layer is deposited. The thin Co layer blocks copper from coming into contact and diffusing into the IMD. During a subsequent anneal, the Mn diffuses to and reacts with the surface of the IMD forming a self-formed $MnSi_XO_Y$ copper barrier layer. The Co diffuses to and adheres to the surface of copper improving electromigration.

Electromigration reliability is improved for copper wires encapsulated with cobalt. There are proposals to replace CVD-W contact plugs and CVD-W local interconnect with CVD-cobalt (CVD-Co) contact plugs and CVD-Co local interconnect. http://www.appliedmaterials.com/products/endura-volta-cvd-cobalt [12]; http://www.appliedmaterials.com/products/cobalt [13] Ruthenium (Ru) is being investigated as a replacement interconnect metal at the 3 nm technology node [14]. Ru does not need to be encapsulated in a high resistance barrier layer. Other metals such as molybdenum, osmium, niobioum, iridium, and rhodium are also being investigated as replacements for copper interconnect to address wire resistance and wire electromigration issues [15]. These metals are being pursued in parallel at semiconductor companies, universities, and consortia to determine which process is the best compromise between electrical properties, reliability, ease of manufacturing, cost of manufacturing, and yield.

Metal wiring technology constantly evolved as integrated circuits scaled and transistor counts ballooned. Typically, a number of approaches were pursued in parallel for thin film deposition, thin film etching, and thin film planarization. Usually only one or at most two of the approaches were found to be manufacturable. The other approaches were abandoned and relegated to the trash heap of history. The companies whose equipment/processes make it into manufacturing were winners and able to recoup their enormous research and development investments. Companies whose equipment/process did not make it into manufacturing lost millions of dollars, sometimes went bankrupt and closed their doors, or were bought out by one of the winners.

References

1. *IBM Delivers World's First Copper Chips*, IBM Press Release, 1 Sep 1998, https://www-03. ibm.com/press/us/en/pressrelease/2486.wss
2. Motorola introduces its first copper processor: PowerPC with AltiVec technology, EE Times, 31 Aug 1999, https://www.eetimes.com/document.asp?doc_id=1188138
3. *Intel Completes 0.13 Micron Process Technology Development*, Intel Press Release, 7 Nov 2000, https://www.intel.com/pressroom/archive/releases/2000/cn110700.htm
4. P. Andricacos, Copper on-chip interconnections a breakthrough in electrodeposition to make better chips, Electrochem. Soc. Interf. (1999), pp. 32–37, https://www.electrochem.org/dl/interface/spr/spr99/IF3-99-Pages32-37.pdf
5. Copper Interconnects The Evolution of Microprocessors, IBM100 Website (2011), https://www.ibm.com/ibm/history/ibm100/us/en/icons/copperchip/. Accessed 4 Mar 2020
6. Process Technology History—Intel, WikiChip Chips & Semi, https://en.wikichip.org/wiki/intel/process. Accessed 4 Mar 2020
7. Intel Microprocessor Quick Reference Guide, WikiChip Chips & Semi, https://www.intel.com/pressroom/kits/quickrefyr.htm. Accessed 4 Mar 2020

8. Low-k films, Silicon Valley Microelectronics (SVM), Website, https://www.svmi.com/custom-film-coatings/low-k-films/. Accessed 4 Mar 2020
9. R. Perry, Producer® Black Diamond® PECVD, Applied Materials, Website, http://www.appliedmaterials.com/products/producer-black-diamond-pecvd. Accessed 4 Mar 2020
10. P. Xu, S. Rathi, A Breakthrough in Low-k Barrier/Etch Stop Films for Copper Damascene Applications, in *Semiconductor Fabtech*, 11th edn. (2000), pp. 239–244, https://www.webpages.uidaho.edu/nanomaterials/research/Papers/General/Bariers%20for%20Cu%20low%20k%20Damascene%20structures%20Fabtech%202011.pdf
11. T. Usui, H. Nasu, S. Takahashi, N. Shimizu, T. Nishikawa, M. Yoshimaru, H. Shibata, M. Wada, J. Koike, Highly reliable copper dual-damascene interconnects with self-formed $MnSi_xO_Y$ barrier layer, IEEE Trans. ED, \, No. 10, 2492, 2006. , https://www.researchgate.net/publication/3075130_Highly_reliable_copper_dual-damascene_interconnects_with_self-formed_MnSisub_xOsub_y_barrier_Layer
12. Endura® Volta™ CVD cobalt, Appl. Mater., Website, http://www.appliedmaterials.com/products/endura-volta-cvd-cobalt, Accessed 4 Mar 2020
13. Cobalt Product Suite, Appl. Mater., Website, http://www.appliedmaterials.com/products/cobalt. Accessed 4 Mar 2020
14. David Manners, "Ru break through for 3 nm interconnect", Electronics Weekly, 10 July 2018; https://www.electronicsweekly.com/news/business/ru-breakthrough-3nm-interconnect-2018-07/
15. Paul McLellany, "IEDM: the World After Copper", Breakfast Bytes Blogs, Cadence, 3 Jan. 2019; https://community.cadence.com/cadence_blogs_8/b/breakfast-bytes/posts/iedm18-interconnect

Chapter 13
Anecdotes

13.1 High-Voltage Transistors for DLP

Early in my career at Texas Instruments I worked on nonvolatile memory EPROMs. At the time these transistors operated at about 2.5 V, but required voltages in excess of 15 V to program and erase. Part of my job as an EPROM transistor engineer was to design and build these high-voltage programming transistors. One day, Larry Hornbeck, the inventor of digital light processor (DLP) technology, asked me if I could help him with the high-voltage transistors he needed to manipulate his digital mirrors. I knew nothing about DLP and asked him how it worked. He told me his DLP was composed of thousands of exceedingly small micro mirrors attached to posts with small strips of metal. He needed high voltage from a high-voltage transistor to attract and bend these mirrors, so they would deflect light in different directions. I thought his idea was crazy. I told him I thought after bending back and forth a few thousand cycles, his metal hinges would get metal fatigue and his mirrors would fall off. He assured me that was not the case. Obviously he was right. Projectors in most theaters today use Larry Hornbeck's DLP technology. In 2015 Larry was awarded an Oscar for his contribution to the motion picture industry (see Fig. 13.1) [2].

13.2 TI Widows and Orphans

I was told by colleagues from other companies that TI was known for running lean and working hard. Since I only worked for TI, I have no comparison. I do know that when TI engineers and their wives got together socially, the TI wives jokingly referred to themselves as TI widows.

When I was running new technology manufacturing flow development projects (It took a couple of years to complete such a project and transfer the manufacturing flow into production), I worked ridiculous hours as did virtually all the engineers

© Springer Nature Switzerland AG 2020
H. Tigelaar, *How Transistor Area Shrank by 1 Million Fold*,
https://doi.org/10.1007/978-3-030-40021-7_13

Fig. 13.1 Larry Hornbeck. Texas Instruments engineer who invented Digital Light Processing (DLP) technology. Dr. Hornbeck was awarded an Oscar in 2105 for his contribution to motion pictures

on my team, and the managers and engineers on associated development teams. Any time day or night, weekdays, weekends, and holidays people would be running equipment in the labs and working in many of their office cubicles. It was impossible to get the job done in a normal workday or workweek. Other than taking a couple weeks vacation (one in summer and one at Christmas), I worked pretty much each and every day. Monday through Friday, I usually attended the morning manufacturing meeting at 7:30 am to get up to speed how processing went on my groups lots the night before and to find out what new problems had come up that needed my groups attention. I'd frequently get to word an hour or so before the 7:30 am meeting and rarely got home before 6:30 pm in the evening. Usually, at least once a week, I would arrive home 10 pm or later—not uncommon to arrive after midnight. On Saturday I would attend the morning meeting at about 7:30 am and work until about 2 pm. I would then meet my wife and her sister and go to a movie, shopping, or? and then we'd go out to dinner. On Sunday after church and lunch, I would go to work for the afternoon. Weekend work was for reviewing data from past week's lots, and planning next week's work. I also used weekends to plan and write manufacturing instructions for design of experiment lots we needed to start. I was not an exception. A great number of engineers and managers spent more time at TI, and more time away from their families than I did. A significant number of marriages did not survive.

13.3   Priority Zero Lots

Before the fab became fully automated and lots were run by humans, normal lots took 2–3 months to complete running 24 h a day, 7 days a week depending upon the number of wiring levels. Priority One lots were tagged and immediately went to the head of the queue when they arrived at a tool to be processed. Priority One lots could be completed in about half the time of a normal lot. Priority Zero (P0) lots were hand carried through the fab by dedicated lot technicians. The next tool in the manufacturing flow was held idle waiting for the P0 lot to arrive. Test runs were run ahead of time to ensure the tool was operating to specifications. A P0 lot usually could be completed in about three to four weeks.

P0 lots were usually reserved to test out new designs for TI's customers. For example, when SUN Microsystems, Inc. sent TI the design for the next-generation servers, TI would work around the clock and rush make the first three reticles, so a P0 lot could be started.

While the P0 lot was running, my group had process integration technicians on three shifts who worked with the manufacturing operators and manufacturing technicians to ensure the P0 lot would wait for nothing. They would notify my engineers a day, a shift, and then by hour before the lot would be going through their process. When a P0 lot was processing through their area of responsibility, the process integration engineers who worked for me would be in the fab, even at 3 am, standing by in case any hiccups occurred. One time a manufacturing supervisor complained to me that my metallization engineer was not in the fab by her machine while the P0 lot was going through metal pattern and etch. I told him I would check into it and get back to him. I knew there must be an explanation since she was one of my best metallization engineers and very conscientious. Turns out the P0 lot was moving so fast through the fab that it was hitting metal pattern and etch about every 4–6 h. My engineer had not been home for over 2 days! Good thing she did not have kids! I got the rules changed to have a metallization process integration technician in the fab when the P0 lot was running and the metallization process integration engineer on call with a pager and with ready access to a phone while the P0 lot was processing through their area. My process integration technicians were exceptionally good and could handle most all metallization issues.

13.4   When Nerds at Play Paid Off Big Time

A couple of Friday afternoons each month in Texas Instrument's (TI's) Semiconductor Process and Device Center (SPDC), the manager of one of the design groups who was planning to use SPDC's next-generation process flow to build the circuits they were designing gave a seminar to inform the SPDC engineers on what circuit they were designing and what were the primary challenges they were facing. The manager of the analog group gave a seminar describing the analog to digital converter (ADC) they were designing to convert the analog signals from hard disk drives into digital signals. These analog signals included music signals so the project was named Mozart. For the music to sound good, the ADC had to convert the analog signal to a minimum number of digital bits. Section 3.10.2 Their challenge was to achieve the required number of bits using the polysilicon/dielectric/polysilicon capacitor in SPDC's next-generation process flow. The problem was that the capacitance of a polysilicon/dielectric/polysilicon capacitor changed with voltage. When a positive voltage is applied to one of the polysilicon capacitor plates, polysilicon grains immediately next to the dielectric become depleted of electrons. These depleted polysilicon grains behave like an additional dielectric thickness causing the capacitance to lessen as the applied voltage increases. This changing capacitance distorts the analog to digital conversion. The designers were having to add very complicated circuitry to accurately measure this nonlinear change in capacitance, so they could apply corrections to the voltages to remove this distortion. A colleague and I were in the audience. We were working on EPROMs and had just added molybdenum silicide ($MoSi_2$) to the EPROM gates to reduce the resistance of the word lines in the EPROM memory. (Without the $MoSi_2$ the EPROM IC was having to wait for the signal to propagate down the EPROM word line). We knew metal/dielectric/metal capacitors have much smaller change in capacitance with voltage than the polysilicon/dielectric/ polysilicon capacitors and thought a MoSi2/dielectric/metal1 capacitor should be much better because $MoSi_2$ behaves as a metal.

These were still early days at Texas Instruments when engineers had a lot of freedom to play and try out new things. My colleague and I decided to build some $MoSi_2$ to metal1 capacitors and give them to the ADC design group to see if these capacitors would work better for them. We knew these custom capacitors could be added to the manufacturing flow we were developing with little added cost. Since we did not know what dielectric (silicon dioxide, silicon nitride, silicon oxynitride, multilayered dielectric) would be best, we ran a couple of wafers with each kind of dielectric. The fab routinely ran capacitor lots to monitor the dielectrics they were depositing in their furnace tubes, so we started a lot using their capacitor reticle set. We told the dielectric process engineers that we wanted them to deposit their dielectric on our $MoSi_2$ capacitor bottom plates. They firmly told us there was no way we were going to contaminate their furnace tubes with metal (Mo) from our wafers. We then asked if they would be willing to run our wafers through their furnaces as the final lot before they shut down their furnace and pulled their furnace tube for cleaning. They said sure, no problem.

We built our $MoSi_2$/dielectric/metal1 capacitors and gave them to the ADC design group for evaluation. We did not hear anything for a couple of months. Later we bumped into one of the ADC design engineers in the hall asked if they had time to evaluate our capacitors. He then questioned if we were sure we could build those capacitors in production without breaking the bank. He said the change in capacitance with voltage on our capacitors was so small they could not measure it using their standard equipment. The had to build custom equipment to measure the capacitance versus voltage repeatedly thousands of times and store these measurements in computer memory. They then took the thousands of measurements and added them together. Since measurement noise is random but the capacitance vs voltage signal is not random, when the thousands of measurements were added together, the random noise signals canceled each other out and the true capacitance vs voltage signal emerged. It was very, very small.

A week or so later, TI's CTO called us together and initiated a program to design and build an ADC using our new MoSi2/dielectric/metal1 capacitor process. It was a super rush job. He wanted it ready for an electronics show in Las Vegas about 6 months later.

We did it and TI's ADC chip with our new capacitor captured the majority of the hard disk drive market. Our capacitor has been used in millions of TI's ICs and is still being used today.

Unfortunately engineers no longer have this kind of freedom to play. Today ideas must be written up and vetted by management teams before permission is given and wafers are allocated to start a lot. I know many good ideas are never tested because of these management and paperwork obstacles.

It took us just a few hours to put together the process flow and start our capacitor lot. We did not have to get approval from anyone. I am rttfairly certain in today's environment we would have figured we did not have the time required to jump through the hoops needed get the lot started and would have dropped it.

13.5   Super Tech Nerds

These guys are rare—very rare. I met two in my career. They know virtually everything about their technology.

In 2000 Jean-Luc Pelloie taught me partially depleted silicon on insulator (PDSOI) technology at Laboratoire d'Electronique des Technologies de l'Information (LETI) in Grenoble, France. He is an expert in SOI manufacturing flows, SOI manufacturing equipment, and SOI processing recipes. He is an expert on SOI design rules. He is an expert in the device physics of SOI transistors and other SOI devices. He is an expert in designing, laying out, and modeling SOI integrated circuits. He is a one stop shop for SOI, for partially depleted SOI, and for fully depleted SOI. This one guy knows more than dozens of Ph.Ds. He designed and characterized a PDSOI cell library of subcircuits for SOI designers. He demonstrated he could save the cost of the expensive SOI starting material because his circuits are smaller, so more chips per wafer. His PDSOI circuits gave higher performance than competitor PDSOI circuits. Other PDSOI design groups relaxed design tolerances, so they could ignore floating body effects and have their circuits still work. Relaxing design tolerances lowers IC performance. Jean-Luc Pelloie exploited floating body effects to switch his PDSOI transistors faster significantly boosting the performance of his PDSOI circuits over that of PDSOI designs with relaxed design tolerances.

I worked with a second super tech nerd while consulting for a major semiconductor company. His expertise is photonics. He is an expert in photonic materials, photonic devices, photonic manufacturing flows, photonic manufacturing equipment, and photonic processing recipes. He is an expert in photonic design rules. He is an expert in the device physics of photonic switches, couplers, lasers, and other photonic devices. He is an expert in designing, laying out, and modeling photonic integrated circuits. He writes is own programs to model photonic devices and to model photonic integrated circuits. He is a one stop shop for photonic technology. This one guy knows more than dozens of Ph.Ds.

He built a photonic processor that in parallel processed over a dozen RF signals at speeds of exceeding 100 GHz before his project was shut down. He designed a process flow and photonic circuit for an optical computer that could operate above 100 THz but has not found funding to build it. (CMOS computers are limited to about 10 GHz).

Because these super nerds are so unique, they are also very rare. It is almost a crime against humanity when their unique talents are ignored and unused. These two super tech nerds appear to be ordinary people. Neither of them is in any way weird (except for how animated they get when talking about their topic of interest.)

13.6   Performance of PDSOI Circuits Versus Conventional CMOS

After I learned PDSOI technology at LETI in Grenoble France, I ran a split lot in Texas Instruments Develop lab with half the lot PDSOI and half the lot conventional CMOS. Except for shallow trench isolation on the conventional CMOS wafers and SOI isolation on the PDSOI wafers, the processing was identical. Performance on delay chains was improved by 25% on the PDSOI wafers when the same power supply voltage was used. When the power supply voltage on the PDSOI wafers was lowered to give the PDSOI and conventional CMOS wafers equal performance, the power consumed by the PDSOI wafers was more than 30% lower.

One barrier to PDSOI is the high cost of SOI starting wafers. This made PDSOI ICs more expensive than conventional (bulk) CMOS ICs. Because WELLs are not needed in SOI ICs, SOI IC area is smaller than bulk CMOS IC area. More IC chips per wafer with PDSOI ICs than bulk ICs for the same circuit only partially compensates for the added starting material cost. A colleague in the PDSOI group designed a low-cost manufacturing flow that ended up making PDSOI ICs lower cost than bulk CMOS ICs. Because PDSOI ICs do not have WELLS and because the turn on voltages of CMOS transistors can be adjusted by implanting through the transistor gates, his flow eliminated all the WELL patterns and implants and all the turn on voltage patterning steps found in bulk CMOS flows. After he formed the SOI isolation, he formed the transistor gates. He adjusted the turn ON voltages of his various transistors by implanting the turn ON voltage (V_T) adjust dopant along with the source/drain extension dopants. Eliminating all the WELL and V_T patterning steps plus the smaller footprint of SOI circuits reduced the cost of PDSOI wafers below that of bulk CMOS. In addition, at the same technology node PDSOI provided a significant boost in IC power and performance.

PDSOI did not catch on at Texas Instruments. TI bulk CMOS designers invented new design techniques to reduce power. (Shut down subcircuits when not in use. Reduce voltage on memory arrays not in use, etc...) One design manager told me he resisted PDSOI because he did not want to take the time nor expend the effort needed to retrain himself and his design team and to come up the PDSOI design learning curve.

13.7   Semiconductor Industry Job Insecurity

The Semiconductor Industry had many ups and downs. They often occurred with little forewarning. Texas Instruments (TI) had a policy that they would not lay off new hires from Universities for 2 years. Before this policy was instituted, TI was having difficulty recruiting. In May of 1982, I was hired into TI's Stafford Plant near Houston, TX, during a hiring burst. By Aug. 1982, because of a business down turn, TI was laying off engineers right and left. Since I was one of the last engineers hired in and was not hired from a university, I was supposed to be one of the first engineers out the door. I had a wife and two small children. Luckily, TI needed a minimum of two metallization engineers and I was metallization engineer number 2. I dodged my first layoff bullet 3 months after I was hired. In Oct. 1982, just two months later, TI decided to shut down the department I was in and move its function to Dallas, TX. I was given 1 month to find another job inside of TI before I was laid off. I had a wife and two small kids. Luckily, the more senior metallization engineer turned down his transfer to Dallas. I jumped at the opportunity, transferred to Dallas and dodged my second layoff bullet within the first year that I was hired.

There were many more up and down business cycles with hiring bursts and lay-off bursts. Luckily I advanced sufficiently to be relatively insulated from layoff. Unluckily, I advanced sufficiently to be the one laying my engineers off. The layoff that still haunts me is the excellent female engineer I laid off. Of all the jobs in my group, hers was the least critical. She cried in my office. She had just found out she was pregnant and her health benefits would run out before she gave birth. Luckily she was able to find another job in TI before her layoff became final.

13.8   A Very Short Career at Texas Instruments

Shortly after I hired into Texas Instruments in Stafford, TX, a gate dielectric problem killed several lots. The gate dielectric engineers were mystified. They checked their gate dielectric furnace tubes daily by building gate dielectric capacitors and testing them. After the failed lots, they built more capacitors, tested them, and found the dielectric from their furnace tube was crap.

They shut down the furnace, pulled the furnace tube, and chemically etched the tube to clean it. They reinstalled the newly cleaned tube, ran a dielectric capacitor test lot, and found the dielectric was excellent.

Next day a couple of additional lots were scrapped due to poor dielectric. The dielectric engineers retested the furnace and again found their dielectric was crap. Panic. Dielectric engineers had no clue as to what was going wrong.

That night on third shift, a newly hired technician called another tech to come over and see something really cool. He then shot a rubber band into the open end of the gate dielectric furnace tube. The furnace tube was heated to 900+ °C. Just a few feet into the furnace tube, the rubber band disappeared in a flash of light and a puff of smoke. Apparently it looked really cool.

Within minutes the newly hired technician was fired and walked to the door. His very short career at TI ended.

13.9   Still Needed Test Equipment

US patent 5410162 [1] for this piece of test equipment was issued in 1995, so the 20-year patent protection has now expired. This equipment is still needed in the semiconductor industry. Someone should build it.

Activation energies are frequently measured to characterize and troubleshoot diode defects in integrated circuits. This is currently done using a heated wafer chuck. The reverse diode leakage current is measured at a series of temperatures from room temperature to 250 °C or more. The slope of reverse leakage current versus the inverse of absolute temperature is the activation energy (E_A).

The current procedure is to place the wafer on a resistively heated/water cooled wafer chuck. Reverse leakage is measured, the chuck heats the wafer to the next temperature and reverse diode leakage current is again measured when the temperature has stabilized. This is repeated for each temperature. One cycle of activation energy measurements at various temperatures and then returning the chuck to room temperature, so the next activation can be measured takes approximately an hour.

USP 5410162 describes using rapid thermal processing (RTP) to heat the wafer to various temperatures and take reverse diode leakage measurements. With an RTP activation energy tool, activation energies could be measured in less than a minute instead of almost an hour. With an RTO activation energy tool, activation energies could be measured inline by inserting an activation energy measurement step in the manufacturing flow instead of having to remove wafers to measure activation energy offline.

13.10 Exploited Technical Artists

There are tens of thousands of highly educated and highly trained technical workers in the semiconductor industry as well as other technical industries such as the chemical, plastics, medical, oil and gas, and automotive. A small subset of these tens of thousands of scientists and technical workers are creative scientists and creative technical artists that invent new materials, new machines, new devices, new technologies, and new toys. Of this small subset, an even smaller subset of creative scientists and technical workers hit a home run and create something that is a hit in the marketplace. A creation that generates millions or even billions of dollars.

One hit song can set an artist up for life. An actor in a hit movie or TV series can be set for life. A creative scientist or technical artist has to retain ownership of their creation and start their own business if they are to reap significant rewards. Few technical inventors are also successful entrepreneurs.

Jack Kilby was a super star technical artist of the highest order. He invented the integrated circuit. With Jack Merryman, in 1967 Jack Kilby invented the hand-held calculator Jack Kilby patented many other inventions (Fig. 13.2). His ICs generated

Fig. 13.2 Jack Kilby. Texas. Instruments engineer who invented the integrated circuit in 1958 and the hand calculator in 1967. Dr. Kilby was awarded the Nobel Prize in Physics in 2000

Fig. 13.3 A 300 mm (12 in.) wafer signed by Jack Kilby. Each small square is a separate integrated circuit

trillions of dollars in wealth, spawned thousands of new technologies, thousands of new industries, and raised the standard of living across the world. With just a small ownership in his creations, he would have become the richest man in the world. Others benefited far more from his creativity than did he. Dr. Kilby (Jack) was kind enough to autograph a 300 mm wafer for me after being awarded his Nobel Prize. (Fig. 13.3).

The scientist or technical star that creates a hit patent may get a bonus or promotion, but seldom is set for life. Employment contracts assign 100% of the patent rights to the company. The justification is that company equipment and company materials are used by the scientist or technical artist to invent. Companies often reap millions or billions, while the scientist or technical star receives the satisfaction of seeing her or his creation being successful in the marketplace plus a token bonus.

Unless things have changed, employment contracts assign ownership of all inventions by the employee to the company—even inventions that have nothing to do with the company business. When I was a relatively new employee, I changed the oil and performed repairs on my own car. When changing oil, I got tired of my hand and wrench getting bathed in oil as I removed the drain plug from the oil pan. I invented a new drain plug with a hollow center so when I loosened the plug oil flowed out through the middle of the drain plug and not over my hand. I asked the company attorney if my employment agreement allowed me to patent my invention and sell it through car parts stores and K-Mart. He first told me I would need a release and then declined to grant me one.

13.11   Antifuse Technology Development Successful and Sold

Texas Instruments had a joint development project with Actel Corporation to develop a metal-to-metal antifuse for field programmable gate arrays (FPGAs). A thin dielectric electrically isolated the intersection between two interconnect wires. A high voltage could be used to breakdown the dielectric forming a permanent electrical connection between the two interconnect wires (antifuse).

I was project leader for a joint antifuse development team of TI engineers and Actel engineers. Although we missed the target delivery date by a few months, in the end we met all the project goals. Just as we were ready to transfer the antifuse process into a TI manufacturing fab, TI decided not to participate in the field programmable gate array (FPGA) business. The project was shut down and sold to Actel. Actel used this technology to manufacture and sell FPGA's for many years. My team had to transition from transferring the technology we had just developed into a TI fab and teaching TI manufacturing how to make it to transferring our technology to Actel engineers and teaching them how to make it. Very disappointing for me and all the TI'ers on the antifuse team.

The reason we missed our target delivery date by several months is that a small percentage of antifuses failed long-term stability testing. The resistance on the small percentage of blown antifuses inexplicably jumped several orders of magnitude at seemingly random time intervals during extended reliability testing. This was a product killer.

One of the Actel engineers on our team noticed that when these failures occurred, usually several antifuses simultaneously failed. Strange. The failures not random, so he looked for other possibilities. He discovered the failures exactly correlated with thunderstorms passing through Dallas. Although we had electrical isolation between our testing equipment and the house power, the isolation was not sufficient for antifuses. Once we beefed up the electrical isolation, our antifuses passed long-term stability testing with flying colors.

13.12   Managing Tech Nerds

Highly educated and really smart tech. nerds are really not all that hard to manage. Most of them work hard and are self-motivated. The major management task is to make sure they clearly understand their goals and clearly understand their projects timetable.

You tell them what to do and when it needs to be done and then step back and give them the freedom to get the job done their way. Along the way you make sure they have the tools and information readily available, so they can focus on doing their task instead of getting distracted and wasting time rounding up resources and information. A wise manager once taught me. Make sure your workers can get their job done as easily as possible. Give your workers as few excuses as possible not to do their job.

Not invented here (NIH) syndrome is a common problem with tech. nerds. They tend to want to do things their way. One admonition I gave to the tech. nerds in my group was, "We have goals to meet and not much time to meet them. If you can steal an existing idea to meet your goal, I want you to steal it and get the job done as quickly as you can. Do not forget to give credit to the person whose idea you stole. I'll not be happy if I find you wasted time re inventing the wheel."

13.13   One Tech Nerd I Failed to Manage

For the anitfuse development program (see Sect. 13.12), I hired a brilliant new graduate from Stanford, Berkley, or MIT. I do not remember which. He was my lead scientist/engineer to develop the metal-to-metal antifuse. The antifuse link was a plug of amorphous silicon between upper and lower metal wires. When we put a sufficiently high voltage across the amorphous silicon, it broke down electrically forming a low-resistance conductive filament. The resistance between the wires changed from many mega ohms to less than 100 Ω.

We had a problem. When we ran current through the antifuse filament for weeks to test long-term stability, the resistance of a few hundred of them out of tens of millions would suddenly switch from a few ohms to a few mega ohms. Not good. My brilliant engineer discovered we were forming a chalcogenide and that chalcogenides could recrystallize under current stress causing the resistance to change. I congratulated him for figuring out the root cause of our problem and then told him his next task was to fix it.

A few weeks later, he came into my office and informed me in no uncertain terms that I was wasting his time and squandering TI's money. He said trying to change a fundamental property of chalcogenides was like trying to change lead into gold. He gave up and stopped trying.

I assigned the problem to another engineer who was not so brilliant and was not from so prestigious a university. She ended up adding a metal layer on top of the amorphous silicon that melted when the antifuse was blown and formed silicide with the silicon in the chalcogenide filament. Her silicided filament eliminated the resistance switching problem. Her solution was used to manufacture millions of field programmable gate array (FPGA) chips.

13.14   Rapid Learning and Change Control

Experiments usually take a month or more to process through the fab and get data. To speed things up, partially processed lots are staged at key processing steps waiting for data to be fed back. For example, if an engineer is working to nail down transistor extensions, she or he will hold a lot just before extension implants and wait for electrical data from a previous lot processed through metal1. As soon as the feedback data is available, new extension implant conditions are defined and the staged lot is released. To start a new lot from scratch would delay learning by 2–3 weeks.

During development changes can be made to the baseline process flow immediately. Once the baseline flow is transferred to manufacturing, extensive data must be collected from multiple full flow lots to verify the baseline change will not negatively impact circuit performance or yield.

During development we would sometimes make baseline changes based upon data from only a couple of wafers in one split lot. Running verification lots prior to making the baseline change would delay process development by weeks. Making the baseline change immediately was the fastest route to obtain feedback data either affirming or rejecting the change. Using this method, I remember of only two times when we got burned and had to go back to the old baseline flow.

Whenever we made a baseline change, our rule was to run at least three split lots where the only split was half the wafers were old baseline and half the wafers were new baseline.

This saved our bacon numerous times. One baseline change we made was to reduce a wet clean time by about 10 s after source/drain extension implant pattern strip to reduce the amount of damaged oxide we were removing. We implanted the source/drain extension dopant through thin deposited amorphous oxide layer. The thin amorphous oxide layer prevented dopant channeling. The implant energy we were using was low to produce shallower extensions. A significant amount of the dopant got trapped in the amorphous oxide layer. We reduced the cleanup time to reduce the amount of dopant atoms we were stripping off with the oxide we were removing. Seemed like a no brainer.

We made additional baseline changes and ran additional split lots between old vs new baseline. Our yield inexplicably dropped 5–7% due to SRAM single bit failures. We had no clue why.

When the lots with old baseline vs new baseline with 10 s. less wet clean came out, there was a clear 5–7% lower SRAM yield on the new baseline with 10 s. less wet clean split. We added back the 10 s. to the wet clean and the SRAM yield went back up. Why? We still haven't a clue.

13.15   Engineer versus Technician

Engineers from product development groups would start and run lots to test out new circuit ideas and calibrate their circuit simulators. They would use the next-generation technology manufacturing flow my group was in the process of developing.

Process integration (PI) technicians in my group (3 shifts, 7 days a week) inspected all lots being processed with our process flow. One of my best PI technicians was a lady with red hair who at one time had been a marine officer. She tended speak her mind.

A product engineer (from the middle east) stormed into my office demanding I immediately fire her for refusing to follow his instructions and for disrespecting him. He said she was insisting his wafers had a problem. He insisted she was wrong. He insisted that it was imposible for his wafers to have the problem she was claiming.

I persuaded them to a truce while we all went into the fab and examined his wafers to determine what if anything was going on. Upon careful examination using a high-power microscope, it turned out his wafers did have a problem but for a different reason than my PI technician thought. Had she followed his instructions, his lot would have been scrapped and over a month of his work would have been down the drain. He and the tech were able to rework and save his lot. He was happy and thankful she had caught the problem. Within a month or so the engineer and PI tech. had become fast friends.

13.16   Customer Crisis Averted

For many technology nodes Texas Instruments developed a custom high-performance manufacturing flow and produced next-generation servers for SUN Microsystems. SUN would send us their reticle set and we would fabricate their chips as quickly as possible, so they could characterize, debug, and tweak their design before sending us their production reticle set.

On one such occasion, the first lots out had zero yield. Failure analysis determined closely spaced aluminum wires on the upper-level metal were shorted.

Turned out SUN designs required thick aluminum to carry high current. Texas Instruments (TI) designs used thinner aluminum with smaller spacing between the aluminum wires. The design rules for SUNs designs were larger than the design rules for TI designs to accomodate the thicker aluminum. A SUN design engineer got permission from someone in TIs design group to use TIs thin aluminum design rules on SUNs thick aluminum designs. No one bothered to check with the process and process integration engineers in my group.

Huge problem—SUNs ICs had zero yield. SUN did not have time to re layout their server chip with the correct design rules—it would have taken months. In addition, the area of SUNs IC chips with widely spaced wires would have been far too large. The TI metallization process and process integration engineers pulled a rabbit out of the hat. They tweaked and retuned their metallization processes and were able to achieve sufficient yield on the second SUN lot so SUN could test and debug their circuit design.

After a huge amount of work, the metallization equipment, process, and process integration engineers were able to develop a metal etching process with satisfactory yield in manufacturing. The aluminum patterns had to be reworked more than was desirable and the metal etch had to be run with tightened specifications, but it worked. A manufacturing process with acceptable yield for SUN's server chips was achieved.

13.17   Reliability Failures on a Qualified Manufacturing Flow

My group had developed a new technology manufacturing flow and transferred the flow into manufacturing. Several products were being manufactured and sold. A new product with a much larger die was manufactured and sold. A few of the chips started failing in customer products and being returned. Customers were very unhappy. The new larger die had millions more vias than previous dies. A via defective rate of a few defective vias per million killed many more of these new dies.

I was traveling to Europe to start work on a new project when the doodoo hit the fan. They tracked my wife and me down in a hotel in Paris and told us get our butts back to Dallas as soon as possible and fix the problem.

A team that met every morning and every afternoon 7 days a week was formed to review data and plan the next steps. We solved the problem in less than 1 month and I went back to Europe to resume my new project.

Patrick Fernandez, a colleague and friend, did an in-depth comparison of equipment and process conditions between the fab that developed and qualified the process and the manufacturing fab into which the process was transferred. He tabulated all differences between the development and manufacturing fabs processes and equipment and helped assess if any of the differences could be contributing to our problem. Two of his key findings were: (1) the base pressure (vacuum level) in the manufacturing fab metal deposition tool was a couple of orders of magnitude higher than the base pressure in the development deposition tool, and (2) the degas temperature in the manufacturing via liner deposition tool spiked higher than in the development tool.

The aluminum wires needed a thin continuous top layer of titanium aluminide (Ti_3Al) for the vias to be reliable. During manufacturing, a thin layer of titanium is deposited on the aluminum and then heated to react the titanium with the aluminum to form the Ti_3Al. Because of the higher base pressure (lower vacuum) in the manufacturing titanium deposition tool, sufficient residual gas remained to form a monolayer film on the aluminum surface. This monolayer prevented a continuous layer of Ti_3Al from forming uniformly on the surface of the aluminum.

An intermetal dielectric layer (IMD) is then formed on the aluminum wires and via holes are etched down through the IMD layer. A titanium/titanium nitride barrier layer is first deposited to coat the inside of the via hole before the via hole is filled with chemical vapor-deposited tungsten (CVD-W). Immediately prior to the barrier layer deposition, the wafer is heated for a few seconds to drive off (degas) any volatile molecules. In the manufacturing tool, because of inadequate control, the temperature overshot the target value for a few milliseconds before settling down to target temperature. Aluminum expands when heated. The high-temperature spike caused the aluminum to expand, break through the thin Ti_3Al layer, and push up into the open via holes. The titanium barrier layer was then deposited directly on aluminum extruded into the open via holes instead of being deposited on the Ti_3Al layer. During subsequent anneals, this barrier titanium reacted with the extruded aluminum forming Ti$_3$Al. Since the titanium plus aluminum shrinks as it forms Ti$_3$Al, this reaction causes the via plug to be under high tensile stress or even form cracks where the Ti$_3$Al is formed. Vias with Ti$_3$Al formed in the via plugs were failing in the customer's electronic equipment. The combination of requiring lower base pressure in the titanium sputtering tool and preventing temperature overshoot in the degas chamber solved our reliability problem.

13.18   Texas Instruments Training Institute and Semiconductors in Asia

A common joke in the semiconductor industry is that TI stands for Training Institute. At some point in their career multiple engineers from a majority of the fabs in the world worked at Texas Instruments. After I was laid off by TI in 2007, I worked for a few years at PDF Solutions, Inc. and visited many fabs around the world. Almost always I would run into multiple ex TI colleagues.

Texas Instruments (TI) was very influential in spreading semiconductor technology worldwide and especially in Asia. TI was the first western company to build a semiconductor manufacturing factory in Japan (Miho in 1968). TI built joint venture manufacturing fabs in Italy (TI Avezzano, now L Foundry), France (TI Niece), Germany (TI Freising), Singapore (TI-Singapore now Tech Singapore), Taiwan (TI Acer, now Acer). In addition, a number of semiconductor wafer fabs such as UMC, TSMC, and Dallas Semiconductor were started by former TI'ers.

It was common for engineers to move from TI to other companies and back again. At one time, I knew of four husband and wife engineers where the husband engineer worked in Taiwan at TSMC or UMC and the wife engineer worked in Dallas at TI. The husband of one of the couples had been one of TI's top technical managers. He left TI to become TSMC's Chief Technical Officer (CTO). His wife stayed in Dallas at TI and managed the group that developed next-generation technology in TI's advanced process development lab.

13.19   GUI Interface to Run Split Lots in Manufacturing

Split lots in which different sets of wafers would be run through different processes were constantly run in Texas Instruments (TIs) development fabs. Engineers using a standard format would write down detailed manufacturing instructions and print out a copy on lint-free paper. This copy, called a traveler, accompanied the split lot as it was processed through the manufacturing fab. Manufacturing operators and process engineers would read the instructions from the traveler, write custom process recipes if needed, and process the lot. Lot technicians would make post process measurements (line width, film thickness, resistance, drive current, etc.) and write the measurements on the lint-free paper traveler.

Using this method, a significant number of lots were processed incorrectly, a significant number of important measurements were missed, and a significant amount of lot information was not recorded correctly. The development fab manager finally edicted that from now on all split lots had to be run using the same online computerized system used for running baseline lots.

This was a huge problem for transistor development engineers. While it took days for engineers to get their proposed experiments signed off by the process engineers and to generate the paper traveler with detailed process instructions, it would have taken many weeks to generate an online traveler. A new process recipe for each new process condition would have to be generated on the manufacturing computer, so it could be downloaded when the lot was logged into that processing step. New data templates would have to be generated on the manufacturing computer to receive measurements from online metrology tools. Downloadable recipes to generate sublots from the parent lot, so the sublots could be processed differently had to be written. Online split lot travelers were an enormous amount of work.

Our process development groups insisted we did not have the resources to generate online travelers for split lots. The manufacturing manager insisted he would no longer process split lots offline using paper travelers.

We went to the IT department and requested they develop a software program system to help transistor development engineers generate online split lot travelers. TI's computerized online manufacturing flow used an SMS system with very restrictive data input. The curser had to be positioned at exactly the correct position on the computer screen and data had to be entered in exactly the correct format. We had a meeting with the IT manager and their top programmers. They declared it could not be done and sent us on our way.

A programmer who was an expert on TI's SMS system as well as several other programming languages was on loan to our development group to help with another software project. I believe his name was Craig Decker. We discussed the problem with him several times. He also told us he did not think it was possible. One day, months later, he came into my office all excited and said, "I figured out how to do it! I can write software to do what you need." We quickly wrote a detailed specification for the split lot generation program. We decided focus on a dopant implant split lot first and then expand the program later to also generate travelers for other process splits (film thicknesses, gate lengths, anneal times, anneal temperatures, etc.)

Craig reported what he was doing to his boss—the IT manager who said it was impossible. The IT manager immediately pulled him back from our group into her IT group and started a new project to generate the split lot software program. The project was to generate generic split lot software for all possible process splits and for all the wafer fabs at Texas Instruments. (No small task since each TI fab had its own customized version of SMS.)

The project took a few years to complete and was a huge success. The travelers for all split lots are now generated using the program made possible by this one programmer. This program has been worth millions to TI in terms of increased engineer productivity, reduced misprocessed split lots, and improved split lot cycle time.

This is but one example of a huge contribution made by one TI employee who saw a need and took it upon himself or herself to address the need. This employee made an enormous impact but received little to no recognition. There are an uncountable number of similar stories.

References

1. H. Tigelaar, M. Moslehi, Apparatus for and method of rapid testing of semiconductor components at elevated temperature, USP 5410162, 1995, https://patentimages.storage.googleapis.com/96/de/7a/d3e50d4bdf191b/US5410162.pdf
2. Texas Instruments Fellow Larry J. Hornbeck, PhD, wins the Oscar, Ciston, PR Newswire, 9 Feb. 2015; https://www.prnewswire.com/news-releases/texas-instruments-fellow-larry-j-hornbeck-phd-wins-the-oscar-300032643.html

Chapter 14
Wrap Up and What Is Next?

14.1 Wrap Up

The ongoing IC revolution changed and continues to change the world. Although transistor scaling has slowed and almost stopped, the demand for ICs continues to expand and grow. New multibillion dollar semiconductor wafer fabs continue to be built for the increased demand.

ICs scaled rapidly for over 40 years. Every 2–3 years, the area required to build a transistor was cut in half. Over 40 years the area required to build a transistor was reduced by over 1 million. Simultaneously the switching speed of transistors increased over 20,000 times. The combination of billions more transistors switching at higher speeds made IC computer chips vastly more powerful. The cell phone in your back pocket has vastly more computing power than your company's mainframe had just a couple of decades ago.

The transistor development and IC design engineers pushed the process equipment and the computers to their extreme limits when developing each new technology node. The process equipment and computer power that was barely adequate to develop the next technology node was totally inadequate to develop the following one. New equipment with new processes plus more powerful computers were absolutely required. Not just a few times, but many times scientists and engineers invented completely new technologies in a wide variety of technical fields to keep scaling going. Not only did they invent these new technologies but they also built equipment that could manufacture these miniscule transistor geometries with precision and uniformity and could be operated by manufacturing personnel, some with little more than a high school education. At each new technology node, it took the successful completion of thousands of tasks, most of them being performed for the very first time, to manufacture the first IC chip for the new node. Thousands of highly technical tasks, all had to be performed and all had to work together flawlessly for the ICs to function correctly. The teamwork required between the thousands of tech nerds to build the first IC on a new technology is mind-blowing. The

© Springer Nature Switzerland AG 2020
H. Tigelaar, *How Transistor Area Shrank by 1 Million Fold*,
https://doi.org/10.1007/978-3-030-40021-7_14

coordination and cooperation between the hundreds of teams and thousands of scientists and engineers to continue to invent and develop new technologies for more than four decades is beyond impressive—it is miraculous.

The enormous profits generated by IC scaling made this miracle possible. Every year hundreds of millions of dollars were reinvested by multiple IC manufacturers to develop the revolutionary new technologies and equipment needed for scaling. Hundreds of millions of dollars were given in grants to professors and Universities to address forseeable problems for upcoming new technology nodes. Typically, multiple approaches were investigated to address each technical challenge. Multiple semiconductor companies and multiple semiconductor equipment companies would invest millions of dollars in their unique approaches. Texas Instruments built some next-generation equipment internally, jointly developed some with semiconductor equipment companies, or upgraded existing tools to meet the more stringent requirements. At each new technology node equipment was purchased from multiple venders to perform the same thin film deposition or etch. Lithography tools from Nikon, Cannon, and ASML were purchased and used throughout development. Usually one vendor was chosen for manufacturing when the baseline flow was qualified.

14.2   What Is Next: AI

A new robotic revolution is underway. Robots with the ability to make intelligent decisions as problems or new situations arise are being developed. Some fear a massive increase in unemployment with a corresponding increase in the welfare rolls.

I am more optimistic. There have been many workplace revolutions in the past. Instead of generating more out of work people, typically these revolutions generated a higher standard of living and generated new unanticipated jobs demanding higher quality workers and paying higher wages.

The agricultural mechanized farm work revolution raised the standard of living by providing less expensive, more plentiful and higher quality food. Displaced farm workers provided the workforce needed in factories to manufacture new cars, appliances, and other labor saving devices as the economy expanded.

The automobile transportation revolution raised the standard of living by providing cost-effective long-range transportation to common people. The horse stables and carriage shops that went out of business were replaced with automobile dealers and service stations. The anticipated horse manure crisis in cities was averted. The auto revolution generated a huge number of new businesses to manufacture and distribute automobile parts and accessories. It gave birth to the trucking industry. It generated thousands of construction jobs to make roads and highways for this new mode of transportation. Refineries and gas stations were spin-offs of the auto revolution as was NASCAR and the Indy 500.

When robots were first introduced to manufacture cars, union bosses fought against them fearing job losses and fewer union workers. Robotic car manufacturing moved to Japan virtually shutting down car making in Detroit and causing massive job losses. Detroit was devastated. Had USA auto manufacturers embraced robotic manufacture, Detroit may have been spared. Robotic automobile manufacture increased the standard of living worldwide. The robotic manufactured cars are higher quality, last longer, are more energy-efficient and are more affordable. Most families in the US transitioned from one car families in the 1950s to two and even three car families.

Many other revolutions have taken place and have devastated some industries but at the same time generated new ones. Digital photography devastated the film industry but increased the standard of living by allowing families and friends to instantly take and share photographs. Digital pictures and videos have generated software programs to edit and enhance. Artists use software programs to generate art, video games, cartoons, and movie special effects. Although the film cameras and film manufacturers took a big hit, the photographic paper industry on which digital pictures are printed is better than ever.

I believe robots with artificial intelligence will put some employees with repetitive jobs out of work, but I also believe many jobs we currently do not anticipate will be created and take up the slack.

14.3   What Is Next: ICs and Electronic Devices

The IC revolution is going strong. IC chips with more and more capability are being manufactured. Multibillion-dollar semiconductor factories are being built at a record pace. The more powerful and more cost-effective ICs enable the design and manufacture of more affordable electronic devices and toys.

On the horizon are medical devices that monitor a person's health real time and immediately warn the person and possibly emergency services of a heart attack, insulin shock, or some other medical emergency. On the horizon are medical devices that constantly monitor a person's health real time and advise the person on what supplements to take, what food to eat, and how to exercise to maintain good health. On the horizon are virtual reality systems that allow sports fans to virtually be in an audience at a game and allow parents to virtually watch their child play sports. On the horizon are virtual reality systems that allow music lovers to virtually attend concerts—alone or next to a friend who also has the virtual reality system.

On the horizon are point-of-care medical diagnostic systems that can diagnose an illness on the spot from saliva, urine, breath, or a drop of blood. On the horizon are inexpensive DNA tests that farmers can use to ensure their crops are salmonella-free or that the red snapper you ordered at a restaurant is really red snapper and not some other fish masquerading as red snapper.

On the horizon are trucks and automobiles that drive themselves.

On the horizon are sensors in the walls of your house to immediately raise the alarm when excessive moisture or heat is detected—sensors on appliances to shut them down when a leak occurs.

On the horizon are wristbands for your young children or elderly parents, so you know where they are and know they are well.

On the horizon are public libraries you can log into from home and download music, videos, movies, concerts, documentaries, and eBooks and get help from a professional librarian in performing literature and patent searches.

On the horizon are businesses that operate from virtual offices with virtual conference rooms. Employees with virtual reality systems will interact with each other as if they are all in the same conference room although they are in their own homes miles or cities or continents apart.

All these and all those not yet thought of need ICs. The demand is not likely to dissipate soon. The IC revolution is likely to keep going and even accelerate as new applications are imagined and implemented.

14.4   What Is Next: Photonic Computers

Virtually all semiconductor companies were started by tech nerds with a vision of what could be. They were willing to gamble fortunes to follow their vision and make it happen. As semiconductor companies grew and matured, the technical leaders with technical vision were replaced by managers with business and financial backgrounds. While these managers continued the scaling that kept the semiconductor industry profitable, they for the most part stopped taking the big gamble on the next breakthrough technologies. Because the gate dielectrics in CMOS transistors are now only a few molecules thick and the gate length of CMOS transistors only a few molecules wide, CMOS scaling has slowed down almost to a halt. Most semiconductor companies have shut down CMOS scaling research and development. It is too expensive—too risky. The chance of getting a decent return on investment is greatly diminished. Only companies who sell huge volumes of one product such as computer CPUs or memory chips can recoup the enormous R&D investments required to continue scaling. R&D still continues at many companies, but has been refocused from making ICs smaller and faster, to reducing the cost of packaging and to increasing functionality (multichip packages), to developing new sensor technologies, to developing Micro Electrical Mechanical (MEMS) devices, and to developing transistors capable of switching higher power and higher voltages.

RC delay and heat generation in digital IC chips now limits how much faster digital computing can be pushed. Digital computing speed is limited to less than about 10 GHz (billion cycles per second). Optical computation technology (photonic computers) with potential for computing speeds over 1000 times faster (100+ THz—trillion cycles per second) has the potential of being the next leap forward. Optical computers are not only feasible but some are already being built. (MIT Optical Chip [1]; Light Matter—Photonic Artificial Intelligence [2, 3]). Photonic computers do not have the heat-generation problem of CMOS computers. Optical computers can process multiple signals in parallel because unlike electrical pulses in wires light pulses in waveguides to not interact with each other.

I believe whoever develops photonic computers first could dominate artificial intelligence. I am concerned universities do not have the resources or expertise required to develop photonic computers and get them to market. While universities may be able to build prototypes and demonstrate feasibility, universities do not have the knowledge, expertise, and financial capability required to build the infrastructure needed to design and manufacture photonic computers in the volume needed to jump start a new market in photonic computing. CMOS semiconductor companies who already have the required resources and infrastructure do not seem to be interested taking the gamble to develop a new breakthrough technology—especially a technology with the potential to displace their current profitable technology. Semiconductor equipment companies who are profitable making and selling equipment for the most advanced CMOS technology nodes do not seem to be interested in funding the development of a new photonic computing technology that can be manufactured using less profitable previous generation technology tools (65, 90, and 130 nm). One problem that could limit optical computing is the size of optical switches compared with the size of CMOS switches. Currently optical switches are about 10 times larger than CMOS switches. I do not know how much effort has gone into scaling optical switches. Maybe optical computer switching speeds over 1,000 times faster than digital switching speed can more than make up for the difference in the size of the switches.

I believe a major photonic computing project funded by the government would be required if the USA is to develop photonic computers first.

I am an optimist. I believe if the right team of engineers from industry and academia were assembled, given access to a 65 nm or 90 nm semiconductor fab, and funded with about $30 million, they could prototype a manufacturable photonic computer in less than 5 years. (Especially if the team included the photonic super tech nerd—Sect. 13.6.) I hope a group of entrepreneurs with the vision for the potential for photonic computing (like the vision the founders of Texas Instruments had for the potential of silicon-based transistor switches) assembles and funds such a team. Alternatively, I hope our government initiates a program to develop a commercially viable photonic computer before our competition does. A marketing manager at a top semiconductor company estimated the potential of optical computing to be in excess of $3 billion.

I believe we can do it with the right team and the right level of funding in 5 years or less. I would love to be proven right. On the other hand, I want to find out if I am wrong.

14.5   What Is Next: 3-D Integration in Silicon

Since transistor geometries have gotten so small that area (2D) scaling has virtually stopped, stacking transistors one on top of the other to pack more transistors on the chip is being aggressively pursued. The most successful implementation of this concept is 3D NAND FLASH nonvolatile memories made by such companies as Intel, Micron, Toshiba, Samsung, and Hynix. Solid State hard drives with one and two terabytes (1 terabyte = 1 trillion bytes = 8 trillion bits = 8 trillion FLASH transistors) can readily be purchased. These 3D NAND memories have 32, 48, and 64 layers of flash memory cells stacked one on top of the other [4, 5] (Figs. 14.1 and 14.2). 3D NAND memories with 128 layers or more of flash memory cells will be available soon.

Fig. 14.1 Cross-sectional
drawing of Samsung 3D NAND
memory array [4]

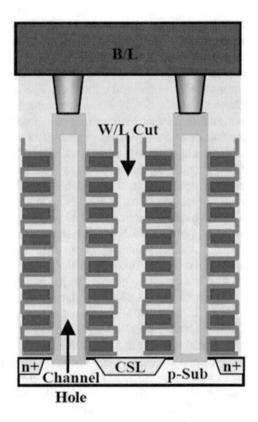

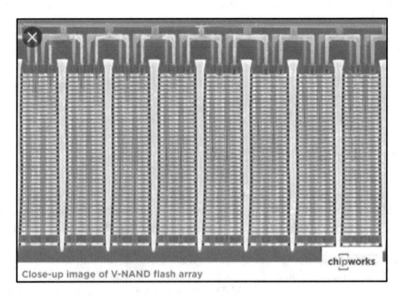

Fig. 14.2 Cross-sectional TEM of Samsung 3D NAND memory array (Courtesy Chipworks)

14.6 What Is Next: More Integration in Packaging— Multichip Packages

Revolutionary new packaging technologies are being introduced to integrate more dies into each semiconductor package, to provide more functionality, to improve performance, to reduce the size, and to reduce the cost of electronic devices that incorporate these packages. Memory dies, plus central processing unit dies, plus various sensors and MEMs are being integrated into single packages. Instead of mounting multiple dies on a circuit board and then mounting the circuit board in the electronic device, multiple dies are integrated into a single package completely eliminating the circuit board (Fig. 2.10).

New technologies are being developed and implemented to package dies on the wafer as the final manufacturing step of the IC wafer. The dies are then singulated (cut into individual packaged dies) and mounted directly on a circuit board.

References

1. M. Heck, Optical Computers Light Up the Horizon, Phys Org, News (2018), https://phys.org/news/2018-03-optical-horizon.html
2. Accelerating Artificial Intelligence with Light, Lightmatter, Inc. blog, https://www.lightmatter.co/. Accessed 4 Mar 2020
3. D. Coldeway, Lightmatter Aims to Reinvent AI-specific Chips with Photonic Computing and $11M in Funding, Techcrunch, join extra crunch, Website, https://techcrunch.com/2018/02/05/lightmatter-aims-to-reinvent-ai-specific-chips-with-photonic-computing-and-11m-in-funding/. Accessed 4 Mar 2020
4. J. Handy, An alternative kind of vertical 3D NAND string, Objective analysis. Semiconductor Market Research, https://thememoryguy.com/an-alternative-kind-of-vertical-3d-nand-string/ Accessed 8 Jun 2020
5. 3D NAND—an overview. Simms, Tech. Talk; https://www.simms.co.uk/tech-talk-2/3d-nand-overview. Accessed 8 Jun 2020

Appendix A: References by Topic

Constructional Analysis: Link to Ice Reports

1. *Constructional Analysis of ICs*, Section 4, Smithsonian Website, http://smithsonianchips. si.edu/ice/icesum.htm. Accessed 4 Mar 2020

Contact Plugs

2. H. Le, K. Banerjee, J. McPherson, The dependence of W-plug via EM performance on via size. Semicond. Sci. Technol **11**, 858–864 (1996)., https://www.ece.ucsb.edu/Faculty/ Banerjee/pubs/Semicond_1996.pdf
3. H. Chang, P. Jeng, Effects of barrier-metal schemes of tungsten plugs and blanket film deposition. Jpn J Appl Phys **39**, 4738–4743 (2000)., https://ui.adsabs.harvard.edu/ abs/2000JaJAP..39.4738C/abstract
4. H. Schmitz, A. Hasper, On the mechanism of the step coverage of blanket tungsten chemical vapor deposition. J. Electrochem. Soc. **140**(7) (1993), https://core.ac.uk/download/ pdf/11465298.pdf

Defects, Failure Analysis, Yield

5. Z. Ming, X. Ling, X. Bai, B. Zong, A review of the technology and process on integrated circuits failure analysis applied in communications products. J. Phys. Conf. Ser. **679**, 012040 (2016)., https://iopscience.iop.org/article/10.1088/1742-6596/679/1/012040/pdf
6. *Yield and Yield Management*, Smithsonian chips Website, http://smithsonianchips.si.edu/ice/ cd/CEICM/SECTION3.pdf. Accessed 4 Mar 2020
7. Prof. R. Leachmann, *Yield Modeling and Analysis*. U. of California, Berkley Course IEOR 130, Methods of Manufacturing Improvement (2017), http://courses.ieor.berkeley.edu/ ieor130/yield_models_rev6.pdf
8. *Defect Inspection and Review*, KLA Tencor Website. https://www.kla-tencor.com/products/ chip-manufacturing/defect-inspection-review. Accessed 4 Mar 2020

© Springer Nature Switzerland AG 2020
H. Tigelaar, *How Transistor Area Shrank by 1 Million Fold*,
https://doi.org/10.1007/978-3-030-40021-7

9. A. Jain, *New Methods of Defect Inspection and Review Improve Yield of Advanced Chips*, Applied Materials blog. (2017), http://blog.appliedmaterials.com/new-methods-defect-inspection-and-review-improve-yield-advanced-chips. Accessed 4 Mar 2020

10. *KLA Announces New Defect Inspection and Review Portfolio*, PRN Newswire, 8 July 2019, https://www.prnewswire.com/news-releases/kla-announces-new-defect-inspection-and-review-portfolio-300880092.html

11. W. Fisk, D. Fulkner, D. Sullivan, M. Mendell, Particle concentrations and sizes with normal and high efficiency air filtration in a sealed air conditioned office building. Aerosol Sci. Technol. **32**, 527–544 (2000)., https://www.tandfonline.com/doi/pdf/10.1080/027868200303452?needAccess=true

12. K. Kuhn, M. Giles, D. Becher, P. Kolar, A. Kornfield, R. Kotlyar, S. Ma, A. Meheshwari, S. Mudanai, Process technology variation. IEEE Trans. Electron Devices **58**(8) (2011)., http://citeseerx.ist.psu.edu/viewdoc/download?doi=10.1.1.707.1390&rep=rep1&type=pdf

Design and Layout

13. T. Yamamoto, N. Ooishi, K. McKay, Using in-design physical verification to reduce tapeout schedules. Des. Reuse Designware Tech. Bull. Synopsis Inc. (2010), https://www.design-reuse.com/articles/24046/in-design-physical-verification.html

Dielectrics: General

14. *Dielectric Materials for Microelectronics*, ed. by R. Wallace Springer Handbook of Electronic and Photonic Materials (2017), https://link.springer.com/book/10.1007/978-3-319-48933-9

Dielectric Deposition: CVD

15. H. Xiao, in *Chapter 10 CVD and Dielectric Thin Film*, Hong Xiao Book, pp. 1–222. http://apachepersonal.miun.se/~gorthu/ch10.pdf

16. M. Abdelgadir, HDP-CVD STI oxide process with in situ post deposition laterally enhanced sputter etchback for the reduction of pattern-dependent film topography in deep submicron technologies. IEEE Trans Semicond Manuf **19**(1), 130–137 (2006)., https://ieeexplore.ieee.org/document/1588870

17. *Improving Electrical Performance Using SACVD Oxide Films*, WebRTC World News, 2 Apr 2008. ;http://www.webrtcworld.com/news/2008/04/02/3363215.htm

18. S. Nguyn, High-density plasma chemical vapor deposition of siliconbased dielectric films for integrated circuits. IBM J. Res. Dev. **3**(1 & 2) (1999), https://ieeexplore.ieee.org/document/5389259

19. L. Xai, S. Nemani, M. Galianmo, S. Pichai, S. Chandran, E. Yieh, D. Cote, R. Conti, D. Restaino, D. Tobben, High temperature subatomospheric chemical vapor deposited undoped silicate glass. J. Electrochem. Soc. **146**(3), 1181–1185 (1999)., http://jes.ecsdl.org/content/146/3/1181.abstract

20. L. Zanotti, S. Rojas, E. Doghiefu, E. Santarelli, Process characterization for LPCVD TEOS-ozone based SiO2 films. J. Phys. IV **3**, 337–343 (1999)., https://hal.archives-ouvertes.fr/jpa-00251403/document

21. J. Foggiato, in *Chapter 3 Chemical Vapor Deposition of Silicon Dioxide Films*, Handbook of Thin-Film Deposition Processes and Techniques, (Noyes Publications/William Andrew Publication, 2002), pp. 111–150, https://booksite.elsevier.com/9781437778731/past_edition_chapters/Chemical_Vapor.pdf

22. R. Curley, T. McCormack, M. Phipps, in *Low-pressure CVD and Plasma Enhanced CVD*, Course ENEE 416 (U. of Maryland, 2011), http://classweb.ece.umd.edu/enee416/GroupActivities/LPCVD-PECVD.pdf

23. A. Tilke, R. Hampp, C. Stapelmann, M. Culmsee, R. Conti, W. Wille, R. Jaiswal, M. Galiano, A. Jain, in *STI Gap-Fill Technology with High Aspect Ratio Process for 45 nm CMOS and beyond*, Proceedings 2006 IEEE/SEMI Advanced Semiconductor Manufacturing Conference, pp. 71–76, https://ieeexplore.ieee.org/document/1638726

24. C. Ching, H. Whitesell, S. Venkataraman, *Improved electrical performance for 65nm node and beyond through the integration of HARP O3/TEOS oxide films for STI, PMD, and thin film applications*. ISSM Paper: PE-P224 (2007), https://ieeexplore.ieee.org/document/4446861

25. eHARP® *(enhanced High Aspect Ratio Process) TEOS/ozone CVD*, Applied Materials Producer® eHARP® CVD, Applied Materials Website, http://www.appliedmaterials.com/products/producer-eharp-cvd. Accessed 4 Mar 2020

Dielectric Deposition: Flowable CVD

26. T. Mandrekar, J. Liang, A. Mallick, N. Ingle, Scaling dielectric gap fill with flowable chemical vapor deposition. Appl. Mater. Nanochip Tech. J. **10**(1), 9–12 (2012)., http://www.appliedmaterials.com/files/nanochip-journals/NanochipTechJournal_Vol10Issue2_2012_1_Security.pdf

27. H. Kim, S. Lee, J. Lee, B. Bae, Y. Choi, Y. Koh, H. Yi, E. Hong, M. Kang, S. Nam, H. Kang, C. Chang, J. Park, N. Cho, S. Lee, *Novel Flowable CVD Process Technology for sub-20 nm Interlayer Dielectrics*, https://ieeexplore.ieee.org/document/6251590

28. Y. Yan, B. Zhang, H. Deng, L. Chen, L. Xiao, B. Zhang, Y. Chen, Flowable CVD process application for gap fill at advanced technology. ECS Trans. **60**(1), 503–506 (2014)., http://ecst.ecsdl.org/content/60/1/503.abstract

29. M. LaPadus, *Applied Flows into Flowable CVD*, EE Times (2010), https://www.eetimes.com/applied-flows-into-flowable-cvd/#

Dielectric Deposition: Low-K

30. C. Chiang, M. Chen, L. Li, Z. Wu, S. Jang, M. Liang, Physical and barrier properties of amorphous silicon-oxycarbide deposited by PECVD from octamethylcyclotetrasiloxane. J. Electrochem. Soc. **151**(9), G512–G 617 (2004)., http://jes.ecsdl.org/content/151/9/G612.abstract

31. Prof. K. Saraswat, *Low-k Dielectrics*, Course EE311 (Stanford University), https://web.stanford.edu/class/ee311/NOTES/Interconnect%20Lowk.pdf

32. P. Xu, S. Rathi, *A Breakthrough in Low-k Barrier/Etch Stop films for Copper Damascene Applications*. Semiconductor Fabtech 11th Edition (2011), pp. 239–244, https://www.

webpages.uidaho.edu/nanomaterials/research/Papers/General/Bariers%20for%20Cu%20 low%20k%20Damascene%20structures%20Fabtech%2011.pdf

33. V. Sekhar, in *Chapter 10 Mechanical Characterization of Black Diamond (Low-k) Structures for 3D Integrated Circuit and Packaging Applications*, Nanoindentation in Materials Science (Intech *Open Science*, 2012), pp. 229–256, https://www.intechopen.com/books/nanoindentation-in-materials-science/mechanical-characterization-of-black-diamond-low-k-structures-for-3d-integrated-circuit-and-packagin

34. *Low-k Films*, Silicon Valley Microelectronics, Website, https://www.svmi.com/custom-film-coatings/low-k-films/. Accessed 4 Mar 2020

35. R. Perry, *Producer®Black Diamond®PECVD*, Applied Materials Website, http://www.appliedmaterials.com/products/producer-black-diamond-pecvd. Accessed 4 Mar 2020

36. S. Beaudoin, S. Graham, R. Jaiswal, C. Kilroy, B. Kim, G. Kumar, S. Smith, An Update on Low=k dielectrics. Electrochem. Soc. Interface, 35–39 (2005). https://www.electrochem.org/dl/interface/sum/sum05/IF08-05_Pg35-39.pdf

37. J. Martin, S. Filipiak, T. Stephens, F. Huang, M. Aminpur, J. Mueller, E. Demircan, L. Zhao, J. Werking, C. Goldberg, S. Park, T. Sparks, C. Esber, *Integration of SiCN as a Low K Etch Stop and Cu Passivation in High Performance Cu/Low-k Interconnect* (Penn State University, AMD, Motorola, 2001), http://citeseerx.ist.psu.edu/viewdoc/download?doi=10.1.1.669.9775 &rep=rep1&type=pdf

Dummy Fill

38. Y. Wei, S. Sapatnekar, *Dummy Fill Optimization for Enhanced Manufacturability*, in International Symposium on Physical Design (2010), http://people.ece.umn.edu/users/sachin/conf/ispd10.pdf

39. *A Novel Approach to Dummy Fill for Analog Designs Using Calibre Smartfill*, Mentor Graphics Website, https://www.mentor.com/products/ic_nanometer_design/resources/overview/a-novel-approach-to-dummy-fill-for-analog-designs-using-calibre-smart-fill-a587f7dc-aadc-466e-a6e8-e31516c8dc92. Accessed 5 Mar 2020

40. *Advanced Process Dummy Fill*, XYALIS Brochure, XYALIS Website, https://www.xyalis.com/wp-content/uploads/GOTstyle-hybrid-dummy-fill.pdf. Accessed 5 Mar 2020

41. D. Payne, *Smart Fill Replaces Dummy Fill Approach in DFM Flow*, Mentor 2011, SemiWiki Website, https://semiwiki.com/eda/670-smart-fill-replaces-dummy-fill-approach-in-a-dfm-flow/. Accessed 5 Mar 2020

Gate Dielectric

42. Prof. K. Saraswat, *Thin Dielectrics for MOS Gate*, Course EE311 (Stanford University), https://web.stanford.edu/class/ee311/NOTES/GateDielectric.pdf

43. P. Nicollian, G. Baldwin, K. Eason, D. Frider, S. Hattangady, J. Hu, W. Hunter, M. Rodder, A. Rotondaro, *Extending the Reliability Scaling Limit of SiO2 through Plasma Nitridation*, IEDM paper 22.5.1 (2000), pp. 545–548, https://www.semanticscholar.org/paper/Extending-the-reliability-scaling-limit-of-SiO%2Fsub-Nicollian-Baldwin/52144bc81f83d5c2d050976 64cc812d879ec062a

44. K. Takasaki, K. Irino, T. Aoyama, Y. Momiyama, T. Nakanishi, Y. Tamura, T. Ito, Impact of nitrogen profile in gate nitrided-oxide on deep-submicron CMOS performance and reliability.

Fujitsu Sci. Tech. J. **38**(1), 40–51 (2003)., https://www.fujitsu.com/global/documents/about/resources/publications/fstj/archives/vol39-1/paper06.pdf

45. R. Geilmkeuser, K. Wieczorek, T. Mantei, F. Graetech, L. Hermann, J. Weidner, *Reliability Response of Plasma Nitrided Gate Dielectrics to Physical and Electrical CET-Scaling*, IEEE International Integrated Reliability Workshop Final Report (2004), pp. 15–18, https://pascal-francis.inist.fr/vibad/index.php?action=getRecordDetail&idt=17806278

46. H. Tseng, Y. Jeon, P. Abramowitz, T. Luo, I. Herbert, J. Lee, J. Jiang, P. Tobin, G. Yeap, J. Alvis, S. Anderson, N. Cave, T. Chua, A. Hegedus, G. Miner, J. Jeon, A. Sultan, Ultrathin decoupled plasma nitridation (DPN) oxynitride gate dielectric for 80-nm advanced technology. IEEE EDL **23**(12), 704–706 (2002). https://ieeexplore.ieee.org/abstract/document/1177959?section=abstract

47. S. Ting, Y. Fang, C. Chen, C. Yang, W. Hsieh, J. Ho, M. Yu, S. Jang, C. Yu, M. Liang, S. Chen, R. Shih, The effect of remote plasma nitridation on the integrity of the ultrathin gate dielectric films in 0.13 μm CMOS technology and beyond. IEEE EDL **22**(7), 327–329 (2001)., https://ieeexplore.ieee.org/abstract/document/930680?section=abstract

48. J. Jeon, P. Yeh, B. En, K. Wieczorek, F. Graetsch, J. Bernard, H. Kim, E. Ibok, C. Olsen, R. Zhao, B. Ogle, *Ultrathin Plasma Nitrided Oxide Gate Dielectrics for sub 100nm Generation CMOS Technology*, https://www.electrochem.org/dl/ma/203/pdfs/0934.pdf

49. F. Cubaynas, *Ultra-thin Plasma Nitrided Oxide Gate Dielectrics for Advanced MOS Transistors*, Dissertation doctor's degree at the Universiteit of Twente, 2004., https://pdfs.semanticscholar.org/ccf6/d8f44d16ca1443e248267361e89928b485cb.pdf

50. M. Bhat, D. Wristers, L. Han, J. Yan, H. Fulford, D. Kwong, Electrical properties and reliability of MOSFETs with rapid thermal NO-nitrided SiO_2 gate dielectrics. IEEE Trans. Electron Devices **42**(5), 907–914 (1995)., https://www.researchgate.net/publication/3062213_Electrical_Properties_and_Reliability_of_MOSFET%27s_with_Rapid_Thermal_NO-Nitrided_SiO2_Gate_Dielectrics

51. Q. Xiang, J.Jeon, P. Sachdey, B. Yu, K. Saraswat, M. Lin, *Very High Performance 40nm CMOS with Ultra-thin Nitride/Oxynitride Stack Gate Dielectric and Pre-doped Dual Poly-Si Gate Electrodes*. IEDM paper 10.8.1 (2000), pp. 860–862, https://ieeexplore.ieee.org/document/904453

52. P. Kraus, K. Ahmed, C. Olsen, F. Nouri, Model to predict gate tunneling current of plasma oxynitrides. IEEE Trans. Electron Devices **52**(6), 1141–1147 (2005)., https://ieeexplore.ieee.org/document/1433107

Isolation: Gap Fill and Planaraization

53. P. Smeys, *Chapter 2 Local Oxidation Of Silicon for Isolation (LOCOS)*, PhD Thesis, Stanford University, 1996, https://web.stanford.edu/class/ee311/NOTES/isolationSmeys.pdf

54. J. Pan, D. Ouma, P. Li, D. Boning, F. Redeker, J. Chang, J. Whitby, *Planarization and Integration of Shallow Trench Isolation (STI)*. VMIC (1998), pp. 1–6, https://pdfs.semantic-scholar.org/ed33/a6084fd959bcd1b4a82ffaffa6beb1563e3b.pdf

55. I. Shareef, G. Rubloff, M. Anderle, W. Gill, J. Cotte, D. Kim, Subatmospheric chemical vapor deposition (SACVD) ozone/TEOS process for SiO2 trench filling. J. Vac. Sci. Technol. B **12**(4), 1888–1892, http://www.rubloffgroup.umd.edu/old/oldresearch/recent_results/publications/publications-pdfs/JVST.95.SACVD_oz-TEOS.paper.pdf

56. S. Nag, A. Chatterjee, K. Taylor, I. Ali, S. O'Brien, S. Aur, J. Luttmer, I. Chen, *Comparative Evaluation of Gap-Fill Dielectrics in Shallow Trench Isolation for Sub-0.25 μm Technologies*. IEDM, 32.6.1 (1996), pp. 841–844, https://www.researchgate.net/publication/224164615_Comparative_evaluation_of_gap-fill_dielectrics_in_shallow_trench_isolation_for_sub-025_mm_technologies

Manufacturing: Machine Control

57. C. Chien, H. Yu, C. Hsu, in *Manufacturing Intelligence for Equipment Condition Monitoring in Semiconductor Manufacturing*, Proceedings of the Asia Pacific Industrial Engineering and Management Systems Conference (2012), https://pdfs.semanticscholar.org/a1b0/dc38e21498bdc9a9ea509d5a4f4f62c0670b.pdf

Metallization: General

58. Prof. Saraswat, *Interconnect Scaling*, Course EE311 Notes (Stanford University), https://web.stanford.edu/class/ee311/NOTES/InterconnectScalingSlides.pdf
59. K. Buchanan, *The Evolution of Interconnect Technology for Silicon Integrated Circuitry.* GaAsMANTECH Conf. (2002), https://csmantech.org/OldSite/Digests/2002/PDF/01c.pdf
60. R. Havemann, J. Hutchby, High-performance interconnects: an integration overview. Proc. IEEE **89**(5) (2001), http://citeseerx.ist.psu.edu/viewdoc/download?doi=10.1.1.461.7813&rep=rep1&type=pdf

Metallization: Aluminum

61. *Metallization: Aluminum Technology*, Semiconductor Technology A to Z, Halbleiter Website, https://www.halbleiter.org/pdf/en/Metallization/Metallization%20-%20Aluminum%20technology.pdf. Accessed 5 Mar 2020
62. K. Gadepally, R. Hawk, Integrated circuits interconnect metallization for the submicron age. J. Arkansas Acad. Sci. **43**, 29–31 (1989)., https://scholarworks.uark.edu/cgi/viewcontent.cgi?article=2277&context=jaas
63. Prof. Saraswat, *Interconnections: Aluminum Metallization*, Course EE311 Notes (Stanford Univ., Spring 2003); https://web.stanford.edu/class/ee311/NOTES/Interconnect_Al.pdf
64. H. Xiao, *Book Chapter 11 Metallization*, http://apachepersonal.miun.se/~gorthu/ch11.pdf
65. T. Doan, M. Bellersen, L. de Bruin, H. Godon, M. Grief, R. de Werdt, *A Double Level Metallization System Having 2 µm Pitch for Both Levels*. V-MIC Conf. 13–14 June 1988, https://ieeexplore.ieee.org/document/14171
66. S. Peng, T. Hsu, G. Ray, K. Chiu, A manufacturable 2.0 micron pitch three-level-metal interconnect process for high performance 0.8 micron CMOS technology. 1990 Symposium on VLSI Technology, 4–4 (1990), pp. 25–26, https://ieeexplore.ieee.org/document/5727450
67. C. Liu, J. Shen, X. Wang, Z. Ji, in *A Technique for Improving Contact Filling with Aluminum*, China Semiconductor Technology International Conference, March 2018, https://ieeexplore.ieee.org/document/8369255

Metallization: Reliability

68. C. Witt, *Electromigration in bamboo aluminum interconnects*, Thesis Von der Fakultat Chemie der Universitat, Stuttgart (2000.); https://www.semanticscholar.org/paper/Electromigration-in-bamboo-aluminum-interconnects-Witt/386550f7d2542d9163b970f396f7df539e4adfee

69. F. Pipia, A. Votta, G. Obetti, E. Bellandi, M. Alessandri, T. Nolan, AlCu metal line corrosion: a case study. Solid State Phenom. **134**, 367–370 (2008)., http://citeseerx.ist.psu.edu/viewdoc/download?doi=10.1.1.429.5748&rep=rep1&type=pdf
70. T. Licata, E. Colgan, J. Harper, S. Luce, Interconnect fabrication processes and the development of low-cost wiring for CMOS products. IBM J. Res. Dev. **39**(4) (1995), https://ieeexplore.ieee.org/document/5389487

Metallization: Copper

71. *Metallization: Copper Technology*, Semiconductor Technology A to Z, https://www.halbleiter.org/en/metallization/copper-technology/
72. Prof. Saraswat, *Interconnections: Copper and Low K Dielectrics*, Course EE311 Notes (Stanford University), https://web.stanford.edu/class/ee311/NOTES/Interconnect_Cu.pdf
73. P. Andricacos, Copper on-chip interconnections a breakthrough in electrodeposition to make better chips. Electrochem. Soc. Interface 32–37 (1999), https://www.electrochem.org/dl/interface/spr/spr99/IF3-99-Pages32-37.pdf
74. *Motorola Introduces Its First Copper Processor: PowerPC with AltiVec Technology*, EETimes 08.31.1999, https://www.eetimes.com/motorola-introduces-its-first-copper-processor-powerpc-with-altivec-technology/#
75. Y. Cheng, C. Lee, Y. Huang, *Chapter 10 Copper Metal for Semiconductor Interconnects*, Copper Metal for Semiconductor Interconnects (2017), https://www.intechopen.com/books/noble-and-precious-metals-properties-nanoscale-effects-and-applications/copper-metal-for-semiconductor-interconnects

Metallization: Copper Barrier Metal

76. H. Yang, F. Zhang, J. Tseng, A. Bhatnagar, N. Kumar, P. Gopalraja, J. Forster, Novel approach extends PVD Ta barrier technology to 32 nm and below. Nanochip Technol. J. **8**(2), 42–45 (2008)., https://www.nxtbook.com/nxtbooks/cmp/nanochip/widget.xml
77. H. Wojcik, B. Schwiegell, C. Klaus, N. Urbansky, J. Kriz, J. Rahn, C. Kubasch, C. Wenzel, J. Barth, *Co Barrier Properties of Very Thin Ta and TaN Films* (2014), pp. 167–169, https://ieeexplore.ieee.org/document/6831854
78. O. van der Straten, X. Zhang, K. Motoyama, C. Penny, J. Maniscalco, S. Knupp, *ALD and PVD Tantalum Nitride Barrier Resistivity and Their Significance in via Resistance Trends.* ECS 226th Meeting, ECS and SMEQ (2014), http://ecst.ecsdl.org/content/64/9/117.short
79. *Applied Says Self-Ionized Plasma System Extends PVD to 100-nm Regime*, EETimes, Nov 2000, https://www.eetimes.com/applied-says-self-ionized-plasma-system-extends-pvd-to-100-nm-regime/
80. X. Liu, Y. Lu, R. Gordon, Improved comformality of CVD titanium nitride films. Mater. Res. Soc. Symp. Proc. **555**, 135–140 (1999)., http://hwpi.harvard.edu/files/gordon/files/conformed_cvd_titanium_matressocsympproc555_135_1999.pdf
81. C. Bryne, *Fabrication and Characterisation of Copper Diffusion Barrier Layers for Future Interconnect Applications*, PhD Thesis, Dublin City University, 2015, http://doras.dcu.ie/20828/1/CBThesis_PostViva.pdf

Metallization: Self Formed Copper Barrier Metal

82. T. Usui, H. Nasu, S. Takahashi, N. Shimizu, T. Nishikawa, M. Yoshimaru, H. Shibata, M. Wada, J. Koike, Highly reliable copper dual-damascene interconnects with self-formed MnSixOy barrier layer. IEEE Trans. Electron Devices **53**(10), 2492–2499 (2006)., https://ieeexplore.ieee.org/document/1705100

Metallization: Cobalt/ Copper

83. M.LaPedus,*DealingWithResistanceinChips*,CobaltInterconnect,SemiconductorEngineering blog, 21 June 2018, https://semiengineering.com/dealing-with-resistance-in-chips/
84. N. Bekiaris, Z. Wu, H. Ren, M. Naik, J. Park, M. Lee, T. Ha, W. Hou, J. Bakke, M. Gage, Y. Wang, J. Tang, *Cobalt Fill for Advanced Interconnects*, https://ieeexplore.ieee.org/stamp/stamp.jsp?arnumber=7968981
85. B. Wang, *Cobalt Can Extend Copper Interconects to the 10 nm Lithography Node for Computer Chips*, EE Times blog, 14 May 2014, https://www.nextbigfuture.com/2014/05/cobalt-encapsulation-can-extend-copper.html

Metallization: Planarization

86. R. Rhoades, in *New Applications for CMP: Solving the Technical and Business Challenges.* NSTI Conference 2009., https://www.entrepix.com/docs/papers-and-presentations/050509_ENTR_NSTI2009.pdf
87. Y. Sun, N. Radjy, S. Cagnina, in *0.8 Micron Double Level Metal Technology with SOG Filled Tungsten Plug.* VMIC Conference (1991), pp. 51–57, https://ieeexplore.ieee.org/document/152965
88. Prof. K. Saraswat, Aluminum Metallization, *Deposition and Planarization*, Course E311 Notes (Stanford University), https://web.stanford.edu/class/ee311/NOTES/Deposition_Planarization.pdf
89. A. Sigov, K. Vorotilov, A. Valeev, M. Yanovskaya, Sol-gel films for integrated circuits. J. Sol-Gel Sci. Tech. **2**, 563–567 (1994)., https://link.springer.com/article/10.1007%2FBF00486310
90. H. Liu, S. Srivathanakul, H. Liu, S. Gaan, X. Cai, X. Rao, J. Shu, S. Kim, PMD and STI gap fill challenges for advanced technology of logic and eNVM. ECS Trans. **52**(1), 397–402 (2013)., http://citeseerx.ist.psu.edu/viewdoc/download?doi=10.1.1.840.2065&rep=rep1&type=pdf

Micoprocessors

91. *Intel Process Technology History, Chips and Semi*, WikiChip Website, https://en.wikichip.org/wiki/intel/process. Accessed 5 Mar 2020
92. *Intel Microprocessors Quick Reference Guide*, Intel Website, https://www.intel.com/pressroom/kits/quickrefyr.htm. Accessed 5 Mar 2020
93. *Intel and Other Manufacturer Microprocessor Transistor Count: Wikipedia*, The Free Encyclopedia, https://en.wikipedia.org/wiki/Transistor_count. Accessed 5 Mar 2020

MMST: Microelectronics Manufacturing Science and Technology

94. M. Moslehi, Y. Lee, C. Shaper, T. Omstead, L. Velo, A. Kermana, C. Davis, *Chapter 6 Single Wafer Process Integration and Process Control Techniques*, Advances in Rapid Thermal and Integrated Processing (Springer-Science + Business, B.V., 1996), pp. 166–191, https://books.google.com/books?id=Y37sCAAAQBAJ&pg=PA166&lpg=PA166&dq=MMST+single+wafer+processing&source=bl&ots=fx85m5hPAj&sig=ACfU3U3DRYRDpigpBxF_LdlBSmkLHn-4tQ&hl=en&sa=X&ved=2ahUKEwj_gtuLhaDjAhXFLs0KHS5FBJ0Q6AEwB3oECAkQAQ#v=onepage&q=MMST%20single%20wafer%20processing&f=false

Moore's Law and Scaling

95. *Chapter 2 The Scaling of MOSFETs, Moore's law, and ITRS*, Sematicscholar.org Website, https://www.semanticscholar.org/paper/The-scaling-of-MOSFETs-%2C-Moore-'-s-law-%2C-and-ITRS-2/c2640e207a23a8262cab7931154145ffbe6a1868. Accessed 5 Mar 2020
96. S. Meoli, *How the Scaling of CMOS Technology is Impacting Modern Microelectronics*, blog, CERN.ch Website, https://meroli.web.cern.ch/lecture_scaled_CMOS_Technology.html. Accessed 5 Mar 2020
97. W. Arden, M. Brillouet, P. Cogez, M. Graef, B. Huizing, R. Rahnkopf, *More than Moore*, White Paper (ITRS, 2013), https://www.yumpu.com/en/document/read/7001330/more-than-moore-white-paper-itrs
98. Kuhn, in *Variation in 45nm and Implications for 32 nm and Beyond*, 2009 2nd International CMOS Variability Conference—London, Intel Website, http://download.intel.com/pressroom/pdf/kkuhn/Kuhn_ICCV_keynote_slides.pdf. Accessed 5 Mar 2020

Photolithography: General

99. P. Silverman, The intel lithography roadmap. Intel Technol. J. **06**(2), 55–64 (2002)., https://www.intel.com/content/dam/www/public/us/en/documents/research/2002-vol06-iss-2-intel-technology-journal.pdf
100. R. Kapoor, R. Adner, Technology interdependence and the evolution of semiconductor lithography. Solid State Technol. 51 (2007), https://go.gale.com/ps/anonymous?id=GALE%7CA178673729&sid=googleScholar&v=2.1&it=r&linkaccess=abs&issn=0038111X&p=AONE&sw=w
101. M. van den Brink, H. Jasper, S. Slonaker, P. Wijhoven, F. Klaassen, *Step and Scan and Step and Repeat, a Technology Comparison*. SPIE Symposium on Microlithography (1996), pp. 1–20, https://www.spiedigitallibrary.org/conference-proceedings-of-spie/2726/0000/Step-and-scan-and-step-and-repeat%2D%2Da-technology/10.1117/12.240936.short

Photolithography: 157 nm

102. S. Dana, P. Ware, A. Tanimoto, *Progress Report: 157-nm Lithography Prepares to Graduate*, Oemaganize (2003), http://spie.org/news/progress-report-157-nm-lithography-prepares-to-graduate?SSO=1
103. A. Bates, M. Rothschild, T. Bloomstein, T. Fedynyshyn, R. Kunz, V. Liberman, M. Switkes, Review of technology for 157-nm lithography. IBM J. Res. Dev. **45**(5), 605–614 (2001)., https://ieeexplore.ieee.org/document/5389065

Photolithography: EUV

104. M. Lapedus, *Why is EUV so Difficult?*, Semiconductor Engineering, Semiconductor Engineering Website (2016), https://semiengineering.com/why-euv-is-so-difficult/. Accessed 5 Mar 2020
105. C. Mbanaso, Fundamental resist exposure mechanisms: a preliminary study based upon mass spectrometer measurements. 2011 International workshop on EUV lithography (2011), https://www.euvlitho.com/2011/P29.pdf
106. S. Moore, *EUV Lithography Finally Ready for Chip Manufacturing*, IEEE Spectrum, 5 Jan 2018, https://spectrum.ieee.org/semiconductors/nanotechnology/euv-lithography-finally-ready-for-chip-manufacturing
107. H. Levinson, T. Brunner, *Current Challenges and Oppportunities for EUV Lithography*, SPIE 10809 (2019), https://doi.org/10.1117/12.2502791

Photolithography: Immersion

108. *Immersion Lithography Explained*, Wikipedia.org, The Free Encyclopedia, https://en.wikipedia.org/wiki/Immersion_lithography. Accessed 5 Mar 2020
109. A. Thampi, *Immersion Lithography*, 3rd Seminar M. Sc. Physics, Cochin Univ. (2006), https://www.slideshare.net/anandhus/immersion-lithography
110. R. French, H. Tran, Immersion lithography: photomask and wafer-level materials. Annu. Rev. Mater. Res. **39**, 93–126 (2009)., https://engineering.case.edu/centers/sdle/sites/engineering.case.edu.centers.sdle/files/immersion_lithography_photomask_and_wafer-level_m.pdf
111. D. Sanders, Advances in patterning materials for 193 nm immersion lithography. Chem. Rev. **110**, 321–360 (2010)., https://pubs.acs.org/doi/full/10.1021/cr900244n

Photolithography: Photo Resists

112. D. Roy, P. Basu, S. Eswaran, Photoresists for microlithography, Resonance, 44–53 (2002), https://www.ias.ac.in/article/fulltext/reso/007/07/0044-0053
113. H. Ito, Chemical amplification resists: history and development within IBM. IBM J. Rees. Develop. **44**(1, 2), 119–130 (2000)., https://researcher.watson.ibm.com/researcher/files/us-saswans/05389371.pdf

114. D. Brock, *Patterning the World: The Rise of Chemically Amplified Photo Resists*, Distillations, The Science History Institute, 2007, https://www.sciencehistory.org/distillations/patterning-the-world-the-rise-of-chemically-amplified-photoresists
115. C. Henderson, *Introduction to Chemically Amplified Resists*, C. L. Henderson Group, Georgia Tech., https://sites.google.com/site/hendersonresearchgroup/helpful-primers-introductions/introduction-to-chemically-amplified-photoresists. Accessed 05 Mar 2020

Plasma Etching

116. Prof. L. Fuller, *Plasma Etching* (Rochester Institute of Technology, 4-13-2017), http://diyhpl.us/~nmz787/mems/unorganized/Plasma_Etch.pdf. Accessed 5 Mar 2020

Premetal Dielectric (PMD) and Intermetal Dielectric (IMD)

117. *Interlevel Dielectric Technology*, EMT 362—Microelectronic Fabrication School of Microelectronic Engineering (University Malaysia Perlis, 2012), https://slideplayer.com/slide/8623819/
118. P. Lee, M. Galiano, P. Keswick, J. Wong, B. Shin, D. Wang, in *Sub-atmospheric Chemical Vapor Deposition (SACVD) of TEOS-ozone USG and BPSG*, VMIC Conference (1990), pp. 396–398, https://ieeexplore.ieee.org/document/127910
119. V. Vassiliev, C. Lin, F. Gn, A. Cuthbertson, *A Comparative Analysis of Pre Metal Dielectric Gap-Fill Capability for ULSI Device Technology*, International Symposium on VLSI Technology, Systems, and Applications, Proceedings, vol. 8 (1999), https://www.research-gate.net/publication/273634944_Comparative_Analysis_of_Pre-Metal_Dielectric_Gap-Fill_Capability_for_ULSI_Device_Applications
120. V. Vassiliev, *Introduction to Pre-Metal Dielectric Gap-Fill Issues and Solutions for Sub-0.25 um ULSI Device Applications*, Invited presentation to semiconbay.com (thin film section) (2000), https://www.researchgate.net/publication/273634800_Introduction_to_Pre-Metal_Dielectric_Gap-Fill_Issues_and_Solutions_for_Sub-025_um_ULSI_Device_Applications_Invited_presentation_to_website_Year_2000
121. V. Vassiliev, in *Void Free Pre-Metal Dielectric Gap-Fill Capability with CVD Films for sub quarter Micron VLSI*, Dielectric for ULSI Multilevel Interconnection Conference (2000), pp. 121–132, https://www.researchgate.net/publication/273634788_Void-free_pre-metal_dielectric_gap-fill_capability_with_CVD_films_for_subquater-micron_ULSI_Invited_paper
122. W. Douksher, M. Miller, C. Tracy, Three 'Low Dt' options for planarizing the pre-metal dielectric on an advanced double poly BiCMOS process, J. Electrochem. Soc. **139**(2) (1992), http://citeseerx.ist.psu.edu/viewdoc/download?doi=10.1.1.907.6538&rep=rep1&type=pdf

Semiconductor Device Physics

123. S. Sze, Semiconductor Devices, in *Physics and Technology*, (John Wiley & Sons, Hoboken, 2012)
124. Y. Taur, T. Ning, *Fundamentals of Modern VLSI Devices* (Cambridge University Press, Cambridge, 2013)

125. B. El-Kareh, L. Hutter, *Silicon Analog Components, Device Design, Process Integration, Characterization, and Reliability* (Springer Science+Business Media, LLC., Berlin, 2015)
126. B. Streetman, S. Banerjee, *Solid State Electronic Devices* (Prentice-Hall Inc., Upper Saddle River, 2015)

Semiconductor Technology

127. Y. Nishi, R. Doering, *Handbook of Semiconductor Manufacturing Technology* (Marcel Dekker Inc, New York, 2007)
128. S. Wolf, R. Tauber, *Silicon Processing for the VLSI Era, Volume 1—Process Technology* (Lattice Press, Huntington Beach, 1986)
129. S. Wolf, R. Tauber, *Silicon Processing for the VLSI Era, Volume 2—Process Integration* (Lattice Press, Huntington Beach, 1986)
130. S. Wolf, *Silicon Processing for the VLSI Era, Volume 3—The Submicron MOSFET* (Lattice Press, Huntington Beach, 1986)
131. S. Campbell, *The Science and Engineering of Microelectronic Fabrication* (Oxford University Press, Oxford, 2001)

Silicides

132. Prof. K. Saraswat, *Interconnections: Silicides*, Course E311 (Stanford University), https:// web.stanford.edu/class/ee311/NOTES/Silicides.pdf
133. Prof. K. Saraswat, *Polysides, Salasides, and Metal Gates*, Course E311 (Stanford University), https://web.stanford.edu/class/ee311/NOTES/Silicides%20&%20Metal%20gate%20 Slides.pdf
134. S. Haiping, X. Qiuxia, Two-Step Ni silicide process and influence of protective N_2 gas. J. Semicond. **30**(9), 096002–096001 (2009)., https://pdfs.semanticscholar.org/c01b/634a9d2 2cb36dc170c3ef6a80bfcf7983582.pdf
135. C. Lu, J. Sung, R. Liu, N. Tsai, R. Shingh, S. Hillenius, H. Kirsch, Process limitation and device design tradeoffs of self-aligned TiSi, junction formation in submicrometer CMOS devices. IEEE Trans. Electron Devices **38**(2), 246–254 (1991)., https://ieeexplore.ieee.org/ document/69902
136. B. Froment, M. Muller, H. Brut, R. Pantel, V. Carron, H. Achard, A. Halimaoui, F. Boeuf, F. Wacquant, C.Regnier, D.Ceccarelli, R. Palla, A. Beverina, V. DeJonghe, P. Spinelli, O. Leborgne, K. Bard, S. Lis, V. Tirard, P. Morin, F. Trentesaux, V. Gravey, T. Mandrekai, D. Rabilloud, S. Van, E. Olson, J. Diedrick, *Nickel vs. Cobalt Silicide integration for sub 80nm CMOS*, 33rd Conv. On European Solid-State Dev. Research (2003), https://ieeexplore. ieee.org/document/1256852
137. J. Lei, S. Phan, X.Lu, C. Kao, K. Lavu, K. Moraes, K. Tanaka, B. Wood, B. Ninan, and S. Gandikota, *Advantage of SiconiT Preclean over Wet Clean for Pre Salicide Applications Beyond 65nm Node*, IEEE International Symposium on Semiconductor Manufacturing (2006), https://vdocuments.site/ieee-2006-international-symposium-on-semiconductor-man-ufacturing-issm%2D%2D58c5fe45923e4.html
138. J. Tang, N. Ingle, D. Yang, *Smooth SiCoNi Etch for Silicon Containing Films*, US Patent 8501629, 2013, https://patentimages.storage.googleapis.com/02/f9/32/ad817ef855bdb4/ US8501629.pdf

139. M. Alperin, T. Holloway, R. Haken, C. Gosmeyer, R. Karnaugh, W. Parmantie, Development of the self-aligned titanium silicide process for VLSI applications. IEEE Trans. Electron Device **ED-32**(2) (1985), https://ieeexplore.ieee.org/document/1484670
140. R. Mann, L. Clevenger, P. Agnello, F. White, Silicides and local interconnections for high-performance VLSI applications. IBM J. Res. Dev. **39**(4) (1995), https://web.stanford.edu/class/ee311/NOTES/IBM_Silicides_Mann.pdf
141. S. Murarka, Silicide thin films and their applications in microelectronics. Intermetallics **3**(3) (1993)

Transistors: Mobility Enhancement

142. Y. Song, H. Zhao, Q. Xu, J. Luo, H. Yin, J. Yan, H. Zhong, Mobility enhancement technology for scaling of CMOS devices: overview and status. J. Electron. Mater. **40**(7), 1594 (2011)., https://link.springer.com/content/pdf/10.1007/s11664-011-1623-z.pdf
143. N. Yu, *Lecture 9, Strained Si—Device Physics*, Course EE290D (U. of Cal. Berkley, 2013), http://www-inst.eecs.berkeley.edu/~ee290d/fa13/LectureNotes/Lecture9.pdf

Transistor and Interconnect: Scaling

144. A. Stillmaker, Z. Xiao, B. Baas, *Toward More Accurate Scaling Estimates of CMOS Circuits from 180 nm to 22 nm*, Technical Report ECE-VCL-2011-4, U. C. Davis, 2011, https://pdfs.semanticscholar.org/8ce1/9ee86b874455bf7b83ae7f3ea5b8dc85e760.pdf
145. P. Bhattacharjee, A. Sadhu, VLSI transistor and interconnect scaling overview. J. Electron. Des. Technol. **5**(1), 1–15 (2014)., https://www.researchgate.net/publication/290883859_VLSI_Transistor_and_Interconnect_Scaling_Overview
146. Prof. K. Saraswat, *Trends in Integrated Circuits Technology*, Course EE311 Handout 2 (Stanford University), https://web.stanford.edu/class/ee311/NOTES/Trends.pdf
147. Prof. K. Saraswat, *Trends in Integrated Circuits Technology*, Course EE311 (Stanford University), https://web.stanford.edu/class/ee311/NOTES/TrendsSlides.pdf

Transistor: Source/Drains and Anneals

148. J. McWhirter, *LXA: Nanosecond Laser Anneal for sub-10nm*, Ultratech, Inc., NCCAVS Junction Technology Group SEMICON West Meeting (2016), https://nccavs-usergroups.avs.org/wp-content/uploads/JTG2016/jtg2016_7mcwhirter.pdf
149. P. Kalra, *Advanced Source/Drain Technologies for Nanoscale CMOS*, PhD Thesis, U. of California, Berkeley, 2008, https://pdfs.semanticscholar.org/e9e3/f0025f0159f91925e13da814faac1a6f292a.pdf

Index

A

Accelerated reliability modeling, 186, 187
Accelerated reliability testing, 186
Activation energy (E_A), 278
Active areas, 76
Active pattern, 75
Active photoresist pattern, 75, 76
Alternating phase shift mask, 167
Aluminum alloy, 131
Aluminum alloy plasma etch, 132
Aluminum bond pad, 73
Aluminum-copper leads, 240
Aluminum forming titanium aluminide
    (Al_3Ti), 245
Aluminum wires, 130
AMAT 5500 Endura^R multi-chamber
    deposition tool, 239
Analog MOS transistors
    180 nm, 45, 46
    current vs. voltage curves, 45
    drain voltage, 45
    I/V curve, 44
Analog signals, 39
    audio, 40
    digitizing, 40–42
    video, 40
Analog-to-digital converters (ADCs), 40, 274
"AND" logic circuit symbol, 38
Anitfuse development program, 282
Antifuse technology development, 281
Antireflection coating (ARC), 248, 260
Apollo 17 mission, 1
Applied Materials, 151
Applied Materials Centura© Gate Stack
    Tool, 84

Ashing, 216
Atmospheric CVD (APCVD), 205
Atmospheric pressure chemical vapor
    deposition (APCVD), 11
Atomic layer chemical vapor deposition
    (ALD), 11
Attenuated phase shift mask, 167, 168
Auger electron spectroscopy (AES), 184
Automated killer particle detection, 183
Automatic teller machines (ATMs), 2
Automobile transportation revolution, 290

B

Bamboo grain structure, 244
Base, 67, 70, 72
Bilayer resist, 165
Bipolar plus CMOS (BICMOS), 48
    NWELL, 49
    STI, 49
    vertical NPN, 49, 50
    vertical PNP, 49, 50
Bipolar transistors
    BICMOS, 48–50
    forward-biased emitter/base diode, 47
    NPN, 46–48
    PNP, 46–48
    reverse-biased collector/base
        diode, 47
Black Diamond II™ (BD), 260
Black Diamond™, 264
Bond pads, 120, 131–134
Boron tribromide (BBr_3), 211
Bottom antireflection coating (BARC),
    164, 259

© Springer Nature Switzerland AG 2020
H. Tigelaar, *How Transistor Area Shrank by 1 Million Fold*,
https://doi.org/10.1007/978-3-030-40021-7

Build self-testing circuits (BIST), 185
Bunny suit, 139, 140
Bytes, 41

C
Capacitance, 55
Central processing units (CPUs), 41
Channel/gate dielectric interface
  CHC, 219
  nitrogen concentration, 219
  PMOS transistor, 219
  self-aligned silicide, 219
  silicides, 219
Channel hot carriers (CHC), 71, 84, 88, 219
Chemical mechanical polish (CMP), 78, 79,
    106, 120, 129, 157, 173–179, 202, 231,
    234, 246, 253
Chemical vapor deposited tungsten (CVD-W),
    209, 238
Chemical vapor deposition (CVD), 10, 11, 85,
    110, 202
Chemically amplified (CA) resists, 162, 163,
    166, 196
Chemical-vapor-deposited tungsten (CVD-W),
    105, 109, 111
Chief Technical Officer (CTO), 286
Clean room attire, 139, 140
Clean room evolutions, 140–142
CMOS IC technology, 137
CMOS inverter, 123–125
  AND logic gate, 38
  circuit symbol, 34
  diagram, 34
  low-power integrated circuits, 33
  OR logic gate, 39
  p-type silicon substrate, 35
  silicon storing logic state, 36, 37
  storing logic state, 34–37
CMOS inverter manufacturing
  active pattern, 76
  cross-sectional view, 74
  IMD1, 73
  metal bond pad layer, 74
  PMD, 73
  STI (see Shallow trench isolation (STI))
  top-down view, 75
Cobalt silicide (CoSi$_2$), 220
Coefficient of thermal expansion
    (CTE), 194
Collector, 67, 70, 72
Columnated sputtering, 13, 239
Complementary metal-oxide-semiconductor
    (CMOS), 7, 8, 14
Computer processing unit (CPU), 210

Computer simulation programs
  complex integrated circuits, 188
  data analysis, 188
  electrical testing, 188
  experiments, 189
  IC processing, 192
  incredible IC scaling, 188
  in-line manufacturing data, 188
  integrated circuit, 189
  integrated circuit design, 192, 193
  photo, 195
  reliability, 194
  sensors, 189
  stress, 194
  thermal heating, 193
  transistor processing, 189–191
  transistors, 188
Conductors, 21
Contact, 53, 56, 57, 61
  fill/metal1, 239–241
  tungsten chemical vapor deposition,
    241, 242
Contact barrier
  material/deposition, 238
  overview, 238
Contact barrier layer, 109, 110
Contact CMP, 111
Contact etch, 108, 110
Contact etch-stop liner (CESL) technology,
    103–105, 108, 114, 209, 237
  deposition process, 223
  NMOS transistors, 223
Contact hole, 103, 104, 106, 107, 109–111
Contact hole metal fill, 111
Contact pattern, 107, 109
Contact plug, 53, 56–58, 105, 112, 114, 116
Contact process
  etch, 237
  overview, 236
  pattern, 236
Copper CMP, 114, 115, 120
Copper damascene, 112
Copper metallization, 178
Copper seed layers, 116, 118, 119, 128
Copper wiring, 7, 8
Coral™, 260
Core transistors, 85
CPU technology development, 254
Creative scientist/technical artist, 279
Crystal defects, 143
Current flow analogy, 51, 52
Customer crisis, 284
CVD-cobalt (CVD-Co), 268
Cyber security, 3
Cybercrime, 3

D

Damascene copper metalization
  dual process, 256
  single process, 256
Data analysis programs, 195, 196
DC sputtering, 13
Decoupled plasma nitridation (DPN), 218
Deep buried diffusions (DUFs), 211
Deep source/drain diode, 100
Defect analysis, 180, 181
Defect detection, 180, 181
Design for manufacturing design rules
    (DFM), 61
Design of experiments (DOE), 189, 191
Design rules, 68, 69
Diamond (single crystal carbon), 23
Dielectric protective overcoat layer (PO), 73
Diffused/implanted wells, 82
Digital light processor (DLP), 3, 271
Digital MOS transistors, 41, 43
  halo (pocket) doping, 43
Digital photography, 291
Digital signal processing integrated circuits
    (DSPs, 40
Digital-to-analog converters
    (DACs), 40
Digitizing analog signals, 40, 41
Diode capacitors, 58, 61, 62
Double-level metal (DLM), 247
Drain extension, 56, 57, 61
Dual damascene, 123, 126–129
Dual damascene copper metalization, 256
Dual inline packaged (DIP), 16
Dual stress liner (DSL) technology, 225
Dummy active geometries, 203
Dummy fill geometries, 173–176
Dynamic random-access memory
    (DRAM), 14

E

Electric fields, 12
Electrical programmable read-only memory
    integrated circuits (EPROMS), 89
Electromigration model, 186
Electromigration reliability, 268
Electronic devices, 291, 292
Electroplated copper, 112, 116, 119, 123, 129,
    256–258, 260, 262, 264, 267
Electrostatic discharge (ESD), 141
Emitter, 67, 70–72
Erasable programmable read only memory
    (EPROM), 15
Etch loading, 168, 169
Extreme UV (EUV) lithography, 157

F

Failure analysis (FA), 207
Fault detection and classification (FDC), 197
Field programmable gate arrays (FPGAs), 281
Figure of Merit (FOM), 53
Film cameras, 291
Film manufacturers, 291
First layer of intermetal dielectric (IMD1),
    7, 8, 114
First robotic revolution, 2
Flash anneal, 179
FLASH memory, 15
Forming gas, 134, 135
Forward-biased pn-diode, 26, 27
    current *vs.* voltage curve, 28
Front opening unified pods (FOUPs), 142,
    145–147, 150
Functional testing, 185

G

Gap fill dielectric, 103, 105
Gap fill layer, 103
Gate critical dimension (CD)/gate length
    parameter, 215
    transistor, 215
Gate dielectric engineers, 277
Gate dielectric scaling
  AMAT, 218
  annealing, 217
  footing, 217
  integrated process, 218
  notching, 217
  PMOS transistor, 217
  RPN, 217
  silicon dioxide, 217
  SiO_2, 218
Gate etch
  bias, 216
  chemistry, 217
  plasma etching, 216
  polymer deposition, 216
  resist ashing, 216
  resist patterns, 216
  transistor, 216
  uniform across, 216
Gate etch vertical transistor gate, 216
Gate poly oxidation, 88
Good electrical dies (GEDs), 16
GUI split lot interface, 286–288

H

Halo, 53, 56–58, 60, 61
HARP™ deposition process (eHARP™), 236

HDP dielectric deposition process
    argon atoms, 205
    choices, 207
    HARP, 205
    high-voltage transistors, 207
    STI isolation voids, 207
    versus technology node, 206
Heavily doped silicon, 100
High aspect ratio process (HARP), 11, 77, 205
High density plasma chemical vapor
    deposition (HDP), 11
High-density plasma (HDP), 205
High-efficiency particulate air (HEPA) filters,
    140, 142
High selectivity slurry (HSS), 204
High-voltage transistors, 271
Hornbeck, L., 271
Hydrogen silsesquioxane (HSQ), 249

I
IBS sputtering, 13
IC factories, 201
IC geometries, 143, 201
IC revolution, 3, 289, 291, 292
IC scaling, 290
ILD2, 257, 265, 267
IMD1 capping layer, 115, 116
IMD2 dielectric stack, 125
IMP sputtering, 13
In situ steam generation (ISSG), 218
Increasing wafer size, 143–145
Industry in transition, 4, 5
Input/output (I/O) transistors, 85
Insulators, 21
Integrated circuit manufacturing process, 8
Integrated circuit technology revolution
    complex real-time computation, 1
    industry in transition, 4, 5
    scaling and technological/workplace, 2, 3
    solid-state transistors, 1
    space program, 1
    standard of living, 4
    universities, industrial labs and
        consortia, 3, 4
Intel computer chips, 138
Intel microprocessor chips, 137
Intel's 2002 Lithography Roadmap, 156, 157
Intermetal dielectric layer (IMD), 53, 73, 130,
    177, 246, 253, 256, 285
Ion beam sputtering (IBS), 238
Ion metal plasma (IMP), 110, 111, 118, 238
Ions, 9, 11–13
Isolation implants, 82

K
Kilby, J., 280
Killer particle analysis, 183, 184
Killer particles, 140, 181, 182
Kodak thin film resist (KTFR), 161

L
Laser anneal, 180
Latch up, 72
Light interference, 168, 169
Light wavelength, 155
Lithography
    IC manufacturing, 151
    Intel's 2002 Lithography Roadmap,
        156, 157
    light wavelength, 155
    157-nm photolithography system, 159, 160
    193 nm Immersion Scanner, 158, 159
    pattern alignment, 152, 153
    photolithography printers, 153, 154
    printing IC dies, 151, 152
    scanners, 157
    steppers, 157
Lithography tools, 290
Lithography wavelength, 155
Local oxidation of silicon (LOCOS), 176, 202
Low pressure chemical vapor deposition
    (LPCVD), 11
Low resistance doped single crystal silicon
    boron, 25
    n-type, 24
    phosphorous, 25
    P-type, 25, 26
Low-k dielectric layer, 114, 115
Low-K Intermetal Dielectrics (Low-K IMDs),
    254, 255
Low-k organo silicate glass (OSG), 260
Low-pressure chemical vapor deposition
    (LPCVD), 75, 104

M
Memory transistors, 14
Metal etch stop layer, 263
Metal replacement gate transistors, 178
Metal wiring, 104
Metal wiring levels, 130
Metal wiring technology, 268
Metal1, 7, 8
Metal1 copper CMP, 120
Metal1 trench barrier layer, 116
Metal1 trench copper fill, 116, 119
Metal1 trench copper seed layer, 118

Metal1 trench etch, 116
Metal1 trench etch stop layer (M1TESL),
    114, 259
Metal1 trench pattern, 115, 118
Metal1 wiring layer, 123
Metal2 barrier, 128
Metal2 copper CMP, 129, 130
Metal2 trench copper fill, 128, 129
Metal2 trench etch, 127
Metal2 trench etch stop layer (M2ESL), 263
Metal2 trench pattern, 127
Metal2 trench plasma etch (M2TESL), 125
Metal2 wiring layer, 123
Metallic nickel, 101
Micro Electrical Mechanical (MEMS)
    devices, 292
Microelectronics Manufacturing Science and
    Technology (MMST)
    program, 149–151
Microscopes, 183
Missed patent opportunity, on halo
    implants, 91
300 mm wafers, 141, 145, 148–150
MnSi$_X$O$_Y$/Ta barrier, 267
Mobile ions, 232
Molybdenum silicide (MoSi$_2$), 219
Moore's Law, 3, 137
Multibillion-dollar semiconductor
    factories, 291
Multilevel aluminum metal
    alloy scaling, 246
    CMP planarization, 250, 251
    IMD1 resist etch back planarization,
        246, 247
    spin-on-glass etch back planarization,
        248, 249
    technology node, 245
    topography, 246
    tungsten filled via plugs, 246, 248
Multi-level metal, 177

N
Negative bias temperature instability (NBTI),
    84, 215, 219
Nickel silicide (NiSi$_2$), 220
    gates and source/drains, 101
    resistance lowering anneal, 102
32 nm dual damascene copper metal1,
    266, 267
32 nm single damascene copper metal1, 266
45 nm dual damascene copper metal2, 264, 265
45 nm single damascene copper metal1,
    263, 264

65 nm dual damascene copper metal2,
    261, 262
65 nm single damascene copper
    metal1, 261
90 nm dual damascene copper metal2, 260
90 nm single damascene copper metal1, 259
130 nm copper multilevel metal, 258
157 nm photolithography, 159, 160
180 nm dual damascene copper metal1, 257
180 nm single damascene copper metal1,
    256, 257
180 nm transistor node, 45, 46
193 nm immersion scanner, 158, 159
NMOS deep source/drain (NSD)
    pattern, 97
NMOS deep source/drain diodes, 97
NMOS transistor, 7, 14, 15, 67, 208
    capacitances, 209
    capacitor, 30
    cartoon, 31
    conductive transistor gate, 30
    dopant atoms, 211
    halos, 44
    ion implantation, 211
    LOCOS isolation, 208, 209
    nickel silicide, 210
    no halos, 44
    n-type channel, 209
    pn-junction diffusions, 211
    power supply voltage, 210
    pre-metal dielectric, 209
    p-type substrate, 208
    resistances, 209
    reverse-biased diode leakage, 209
    scaling, 210
    source /drain extensions, 209
    symbol, 30
    turned OFF, 30
    turned ON, 30, 31
NMOS transistor cartoon, 54, 57
NMOS transistor source/drain extension
    arsenic, 89
    carbon implant, 89
    dopant atoms, 89
    halo dopant implants, 89, 91, 92
    wafer, 91
Nonfunctional and partially functional dies
    (non GEDs), 16
Not invented here (NIH)
    syndrome, 282
Novolak positive photoresist, 162
Novolak resist, 162
NPN bipolar transistor, 213
N-type dopant, 79, 83

NWELL formation
  implantation, 79, 80
  pattern, 79
  photoresist pattern, 79, 80

O

Online metrology tools, 287
Optical computers, 292
Optical proximity correction (OPC), 158,
  168, 170–173
Organo silicate glass (OSG), 114, 115, 264

P

Packaged integrated circuits, 135
  IC dies, 16
  multiple dies, 18
  wafer, 16
Pad oxide, 75
Parasitic bipolar transistors, 70–72
Parasitic MOS transistors, 67–70
Parasitic MOS transistors between core MOS
  transistors, 69
Parasitic MOS transistors between core MOS
  transistors and WELLS, 70
Partially depleted silicon on insulator
  (PDSOI), 275
  vs. conventional CMOS, 276
Pattern alignment, 152, 153
Pelloie, J.-L., 275
Personal digital assistants (PDA), 3
Phase shift masks, 167
Phosphine (PH$_3$), 210
Phosphorus doped silicate glass (PSG), 107,
  232, 233
Photo acid generator (PAG), 163
Photo resist, 8, 9
Photo simulation programs, 195
Photolithography, 9
Photolithography printer, 153, 154
Photomasks, 166, 167
Photonic computers, 292, 293
Photonic processor, 276
Photoresist
  BARC, 163, 164
  bilayer resists, 165
  capping molecule protects, 161
  chemically amplified, 162, 163
  coating wafers, 161
  KTFR, 161
  novolak positive photoresist, 162
  organic polymers, 161
  TARC, 164

  ultraviolet hardened resists, 166
  wavelength, 161
Photoresist mask, 76
Physical science departments, 3
Physical vapor deposition, 12, 13
Planarization, 138, 173, 177
Plasma enhanced chemical vapor deposition
  (PECVD), 11, 95
Plasma etching, 8, 9, 76
Plasma-enhanced chemical vapor deposition
  (PECVD), 107, 114
PMD1 planarization, 106
PMOS deep source/drain (PSD) pattern, 99
PMOS transistor, 7, 67
  cartoon, 31, 32
  electrical symbol, 32
  turned OFF, 32
  turned ON, 32, 33
PMOS transistor source/drain extension
  photoresist
  dopant implant, 93
  halo implant, 93, 94
  reticle, 93
PN-junction diode, 26
  cartoon, 26
  forward-biased, 26, 27
  reverse-biased, 28
  symbol, 26
PO etch, 131
PO pattern, 131
Polysilicon gate, 215
Positive charge carrier (p-type), 25
Pre-metal dielectric (PMD), 7–9, 53, 61,
  103–111, 114, 209, 241
  challenge, 230
  CMP, 231, 234
  CMP planarization, 230
  CVD technique, 236
  gap fill, 235, 236
  phosphorous, 231
  REB, 234
  reflow, 231–233
  stack, 229
  technology node, 233
  transistor gate topography, 229
  transistor gates, 230
Pre-metal dielectric BPSG layer (PMD2), 235
Pre-metal dielectric planarization, 177
Pre-metal TEOS dielectric layer (PMD1), 235
Printed wafer, 153, 167, 168, 170
Priority zero lots, 273
Process flows implant, 224
Process flows source/drain anneal, 223
Process integration (PI) technicians, 283

Protective overcoat (PO), 7, 16, 120, 131
P-type dopant, 80, 82
P-type substrate, 76
Punch through implant, 82, 83
PWELL formation
    implant, 81
    photoresist pattern, 80, 81
    RTA tools, 80

Q

Qualification reliability testing, 187

R

Rapid learning/change control, 282, 283
Rapid thermal anneal (RTA), 80, 101, 179, 218
Rapid thermal oxidation (RTO), 218
Rapid thermal processing (RTP), 10, 150,
    179, 278
    flash anneal, 179
    laser anneal, 180
    RTA, 179
    spike anneal, 179
    transistor source/drain junctions, 179
RC delay, 254
Reactive sputtering, 13
Reliability failures, qualified manufacturing
    flow, 285
Reliability simulation, 194
Reliability testing, 185
Remote plasma nitridation (RPN), 217
Research and development (R&D), 4
Resist etch back (REB) planarization, 234, 246
Reticles, 166, 167
Reverse active patterning and oxide etch back
    processing steps (RPOE), 203
Reverse-biased diode capacitor, 58, 59
Reverse-biased pn-diode
    avalanche breakdown, 29
    current *vs.* voltage curve, 29
    depletion region, 27
    NMOS/PMOS transistors, 29
Reverse leakage, 278
RF sputtering, 13
Robotic automobile manufacture, 291
Robotic revolution, 290
Rubber band, 278

S

Scaling source/drain junction diodes
    extensions, 212, 213
    implant angle, 211

Scanners, 157
Scanning electron microscopes
    (SEMS), 53, 183
Scanning tunneling electron microscopes
    (STEMs), 183
Second layer of intermetal dielectric
    (IMD2), 7
Second robotic revolution, 2
Self-aligned silicide, 219
Self-formed copper diffusion barrier, 265
Self-ionized plasma (SIP) process, 13, 118
Semiconductor, 21
Semiconductor companies, 292
Semiconductor industry, 286
Semiconductor industry job insecurity, 277
Semiconductor Process and Device Center
    (SPDC), 273
Semiconductor wafer, 151
Shadow mask sputtering, 239
Shallow trench isolation (STI), 49, 67, 174,
    176, 209
    active photoresist pattern, 75, 76
    challenges, 202
    CMP, 78, 79, 202
    CVD, 202
    design rules, 203
    dummy active geometries, 203, 204
    etch, 76, 77
    HSS process, 204
    isolation, 203
    planarization process, 204
    process, 202
    reverse active photoresist pattern, 204
    RPOE, 204
    trench dielectric fill, 77, 78
Short channel effects, 88
Shrinking transistor isolation
    APCVD, 205
    HDP-deposited silicon dioxide, 206
    LOCOS, 202
    SACVD-deposited silicon dioxide, 206
    STI, 202
    technology, 201
    void-free gap fill, 205
Siconi© preclean process, 220
Signal time delay, 254
Silicide, 53, 56, 57
Silicon-based integrated circuit, 21, 22
Silicon carbon nitride (SiCN), 259
Silicon-controlled rectifier (SCR), 72
Silicon dioxide (SiO_2), 7, 9, 10, 60, 104,
    176, 202
Silicon germanium (SiGe), 94
Silicon nitride (SiN), 104, 131

Silicon oxynitride (SiO_XN_Y), 131
Silicon-to-silicon covalent bonds, 24
Silicon wafers, 80
Simulated Program with Integrated Circuit
    Emphasis (SPICE), 192
Single contact etch-stop liner stress
    technology, 224, 225
Single crystal gallium arsenide (GaAs), 21
Single crystal gallium nitride (GaN), 21
Single crystal indium phosphide (IP), 21
Single damascene copper metalization, 256
Single-level metal
    aluminum alloy metal, 243
    CVD-W, 242
    electromigration, 243, 244
    hillocks, 245
Single-wafer processing equipment, 150, 196
Sinter, 134, 135
SIP sputtering, 13
Snapback, 71
Social media, 3
Source extension, 56, 57, 61
Source/drain extension implants, 100
Source/drain extension resistance and
    capacitance, 213, 214
Space program, 1
Spike anneal, 179
Spin-on-glass (SOG), 237
Sputter deposition, 12, 13
Static random-access memory (SRAM), 14
Steppers, 157
STI-CMP, 173
Stress, 194
Stress simulation, 194
Sub atmospheric pressure chemical vapor
    deposition (SACVD), 11
SUN microsystems, 172, 190, 284
Supersaturation
    arsenic /phosphorus atoms, 214
    equilibrium condition, 214
    implanted dopant, 215

T
Ta/CuMn copper diffusion barrier
    layer, 267
TaN/Ta barrier, 260, 261
Tantalum barrier layer, 116
Tantalum nitride (TaN), 116, 118
Tech nerds, 281, 282
Technology Computer-Aided Design
    (TCAD), 189–191

Tensile memorization layer, 224
Tensile stress memorization, NMOS
    transistors
    implant, 223
    process flows, 223
Tetraethyl orthosilicate (TEOS), 10, 77, 205
Tetraethysiloxane (TEOS), 233
Tetramethylammonium hydroxide
    (TMAH), 162
Texas Instruments, 137
Thermal heating, 193
Thermal oxidation, 10
Thermal simulation, 193
Third level of metal (TLM), 246
Thyristor, 72
TI metallization process, 284
Titanium adhesion layer, 110
Titanium aluminide (Ti_3Al), 285
Titanium forms, 241
Titanium nitride (TiN), 132, 267
Titanium silicide, 220
Titanium/tungsten (Ti/W) barrier layer, 132
Top antireflective coating (TARC),
    164, 165
Traditional capacitor, 58
Transistor capacitance, 57, 59
Transistor channel, 53, 55–57, 61
Transistor conventional capacitors, 59–61
Transistor count, 137
Transistor deep source/drain, 53, 56, 57, 61
Transistor dielectric sidewalls, 95
    etch, 96
    nitride, 96
Transistor gate
    dielectric/channel interface, 84, 86
    etching, 86, 87
    patterning, 86, 87
    polysilicon layer, 85, 86
Transistor geometries, 137
Transistor processing, 189–191
Transistor resistance, 55
Transistor series resistance, 55–57
Transistor source, 53
Transistor switching speed, 137
Transistors, 289
Triethylborate (TEB), 233
Trimethylphosphate (TMPO), 233
Triphenyl sulfonium hexafluoro antimonate
    (TPSHFA), 163
Tungsten hexafluoride (WF_6), 241
Tungsten silicide (WSi_2), 219
Tunneling electron microscopes (TEMs), 183

Turn ON Voltage (V_{TN})
    NMOS transistors, 82, 83
    PMOS transistors, 83

U
ULK dielectric, 264
Ultraviolet hardened resists, 166
Undoped silicate glass (USG), 107

V
Vacuum wand, 147–149
Via defect rate, 285
Via1, 123, 128, 129
Via1 etch, 125, 127

Via1 etch stop layer (V1ESL),
    260, 263
Via1 pattern, 125, 127
Via1 self-aligned, 266

W
Wafer carriers, 145, 146
Wafer handling tools, 147, 148
Wafer tweezers, 147, 148
Water flow, 51, 52
Wireless revolution, 3

X
X-ray energy spectroscopy (EDX), 184

Printed in the United States
by Baker & Taylor Publisher Services